BUILT IN BRITAIN
The Independent Locomotive Manufacturing
Industry in the Nineteenth Century

BUILT IN BRITAIN

The Independent Locomotive Manufacturing Industry in the Nineteenth Century

Dr. Michael R. Bailey MBE

RAILWAY & CANAL HISTORICAL SOCIETY
2021

Front cover: Export to the Imperial Railways of Japan. Nasmyth Wilson & Co. Ltd. 1888-built 2-6-2T (NW 336/88) [See Fig. 8.4 for side view].   *P. Wardle collection.*

Back cover: Beyer Peacock & Co. 1857-built 2-2-2 (BP 56/57) [See Fig. 3.21 for details].

First published 2021
by the Railway & Canal Historical Society

www.rchs.org.uk

The Railway & Canal Historical Society
was founded in 1954 and incorporated in 1967.
It is a company (No. 922300) limited by guarantee
and registered in England as a charity (No. 256047)
Registered office: 34 Waterside Drive, Market Drayton TF9 1HU.

© Michael R. Bailey 2021

All rights reserved.
No part of this publication may be reproduced or transmitted in any form
or by any means without the prior written permission of the publisher.

ISBN 978-0-901461-70-4

Designed by Stephen Phillips
Printed and bound in Great Britain by
Short Run Press, Exeter

Contents

	Foreword	7
	Preface	8
Chapter 1	Introduction	9
Chapter 2	Development of the Locomotive Market	17
Chapter 3	Marketing and Sales	38
Chapter 4	Technology and Design	67
Chapter 5	Manufacturing	99
Chapter 6	Management, Employment and Industrial Relations	131
Chapter 7	Strategy and Administration	160
Chapter 8	Conclusions	184
Appendix	List of Independent Locomotive Workshops	195
	Bibliography	208
	Index	215

*Dedicated to
The Proprietors, Directors, Managers, Superintendents, Foremen,
Craftsmen, Tradesmen, Draughtsmen, Clerical Staff, Apprentices, Labourers
and Agents of the British Independent Locomotive Industry.*

Foreword

Great Britain's 'Industrial Revolution' during the Nineteenth Century was led by its prowess in manufacturing for the marine, mining, textile and railway industries. Locomotive building was one of the largest of the manufacturing activities, helping to expand growing railway networks, both at home and overseas. Peculiar to Britain was the division of the locomotive manufacturing industry between the mainline railway companies, with their own workshops, and the independent industry catering for the residual needs of the mainline railways, and a large industrial market, together with a major expanding market for overseas railways networks.

The story of mainline railway workshops has attracted the attention of a number of authors over the years, but, surprisingly, the independent manufacturers have received less attention, other than occasional company histories which provide case-studies of an otherwise un-assessed industry.

Drawing upon his extensive knowledge and deep research, Dr Bailey outlines the unknown story of the independent locomotive manufacturing industry in nineteenth century Britain. Michael Bailey highlights the ever-changing political and economic backdrop under which these locomotive builders carried out their business. He shows how these independent manufacturers marketed their products, competing both with one another and with overseas competitors. The firms innovated in both product design and manufacturing procedures and practices, although craft traditions sometimes persisted alongside advanced mechanisation. They had to manage the need to train and retain skilled craftsmen in an economic environment where they were constantly in short supply; to have a workforce available in sufficient numbers to supply often volatile markets; and to cope with evolving employment practices and campaigns for improved conditions of employment. Michael shows how the builders dealt with these challenges, while pioneering new forms of business organisation based on management information and accounting procedures.

The author has done a remarkable job in providing this first industry-wide assessment of independent locomotive builders. Dr Bailey provides a frank assessment of the strengths and weaknesses of the sector, as the market grew during the century. He offers cogent insight into the conduct of the proprietors, and their successes and failures under evolving market opportunities. Michael Bailey's '*Built in Britain*' is a triumph of business and engineering history, beautifully illustrated with telling drawings and photographs, well supported with statistical evidence and scrupulously referenced. This book is a breakthrough in our understanding of a crucial part of the Victorian economy.

Dr. Jonathan Aylen   President of the Newcomen Society
Matthew Searle   President of the Railway & Canal Historical Society
Brian Dotson   Chairman, The Stephenson Locomotive Society

Preface

The growth and development of railways was dependent upon the skills and initiatives of the engineering companies who provided the motive power. Only through their endeavours was it possible for railways to demonstrate reductions in time and cost for the movement of passengers, goods and minerals. Once demonstrated, investment flowed and networks grew at an astonishing rate. Expectations of speed and load haulage grew, further stimulating progress of the designs, materials and manufacturing processes of the locomotive industry.

Surprisingly, the story of these pioneers has thus far gone unrecorded, save for the occasional volumes outlining the particular story of individual factories who participated in the industry. Just over one hundred independently-owned factory sites were engaged in locomotive construction at times through the nineteenth century. These were privately and publicly owned businesses independent of railway-owned sites, such as Swindon and Crewe, that provided for the particular needs of their company locomotive fleets.

The industry, always an integral part of Britain's heavy manufacturing sector, grew to become one of its leading activities through the century. It served not only the needs of the domestic railway companies, but the rapidly expanding railway networks of almost every country in the world. Furthermore, it provided for the requirements of so many other industries at home and overseas that were engaged in mining, processing, manufacturing, construction, agriculture and shipping activities, together with defence requirements.

Did the independent locomotive industry perform well with this all-important manufacturing activity, or could changes have been made resulting in a more efficient and profitable outcome for the benefit of the proprietors, and their staff and customers? This large enquiry has been conducted with reviews of the locomotive market, the companies' sales and marketing activities, the advancement of technological progress, the improvements made in manufacturing techniques, the evolution of employment and industrial relations, and the advances made in management information and the resulting strategies that were pursued.

Although the industry was the world's largest through much of the nineteenth century, the growth of the American and German locomotive manufacturing industries towards the end of the century allows important comparisons to be made with the British industry.

This enquiry has only been made possible through the assistance of many people including, and especially Professor Colin Divall, formerly of the Institute of Railway Studies in York, who guided me through the academic expectations for my initial thesis on this topic. This work has now been re-written, and illustrated for a general, but knowledgeable readership.

Particular thanks are due to my late good friend, James (Jim) W. Lowe, whose major work, *British Steam Locomotive Builders*, first published in 1975 and re-published in 2014, was my starting point in assembling the many sources of information about the independent locomotive industry. Thanks are also due to the many locomotive historians who have laboured over so many years to produce classified lists of the outputs from each manufacturing company. The Stephenson Locomotive Society has a comprehensive library of these lists which I have liberally referred to in building up an overall assessment of the industry's output in the nineteenth century, expressed by type and by destination country. There are, of course, many queries that remain to be resolved, but the accuracy of the data that I have adopted may be taken at between 95 and 98 per cent.

Further thanks are due to the librarians, archivists and other staff members of the National Railway Museum in York, and the staff of the many other libraries and record offices around the country. Also to fellow members of the Newcomen Society for the study of the history of engineering and Technology, the Stephenson Locomotive Society, the British Overseas Railways Historical Trust, and the Railway & Canal Historical Society.

I would also like to thank the Publications Committee of the RCHS and, in particular, David Joy and Stephen Phillips, who have made possible the publication of this volume. David and Stephen have, quite rightly, sought to correct me on matters of consistency and clarity, and I am most grateful to them for all their endeavours.

Above all, I am pleased to acknowledge, with so much gratitude, the enormous help and encouragement of my wife, Jennifer. Her forbearance, for many extended periods over so many years, has far exceeded the bounds of normal marital tolerance, which has made possible the completion of this volume.

Chapter 1
Introduction

Great Britain's heavy manufacturing industry was a major part of its developing economy during the 19th century. Supplying heavy equipment for the process, mining, shipping and transport sectors, the industry boosted the country's output both for the expansion of national economic activity and through major export trading. The manufacture of railway locomotives formed a large part of this industry for the home and overseas markets, in turn stimulating economic activity around the world through expansion of railway and industrial networks.

The British locomotive industry started in 1825 and underwent such an extraordinary growth that, by the end of the 19th century, it had become the country's third largest manufacturing activity, after textile machinery and railway carriage and wagon building, with a gross annual output of over £12million.[1] Nearly two-thirds of this activity was undertaken by the railways' own workshops. The subject of this book, however, is the sector which was independent of railway ownership, manufacturing locomotives for main-line railway and industrial customers, both at home and overseas. This sector, which itself grossed annual earnings of £4.5 million at the end of the century, was the world's largest locomotive export industry.[2]

An interesting subject for study has been the growth of entrepreneurship, business performance and the evolution of firms from their 18th century 'sole proprietor' or 'family' origins to latter-day 'managerial' enterprises.[3] The heavy manufacturing sector developed from the late 18th century as factory-based activities employing multiple craft skills and metal forming techniques to produce robust machines and structures for the marine, colliery, iron, machine tool, textile and other industries. Much consideration has been given to the quality of the management in the attempt to provide a better understanding of the origins, growth and, sometimes, death of firms.[4] This enquiry seeks to consider the decision-making capabilities of the proprietors of the independent locomotive manufacturing industry as they faced the ever-changing opportunities and threats to their markets through the century.

The histories of several of the one hundred or so companies involved in locomotive manufacture in the 19th century have been considered by authors over the years. However, the development of the industry as a whole has not previously been studied to provide an understanding of the opportunities and problems that faced the proprietors of these companies. Indeed, this provides a case-study of Britain's heavy manufacturing industry generally as the companies concerned were, to a greater or lesser extent, also providing other equipment to several other industrial sectors.

Fig. 1.1 4-4-0 passenger locomotive built in 1887 by Kitson & Co. (K 310/87) for the Manchester, Sheffield & Lincolnshire Railway (No. 561). *Kitson collection, The Stephenson Locomotive Society.*

In contrast to American in-depth assessment of business history,[5] previous British studies have recognised the importance of a more specific and analytical assessment of British manufacturing history, highlighting the development of business from 'personal' to 'managerial' capitalism. These studies demonstrate that the British model was a rational response to the growth of the country's industry.[6]

'Sole proprietor' or 'family' capitalism was the form of entrepreneurial enterprise adopted by individuals and families who risked investment to innovate and exploit new technologies and services.[7] Sole proprietors, such as Timothy Hackworth of Shildon in County Durham, depended on their total understanding of an industry in its infancy with the ability to design and manufacture machines that his customers wanted, with a knowledge of costing and material supply, engaging men with the right skills supervised by experienced foremen.

'Family firms', such as R. & W. Hawthorn of Newcastle upon Tyne, refer to those businesses in which the founders or their heirs have gone on to engage managers, but have continued themselves to hold executive positions, and who exercised a decisive influence on policy matters.[8]

Family firms provided kinship networks and personal connections which offered mutual trust and helped to offset the uncertainties and risks of their developing markets. The rapid expansion of family firms in the early-mid 19th century contributed an extraordinary dynamism, particularly to the manufacturing sector, such firms being successful in both scale and structure.[9] The Quakers' regular Meeting House gatherings, for example, became the forum for discussions on investment and joint ventures, exploiting geographically dispersed pools of capital.[10] Such business networks, including manufacturing firms, were successfully established by these Quaker 'dynasties', being extended family enterprises whose beliefs developed a strong business culture.[11] Robert Stephenson & Co. of Newcastle upon Tyne was an example of Quaker capital invested in a locomotive manufacturing company.

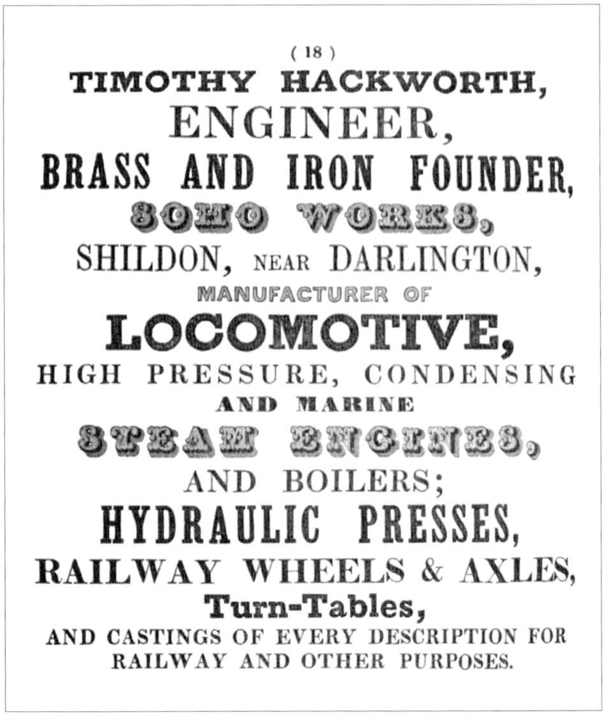

Fig. 1.2 Advertisement for Timothy Hackworth's Soho Works in Shildon, County Durham.
Slater's Trade Directory, Teesside, 1848, p.18 – digitally enhanced.

Family firms maintained a longer-term perspective on their business than did managerial enterprises, and they developed strong corporate cultures which yielded powerful competitive advantages. The crucial factor in their on-going prosperity, however, was the generational transition from the entrepreneurial originators to their offspring. There was a perceived advantage in young members of a family developing an extensive knowledge of their firm, providing them with valuable experience when they themselves came to take decisions.[12]

Manufacturing firms, such as William Fairbairn & Sons of Manchester, were developed in the expectation of the proprietorship passing to the junior generation. However,

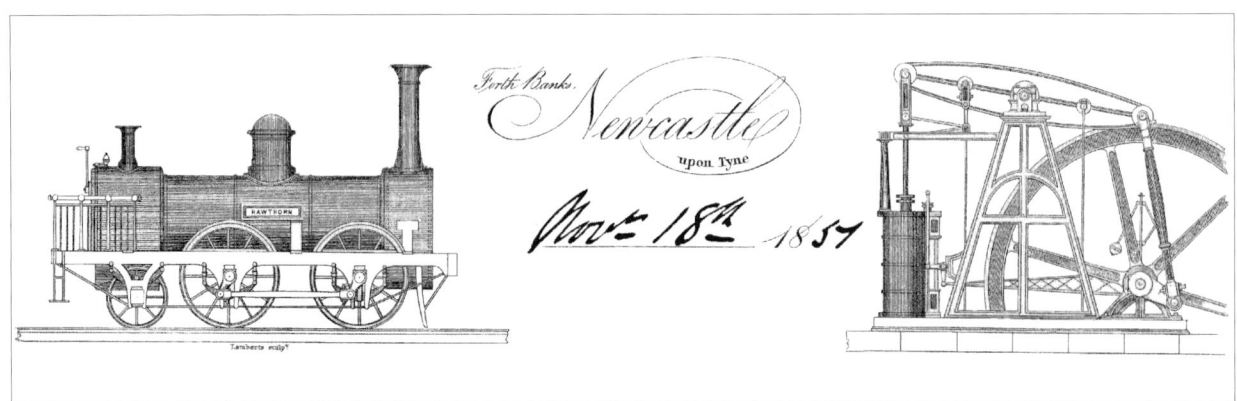

Fig. 1.3 Letterheading of R. & W. Hawthorn, the Newcastle-upon-Tyne manufacturing partnership. Dated November 18th 1851.
Author's collection.

Fig. 1.4 Business card for Robert Stephenson & Co. of Newcastle on Tyne from the mid-19th century. *Warren, 1923, p.107.*

subsequent generations were rarely able to respond adequately to the challenges of technical change and, regardless of relevant career credentials, enterprise was stifled.[13] Their individualistic culture led owner-managers to take decisions containing a firm's growth within the limits of existing resources.[14] Success in business often resulted from these strong personalities, for whom retirement was perceived as a personal defeat.[15]

However, many of Britain's larger manufacturing enterprises in the 19th century were multi-skilled partnerships, which went on to become the organisational building-block of British business.[16] By their nature, partnerships shared the responsibility for the growth and health of their firms between their partners. They provided greater capital-raising potential to meet higher levels of investment and were flexible enough to provide for the withdrawal and recruitment of partners as age, experience and financial circumstance determined.

The repeal of the 1719 'Bubble' Act in 1825 had allowed firms to have more than six partners, stimulating the raising of capital and expansion of manufacturing industry.[17] However, until 1856 British company law encouraged a highly individualistic business culture.[18] Entrepreneurs were forced primarily to form partnerships, as it was prohibitively expensive to set up joint stock companies. Much of the capital employed was working capital, which created an extensive 'web of credit' between industrialists, merchants, banks and acceptance houses, reinforcing the tendency to re-invest most profits.[19]

With its need for high levels of investment and risky markets, Britain's 'heavy' manufacturing industry in the 19th century was primarily formed of partnerships. The industry produced machinery, particularly steam engines, in small batches according to customer specification, and was composed of firms with a vertically-integrated workshop structure, quite unlike the small, repetitive production of the 'light' manufacturing sector. Partners had technical, production and commercial responsibilities, as well as sharing strategic planning with their non-executive colleagues. Their senior clerks and foremen were themselves potential partners, and this business structure generated novel problems of management.

The management progression of heavy manufacturing industry may be compared with that of the other two countries that followed the British lead, namely the United States and Germany.

With the growth of 'competitive managerial' capitalism in the United States, three-tiered managerial hierarchies with pre-determined responsibilities sprang up in large numbers from the mid-19th century, and from the late-century with the growth of 'co-operative managerial' practice in Germany.[20] Other business structures may have been just as effective as those of the independent locomotive manufacturers.

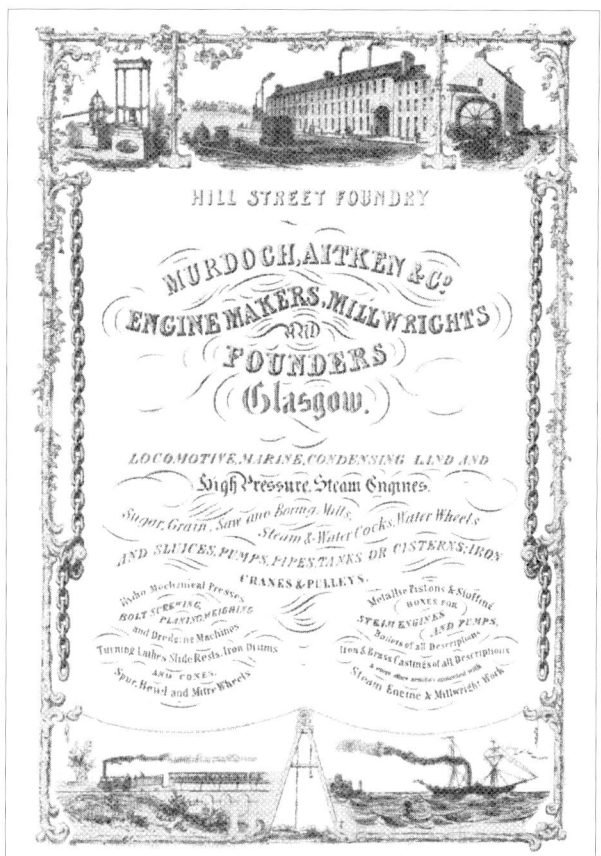

Fig. 1.5 An early example of a Partnership enterprise. Advertisement for Murdoch Aitken & Co. from the 1830s showing depth of involvement in manufacturing, including locomotives. *Mitchell Library, Glasgow.*

Fig. 1.6 End of the century advertisement by Nasmyth, Wilson & Co. Ltd. of Patricroft. *Author's collection.*

In Britain, the introduction of specialist managers into firms was not dependent upon a change in the law, and there were examples in the early 19th century locomotive industry, such as Sharp Brothers of Manchester, where managers were recruited to manufacturing by partnerships. However, the opportunity for developing this policy and creating full managerial enterprises only really came with the Companies Act of 1856, and the consolidating Companies Act of 1862, which provided for the establishment of joint-stock companies without the necessity of an enabling Act of Parliament, and with limited liability status. The intention of the legislation was to make possible the attraction of additional investment from shareholders with no involvement in the enterprises, building on the profitable experience of the railway operating companies.

In practice, however, the investments of the majority of the large number of firms which were incorporated in the years following the Acts were provided by their owner-proprietors seeking greater security from the hazards of cyclical markets.[21] This development of 'private limited companies', such as the Vulcan Foundry Co. Ltd. of Newton-le-Willows, was a British compromise, which, in perpetuating their ownership, may actually have discouraged wider investment.[22] Conversion to public companies was very slow, and largely remained the preserve of the large banking and utility organisations.[23] A small number of the larger manufacturing firms, such as the Yorkshire Engine Co. Ltd. of Sheffield, took advantage of public status, but most proprietors sought to maintain their involvement with their companies, either through existing partnerships or through the new private companies.

Britain's relative decline in the late 19th century, compared to the United States and Germany, has been attributed to entrepreneurs who kept their family firms and passed them on to their heirs, such as locomotive builders Kitson & Co. of Leeds. Some historians have suggested that in the latter part of the 19th century this was a 'gentrification' of industrialists who invested their wealth in landed estates rather than expanding and modernising their capital equipment.[24] This delayed the introduction of 'managerial capitalism' which would have provided improved organisational capabilities.[25]

Furthermore, Britain's 'proprietary capitalism' worked well until faced with the technological complexities and high fixed costs that developed from the late 19th century.[26] In particular, in personally-owned and managed enterprises, such as Nasmyth Wilson & Co. of Patricroft, Manchester, the proprietors sought to maintain an assured income rather than appreciate their assets, and their dividends correspondingly depleted the level of investment available for long-term growth.[27] On-going investment in capital equipment was necessary to exploit fully the economies of scale of improving production processes, and thus a policy of pursuing long-term profits based on long-term growth should have become more important.

The retardation of British industrial development was due to the wider issue of 'business culture', which was one of the principle 'institutional rigidities' which hindered the decision-making process and provision of management information and, in turn, Britain's competitiveness.[28] These studies move beyond the simple concept of profit maximisation towards an appreciation of behavioural rationality, both individually determined, and reflecting wider social forces. However, much remains uncertain, and business historians' knowledge of how businessmen undertook decisions, and the framework within which they were made, whether motivated by economic, cultural or social considerations, remains limited.[29]

Consideration of the evolution of proprietorial responsibilities, can be pursued through an understanding of strategic decision-making relating to the development of marketing, sales, technology, design, manufacturing, management, skills and employment, and the tactical issues governing their implementation. This evolution also spotlights the benefits and drawbacks of incorporation as private and public companies, and the objectives and motivations of their proprietors. It includes an assessment of management information systems, including accounting, which contributed to be the process by which decisions were made, together with the economic outcomes of those decisions.

Locomotive production was largely undertaken by firms, such as William Fairbairn & Sons of Manchester, which pursued several markets with varying levels of specialisation. Similar to the textile machinery firms in this same period, it began through the diversification of early manufacturing firms, which already had vertical integration of manufacturing processes and the administrative experience to take on this new market opportunity.[30]

Engines and other equipment were usually manufactured in small quantities, the many component variations for particular applications limiting batch production opportunities. The in-house development of machine tools, handling equipment and steam power extended the manufacturers' capabilities, both in terms of new product development and organisational efficiency. By its very nature, each item of equipment was designed, manufactured and erected by skilled craftsmen, particularly millwrights, foundrymen, and boiler-makers, whose one-time independence and discretion over the labour process was being subsumed into the collective activities and hierarchical subservience of manufacturing firms.[31] Machinery manufacturers had the option of in-house production or buying-in their interchangeable components on a 'hub and spoke' system, the latter often being more cost effective.[32]

The output of heavy industry contrasts with the light manufacturing sector in the 19th century. The characteristic of four levels of batch size, namely custom, batch, bulk and mass production provide growing economies of scale that is inherent in these processes.[33] The latter undertook quantity production of domestic ware, agricultural implements, firearms and, particularly, components for sub-assembly into larger industrial machines. The economics of repetitive component production by specialist producers, using unskilled labour, were very different from the production and erection of machines. Locomotive production had been born into a custom industry and developed, with varying degrees of success, into a batch industry.

The independent locomotive sector is an example of the relationship between entrepreneurial and behavioural characteristics in the heavy manufacturing industry. With our lack of understanding of how decisions were made, the conventional assertion that the sole motivation was profit maximisation will, for the locomotive industry, need to take account of the varied cultural as well as professional backgrounds of the proprietors. In the industry's early years, this culture, combining enthusiastic engineer, anxious to improve upon successful innovation, and entrepreneurial businessman, with the acumen to create wealth and provide employment, was rarely found in one person, and partnerships were inevitably the way to combine these attributes.

The British independent locomotive industry, which at any one time had between twenty-five and thirty-five partnership firms, was characterised by a wide divergence of entrepreneurial and managerial skills. At one extreme were 'progressive' firms, such as Beyer, Peacock & Co. of Manchester, which encouraged a large amount of equity and loan capital for investment, pursuing increases in productivity through improving capital equipment, employment and production procedures. At the other were firms without such attributes, such as Fletcher Jennings of Whitehaven, that instead retained their traditional craft-dependent working practices. In considering the decision-making attributes of the industry, it is therefore necessary to determine whether this diversity was generally symptomatic of partnership enterprises, and applied equally to all policy areas, or whether there were different motivations on some issues that gave rise to the diverging strategic decisions.

The establishment and development of large workshops in the 19th century called for personal attributes among proprietors which would encourage investors to provide sufficient equity and loan capital. Only by demonstrating sufficient return on that capital could proprietors of firms, such as the Yorkshire Engine Co. Ltd. of Sheffield, stimulate further investment for expanded and modernised manufacturing facilities. From the mid-century, however, partnerships were limited in their ability to maintain sufficient levels of investment; these were heavily dependent upon sustained confidence, good profitability records and high collateral value of sites and capital equipment, not all of which could be guaranteed. The entrepreneurial flair of many of the first partnerships had to be renewed as their older members retired and were replaced with new partners, either from within family circles or by promotion through talent.

The opportunity to adopt limited company status, and thereby encourage further investment opportunities, followed the Companies Acts of 1856 and 1862, which not only allowed proprietors to attract further capital, but also provided the opportunity to attract new entrepreneurial talent. Understanding the changes in proprietorial culture helps to explain the evolution of the industry during the 19th century, especially the extent to which proprietors sought to maintain control over their firms, either through continuing partnerships or through private limited companies, or opted for public company status.

In addition to their entrepreneurial and engineering attributes, proprietors and their managers needed to draw on a third quality, namely strength of character coupled with sensitivity, with which to earn and maintain the respect of the labour force. This was a particular requirement of the locomotive firms as craft skills were eroded and repetitive tasks passed to un-skilled men. Too harsh an approach would lead to industrial strife, too soft an approach could engender such loyalty to the workforce that the motivation for maintaining employment levels in the short-term, and even remaining in business, became stronger than profit incentive alone. With some of the locomotive-building firms, such as R. & W. Hawthorn Leslie & Co. Ltd. of Newcastle upon Tyne, this third quality notably provided alternative motivation to profit maximisation.

Employed managers had been engaged in manufacturing since the mid-18th century, and by the early 19th century they were in great demand as owner-managers struggled with the challenges of growing businesses.[34] From the beginning of the railway era managing partners were one of the more enduring solutions in the compromise between individualism and economic reality.[35] As the businesses grew, and delegated decision-making responsibilities became quite varied, the introduction of specialist managers, in firms such as Robert Stephenson & Co. of Newcastle upon Tyne, took over the responsibilities for strategic and administrative changes of the vertically-integrated operations, which became a significant step forward.

Each of these operations could be regarded as a cost centre, and the principal decision faced by proprietors, as the scale and scope of their operations expanded, was how to develop their administration through introduction of a management hierarchy. The managerial responsibilities included interpretation of, and response to, cyclical market changes, raw material price movements, and the corresponding effects on employment policies and industrial relations. The locomotive industry, however, unlike the quantity-production light industries, had little opportunity to achieve the full benefits of administrative co-ordination. The manufacturers' managerial responsibilities had to become increasingly technical and specialised in order to deal with demanding tactical and strategic decisions, on investment and use of assets.

By the end of the century, the diversity in the locomotive industry was very marked. The progressive firms, such as Dübs & Co. of Glasgow, employing up to 3,000 men with advanced managerial and manufacturing procedures and a high degree of specialisation in locomotive production, contrasted sharply with the craft firms, such as Richard Walker and Brother of Bury, Lancashire, employing several hundred men, and pursuing a broad market base of higher cost, small-batch orders for capital equipment. Throughout the century, the emphasis for all firms was on survival, the pursuit of which resulted in far more diverse organisations than was the case with their competitors in the United States and Germany.

Fig. 1.7 Locomotive built by a craft manufacturer. A 2-2-2 locomotive built by R. Walker & Bro. of Bury in 1846/7 for the East Lancashire Railway No. 10, *DIOMED*. *Courtesy, Lancashire & Yorkshire Railway Society*.

By 1900, twenty-six large and medium sized independent firms were regularly making locomotives for the home and export markets, compared to less than half that number in the United States making three times the output of the British industry. The cultural differences between the British proprietors and their counterparts in America affected corporate decision-making. The failure rate of American locomotive firms was quite high, being unable to remain solvent during the extraordinary periodic downturns in the market.[36] Those that survived through superior decision-making strengths, were mostly incorporated firms, but the Baldwin works, the largest by far, remained a partnership. The company's progressive culture and ability to survive the market swings was due to a distinctive business strategy that possessed its own internal coherence and logic. It minimised risks whenever possible while capitalising on opportunities for growth.[37]

A generation later, however, the American industry itself was faced with the cultural challenges of a major change in technology, during the transition from steam to diesel between 1920 and 1955. Decision-making behaviour in the American steam locomotive industry during this period have parallels with some of the attitudes in Britain's 19th century locomotive industry.[38] The manufacturers' managerial objectives were not directed towards profit maximisation alone, but were combined with diverse preferences for status, pecuniary awards and the steam technology itself. These preferences reflected vested interests in established production and marketing methods, including security and achieved status.

These contrasts within the American industry had their parallels in Britain. In spite of the relatively high 'survivability' of British locomotive firms during the 19th century, a cultural gulf developed between the 'progressive' firms and the 'craft' firms that retained their traditional working practices. The greater specialisation in locomotive production practised by the former firms, such as the Hunslet Engine Co. of Leeds, gave potential benefits of larger batch production, which had to be balanced against the risks of being committed to an uncertain market. The move towards specialisation would have been a conscious decision which became increasingly irreversible, as the growth of special mercantile relationships, highly skilled labour forces and the evolution of particular types of managerial talent made any return to an earlier, more flexible, position more expensive and difficult.[39] The 'craft' firms, on the other hand, were reliant on tactical decisions in order to survive, including diversification into alternative markets and the development of alternative employment policies. It is therefore germane to consider their motivations and attitudes to changes in production techniques, employment terms and marketing.

There were no moves towards amalgamation of the many locomotive firms still in production at the end of the century, which suppressed the opportunity for further production economies in the way that the American industry had evolved. Mergers were seen to be beneficial to certain sectors of British industry from the late 1880s and an average of sixty-seven firms were merged with others in each year between 1888 and 1914, although it did little to create an oligopolostic market structure in the country.[40] Indeed, almost all were horizontal combinations, essentially defensive measures by proprietors seeking the continuation of their businesses, and quite unlike the vertical mergers of the United States which gave closer harmonisation of industrial and financial undertakings.[41] Consideration of mergers within the locomotive industry will, therefore, help to illustrate further the cultural importance in decision-making in the manufacturing sector.

Fig. 1.8 End of century advertisement by the Hunslet Engine Company of Leeds. *Author's collection.*

Notes – Chapter 1

1. Saul (1968), Table 1, p 192.
2. *ibid.*
3. The standard work has been Coase (1937), pp 386-405, developed by Williamson (1985); also, Williamson (1986). The subject has also been covered by Devine, in Devine, Jones, Lee and Tyson (Eds) (1976); and George, Joll and Lynk (1991).
4. For example, Jenkins and Ponting (1975); Lloyd-Jones and LeRoux (1982), No. 2, pp 141-155; and Boyns and Edwards (1995), pp 28-51.
5. Chandler Jr. (1977); Chandler Jr. (1990).
6. Supple (1977); Brown and Rose (1993).
7. Chandler (1990), Part III, pp 235-392.
8. Chandler (1990), Part III, p 240; Church (1993), pp 17-43; Jones and Rose (1993), pp 1-16.
9. Church (1993), p 19.
10. Prior and Kirby (1993), p 67.
11. Prior and Kirby (1993), pp 66-85.
12. Jones and Rose (1993), p 4.
13. Lazonick (1991), p 49.
14. Payne (1988), pp 40-43.
15. Church, (1993), p 30.
16. Payne, in Mathias and Postan (Eds) (1978), p 192. Also, Cottrell (1980), pp 39-75.
17. Lee (1978), p 237.
18. Wilson (1995), p 56.
19. S. Pollard, in Creuzet (1972), p 154.
20. Chandler (1977), and Chandler (1990). Also, Chandler Jr. and Daems (1980), p 3.
21. Wilson (1995), p 120.
22. Payne (1967), p.520.
23. Cottrell (1980), pp 39-45.
24. For example, Wiener (1981), p 137, provides the example of Marshalls, the Leeds firm of flax spinners. Also, Fitton (1989), pp 182-184, refers to the acquisition of estates by the Arkwright family.
25. Chandler (1990), p 286.
26. Lazonick (1991), pp 25-27, 45-49.
27. Chandler (1990), pp 594/5.
28. Elbaum and Lazonick, in Elbaum and Lazonick (Eds), 1986, pp 1-15. Also Boyns and Edwards (1995), pp 28-51.
29. Boyns and Edwards (1995), pp 28-51, reporting the Post-Chandlerian Business History Seminar, University of Reading, 4th March 1994.
30. Jenkins and Ponting (1975), p 302.
31. For example, discussed by Chandler (1977), pp 269-272; Musson (1975), pp 109-149; Saul, in Aldcroft (Ed.) (1968), pp 186-237; and Drummond (1995), pp 40-132.
32. Cookson (1997), p 4.
33. Scranton 1997), p 10.
34. Wilson (1995), p 27.
35. Wilson (1995), p 27.
36. Brown (1995), pp 31-35.
37. Brown (1995), p 235.
38. Marx (1976), p 19.
39. Payne (1967), p 525.
40. Hannah (1983), pp 21/2.
41. Hannah (1974), pp 1-20.

Chapter 2
Development of the Locomotive Market

Introduction

As railway networks expanded through the 19th century in Britain, Europe and the wider world, the market for locomotives grew considerably. The development of those railways and, hence, the British locomotive industry's domestic and foreign markets, was determined largely by economic and political considerations, but deflected by the expanding manufacturing aspirations of the British main line railway industry. The way in which the manufacturers interpreted these market developments, and acted upon them, was a major determinant in their profitability or failure. Their strategic decisions on investment, product diversification and employment were based on the interpretation, not only of long-term market trends, but also of the major fluctuations in demand that affected each geographic and economic region, which were determined by events quite outside the influence of the manufacturers. Long-term economic cycles and short-term market variations made profitable production difficult to maintain, and, with limited opportunities for scale economies through batch production, interpretation of market growth made investment decisions risky.

The locomotive industry's main customers in the 1830s and 1840s, the British main line railway companies, went on to become the largest companies in the country through to the First World War.[1] They were both capital-intensive, subject to the variations of the capital market, and major transport utilities, subject to national and, increasingly, international economic health and political stability. With these fluctuations having such an important influence on their affairs, the manufacturers' interpretation of the U.K. market, and their strategic and tactical response to the changes, became important elements in their policy making. The economic cycles of each world region in turn greatly affected the overseas market for locomotives, which followed an, often unpredictable, demand pattern.

The manufacturers' investment decisions had to anticipate the likely capacity requirements when demand was high, without over-providing when demand was low. The latter usually led to cash-flow problems and

Fig. 2.1 Locomotive manufactured for export in the 1860s. Sharp Stewart & Co. 1862-built 0-6-0 (SS 1350/62) *MANCHESTER* for the standard gauge Egyptian Railways. *Peter Wardle Collection*.

manpower reductions. They were thus required to monitor each changing market to predict potential demand, and, towards the end of the century, the external influences that could divert that potential to their foreign competitors. The industry's use of representative agents to keep them informed of market potential followed practices developed since the 18th century. However, with the growth of foreign direct investment based in London, market intelligence in this major economic sector became easier for the proprietors.

Figure 2.2 illustrates the annual production for Britain's independent locomotive industry during the century, demonstrating its swings between periods of growth, followed by three or four years of (sometimes substantially) reduced output, before resuming its growth.[2]

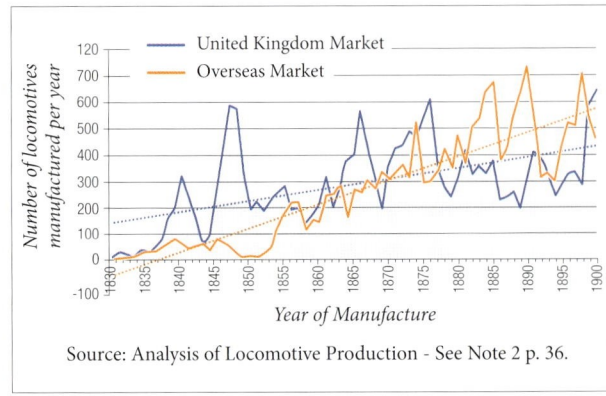

Source: Analysis of Locomotive Production - See Note 2 p. 36.

Fig. 2.3 Annual Locomotive Procuction for the United Kingdom and Overseas Markets 1830-1900.

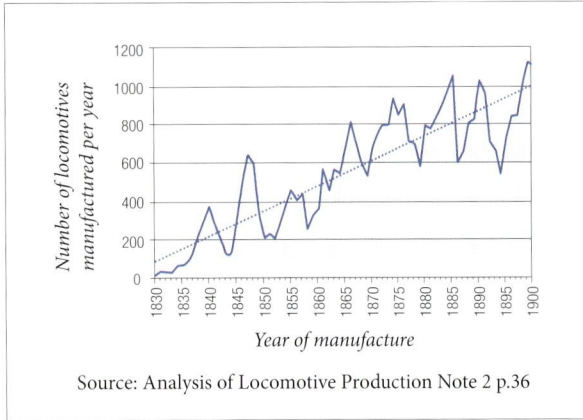

Source: Analysis of Locomotive Production Note 2 p.36

Fig. 2.2 Annual Locomotive Production 1830-1900.

From its start-up in 1830, the industry's initial growth was to meet the demand of the new railways in Britain but, as the market grew to fulfil the needs of railways around the globe, the industry's output grew to over 1,100 locomotives a year by the end of the century.

The market could have been considerably larger during the century, but was curtailed by most of the British main-line railway companies commencing their own locomotive manufacture from the 1840s. This loss of a large part of the market was therefore a major blow to the independent industry, which nevertheless went on to benefit both from new overseas and industrial markets.

Figure 2.3 illustrates the home and overseas markets during the century. The growth in output was interrupted in each decade, as national, regional and world events influenced growth in rail transport, resulting in considerable fluctuations in demand. Whilst the residual requirements of the British main line railway market showed only slow growth in the second half of the century, the industry was boosted by a significant and growing number of locomotives required for the rapidly expanding industrial sector.

The overseas market on the other hand showed a significant growth through the century, and exceeded the home market from the 1870s, although this growth was also punctuated by major fluctuations in seven to ten-year cycles. By the end of the century, the industry was manufacturing up to 700 locomotives annually for overseas markets, compared to more than 400 for its domestic industrial and residual main-line markets, although the major demand at the very end of the century temporarily brought them nearly back into balance.

The overseas market was, to quite a large extent, bolstered by the export of British capital which grew considerably in the second half of the century. The pattern of capital exports after 1855, itself closely following Britain's current account balance, reveals a reasonable correlation with these cyclical movements (Fig. 2.4).

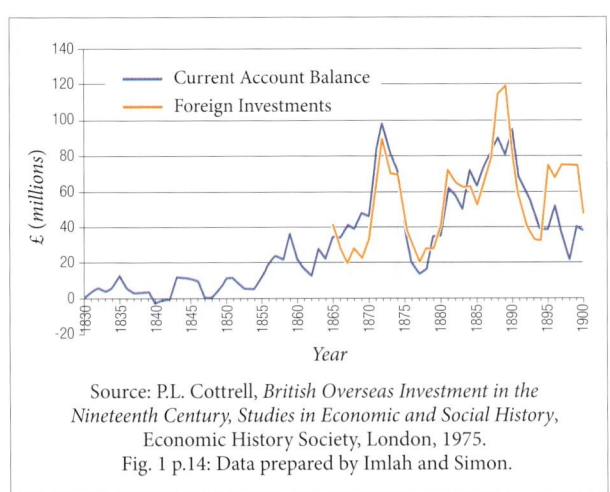

Source: P.L. Cottrell, *British Overseas Investment in the Nineteenth Century*, Studies in Economic and Social History, Economic History Society, London, 1975.
Fig. 1 p.14: Data prepared by Imlah and Simon.

Fig. 2.4 Britain's Current Account Balance and Volume of Capital Exports.

This pattern of British trade and overseas investment followed an eight-twelve year boom/depression cycle, the so-called 'long swing', which was largely followed by the overseas locomotive market.[3] The health of British trade and overseas investment, in turn, accelerated or suppressed the British domestic economy. As the manufacturing sector was wholly dependent on railways for raw material and finished product movements, the domestic railway industry was a barometer of the state of the economy, its rising and falling traffic patterns in turn determining domestic locomotive requirements.

At its start-up in the 1830s, locomotive technology was dominated by British progress, although there was a rapid diffusion to America and continental European countries, which soon developed their own industries. After significant exports of locomotives from Britain, manufacturers in the United States took over their home market from the late 1830s, whilst manufacturers in Germany, France and Belgium dominated their respective home markets from the 1850s, and in Russia, similarly, from the 1890s.

However, the main-line market expanded considerably in the second half of the century as railways were developed in other European countries, such as Scandinavia, the British Empire (particularly India and Australasia), Latin America, the Middle East and the Far East, in addition to Britain itself. These locomotive markets were dominated by the British locomotive industry, albeit from the 1890s facing increasing competition from the American and German locomotive industries. By the end of the century, the American industry manufactured 3,000 locomotives annually, largely for its domestic market, but including export totals, of which the Baldwin company alone made 300.[4]

Evaluation of the growth of the locomotive market, and its evolution through the demand fluctuations, allows an understanding of the context within which manufacturers' decisions were made. To provide sufficient capacity to meet peak demands, without surplus capacity when the market was low, required strategic decisions on investment and employment to preserve scarce labour skills whilst maintaining profitability. This understanding of the market influences led the industry towards structural and corporate changes, particularly in the second half of the century. The changes ranged between progressive specialisation through investment, to complete failure and closure.

Market Evolution 1830-1850

Interest in railways had grown quickly in the 1830s after high dividends were awarded by Britain's first main line railways. The volume and strength of British capital in the development of railway projects led to major investment in railways in Britain, the United States, Belgium, France, Holland and the German states.[5] This assisted the locomotive industry to develop and maintain a dominant

Fig. 2.5 The early years of locomotive manufacture (1). 2-2-0 *Planet* type locomotive built by R. & W. Hawthorn in 1837 (RWH 230/37) for the Stockton & Darlington Railway No. 43, *SUNBEAM*. *National Railway Museum, R.H. Bleasdale collection.*

position in the world market, which it went on to maintain through the 19th century. Until the mid-1840s, almost all locomotives for Britain, Ireland and overseas were made by the independent manufacturers, whose factories in the northeast of England, Lancashire and Yorkshire were well placed to meet the requirements of the new market.

The rapid growth of the locomotive market initially surprised the manufacturers' proprietors. Several of them were already engaged in the heavy manufacturing sector, especially those producing industrial and marine engines, textile and other machinery, who diversified into locomotive manufacture (Appendix). They quickly diverted capacity which allowed an increase in annual production to 100 locomotives by 1837. The dominance of British exports in the 1830s and 1840s was entirely due to Britain's technological and manufacturing lead.

Fig. 2.6 The early years of locomotive manufacture (2). 2-2-2 built by Charles Tayleur & Co. in 1837 (VF 51/37), *VULCAN*, for the broad gauge Great Western Railway. Shown as rebuilt to a 2-2-2BT in 1846. *National Railway Museum, R.H. Bleasdale collection.*

This increase in demand, however, masked early variations in national markets. The market for locomotives for the United States of America, grew quickly from 1831. However, in spite of the dominance of British railway finance,[6] British locomotives, designed and built for the well-made European railways, were less suited to the cheaper, lightly-laid American track. From the early 1830s therefore, a locomotive industry developed in Philadelphia and New York, which from the late 1830s, dominated the United States' domestic market. The American financial crisis in 1837, primarily due to its accumulated indebtedness to London, caused railway building and, hence, demand for locomotives to reduce substantially. When financial stability returned and railroad building recommenced from 1838, the growing American domestic locomotive industry was able to meet almost all the country's requirements for motive power.[7] Indeed, they subsequently went on to compete strongly against British manufacturers for the Canadian market, the characteristics of which were similar to those of the United States.

The technical differences with British-built designs could however act against American manufacturers in European markets. A short-lived venture by the American Norris company to sell locomotives in Britain and Europe in the late 1830s and early 1840s failed to secure sufficient rewards to justify its continuation.[8] Although some of the Norris design characteristics were adopted by Austrian and German manufacturers, the early Europeanisation of the designs served to emphasise the distinctions between the American and European markets. A manufacturing concession to the American Eastwick & Harrison company in Russia from the 1840s was more successful, however, and several hundred locomotives were manufactured there over twenty years.[9]

The growth of manufacturing capability in the United States from the 1830s, and in the main economic regions of Europe in the 1840s, coupled with the growth of the European capital markets, gave rise to increasing competition from manufacturers in those regions. Until the 1850s, the investment in the growing railway networks in the United States and Europe was largely portfolio capital. There was, however, no direct link between this investment, mostly administered by the City of London commercial banks, notably Hambros, Barings, Schröeders and Rothschilds, and the locomotive market.[10]

The loss of the American market coincided with the market expansion in Britain and Europe.[11] The home market, in particular, was strong after the first railway 'mania' of 1836, when more than 1,000 miles of new railway were promoted in Britain alone.[12] Government and 'concession' railways in Belgium, France and the German states, which were largely dependent upon Britain for capital,[13] gave rise to a rapid market expansion for the British manufacturers, which canvassed hard for orders around the European capitals.

Fig. 2.7 The early years of locomotive exports. Robert Stephenson & Co. 1836-built 2-2-0 *PLANET*-type locomotive (RS 143/36) for the standard gauge Bangor, Piscataquis Canal and Rail Road Company of Maine, USA, the *PIONEER*. Railroad absorbed into the Bangor Old Town and Milford Railroad, in 1849, on which line the photograph was taken after withdrawal from service in c.1867.
Author's collection – Original retained in the Bangor, Maine, Public Library, Whipple Lantern Slide Collection.

CHAPTER 2 – DEVELOPMENT OF THE LOCOMOTIVE MARKET

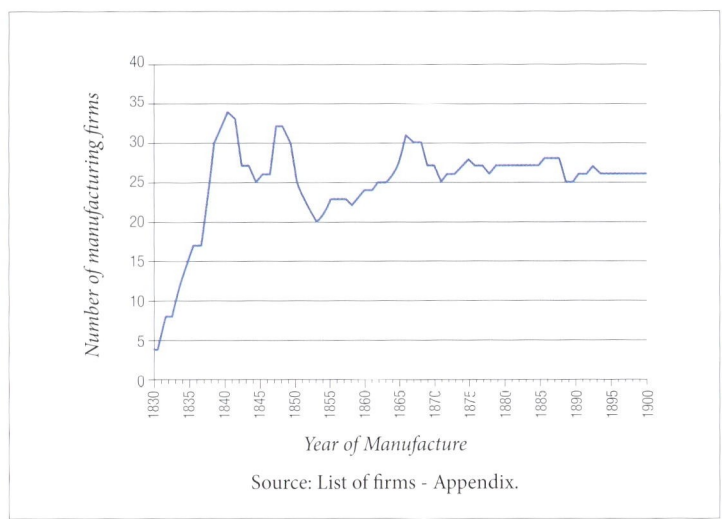

Fig. 2.8 Number of independent firms manufacturing locomotives 1830-1900.

As the British and European market continued to rise, further manufacturing capacity was provided by firms in Scotland and Bristol, as well as the northwest and northeast of England. By 1840, the industry had grown to thirty-four firms (Fig. 2.8) with annual production rising to 380 locomotives. This large increase in production was possible through capital investment in new self-acting machine tools and employment of unskilled labour, to overcome the shortage of craftsmen.

However, in later years the market fluctuated year on year, the effects of which significantly affected the industry as firms started-up, diversified into, or withdrew completely from locomotive manufacture. In the 1840s the market fluctuation was affected by several diverse influences with which the manufacturers were initially unfamiliar. The rate of railway growth in each country varied widely between the un-regulated network growth in Britain to the planned Government systems, including 'concession' networks, in several European countries.

The decisions made by the locomotive manufacturers on investment, employment and diversification were determined by their interpretation of this evolving market. In the first twenty years of main-line railway operation, the industry dominated the home and European markets, and largely determined the pace of technological, design and manufacturing progress. Demand fluctuated widely, however, and the industry realised that it could not control what it soon saw as a high-risk market, in which it would have to participate only with caution, whilst maintaining involvement with other capital goods markets including stationary and marine engine production.

There was an abrupt decline in railway projects in the early 1840s as Britain moved into recession.[14] Trade with the near European countries declined both absolutely and relatively, and there was a cut-back in investment by the capital market, with dividends from continental securities returning to London rather than being re-invested in Europe. The decline was given momentum by the imposition of the 'six hostile tariffs' by Russia, France, Belgium, Portugal, the United States and the 'Zollverein' (German states' customs union). With this substantial reduction in new railway projects,[15] demand for locomotives declined and the locomotive industry experienced the first downward movement of the market cycle it was to experience periodically for the remainder of the century.

Fig. 2.9 During the recession. Sharp Roberts & Co. 1842-built 2-2-2 locomotive (SR 168/42) for the standard gauge Wien Raaber Eisenbahn (Vienna Raab Railway) in Austria (No. 29, *SEMMERING*). *The Locomotive Magazine*, Vol. XVI (1910), p.153.

As new railway routes previously sanctioned were completed and equipped with locomotives, production fell by 1843 to less than a third of its 1840 peak. Nine firms, such as Fenton Murray & Jackson of Leeds, either failed at this time, or withdrew from locomotive manufacturing, such as G. & J. Rennie of London, and returned to their original markets. Other firms, such as the Neath Abbey Ironworks, also sought to increase their involvement in other heavy manufacturing markets, whilst persevering with the limited locomotive market.

The recession gave rise to £20-25 million of inactive capital in the City of London.[16] Investors continued to receive good dividends from railway companies, however, reviving interest in railway investment from 1844, which thus increased rapidly to become the extraordinary railway 'mania' between 1845 and 1847. Unlike the European centrally planned railway networks, the British Government permitted a scramble for new routes to go unchecked.[17] Fig. 2.10 quantifies the mania years:

Year	Acts of Parliament for railway construction	Capital authorised £million	Total actual capitalization £million	Miles in operation
1844	48	17.8	72.3	2,148
1845	120	60.8	88.4	2,441
1846	270	136.0	126.1	3,036
1847	190	40.3	166.8	3,945
1848	85	4.6	200.4	5,127
1849	34	3.1	230.0	6,031

Fig. 2.10 Railway Development in the U.K. 1844-1849.
Source: Jenks (1927), p.129.

This surge in British railway projects coincided with an expansion of railways in France, the Benelux countries and the German states, to which were added further routes in Scandinavia and the Mediterranean Kingdoms. British investment in French railways increased at an extraordinary rate in 1845, fifty companies being authorised in that year, towards which at least £80 million was promised from London.[18] Also in that year, eight British-owned companies, with London head offices, were formed with concessions to build and operate railways in Belgium.[19] In 1847, forty foreign railways were quoted on the Liverpool Stock Exchange, the second largest group after domestic railway companies.[20]

This growth of railway schemes led to a sharp increase in demand for locomotives, production of which nearly trebled to 300 in 1845, and doubled again in 1847 to a peak of more than 600 in the year. This led to profound and lasting effects on the locomotive industry, which was not to see a return to such high production levels for another twenty years. More manufacturing firms, such as Neilson & Mitchell in Glasgow, were attracted to locomotive manufacture by the prevailing high prices, and existing firms, such as Kitson, Thompson & Hewitson in Leeds, sought to introduce or divert additional capacity. However, the industry as a whole, even with over thirty firms engaged in locomotive construction, was unable to provide sufficient capacity in such a short time. The shortage of craftsmen further stimulated the introduction of new capital equipment that was less reliant on skilled labour, and significant improvements were made to production procedures that recognised the benefits of batch production.

The extraordinary demand peak could not have been predicted by the manufacturers, but even when the full extent of the demand was realised, the manufacturers were unable to increase their capacity quickly, and most were reluctant to invest in too much further capacity in anticipation of a subsequent market decline. They increased their prices and demanded down payments to secure priority places in the lengthening production queues. Large orders were spread over several months, sometimes years. Delivery times lengthened unacceptably and prices rose substantially.[21]

The London & Birmingham Railway was so concerned to overcome the problem, that it sought an exclusive manufacturing agreement with the then largest

Fig. 2.11
Manufactured during the railway 'mania'. Gilkes Wilson & Co. 1847-built 2-4-0 for the Stockton & Darlington Railway (No. 66, *PRIAM*) later running on the North Eastern Railway as No. 1066.
National Railway Museum, R.H. Bleasdale collection.

manufacturer, Robert Stephenson & Co. The railway's Secretary wrote: 'I am desired to say, that our Company are prepared to deal with you for a supply of Engines to an extent that would probably make it worth your while to devote your Establishment to the execution of our orders exclusively, & with a view to a more prompt delivery of them than might under other circumstances be thought convenient.'[22] With a very large order book, high prices and advanced payments, the Stephenson Company declined the proposal.

Due to the large increase in demand arising from the railway 'mania', the established railway companies thus became irritated at finding themselves in a long queue for locomotives alongside new railway companies which had yet to establish their services. To mitigate high prices and delayed delivery times they embarked on programmes for building their own locomotives, and their workshops were correspondingly expanded to undertake this new work. This therefore led directly to the long-term reduction of much of the British main-line locomotive market for the independent industry. Using their existing maintenance workshops railways, such as the Great Western with its Swindon workshop site, invested in new facilities and skills, such as boiler-making. This provided economies of scale through wider use of their workshop resources, but carried the risks of demand fluctuation without the opportunity to diversify into alternative markets when few new locomotives were required.

In addition to the Swindon site, the Grand Junction, London & South Western and Glasgow, Paisley & Greenock Railways were amongst the first to commence locomotive manufacture. Joseph Locke, Engineer of the Grand Junction Railway, later wrote: 'Then arose the question, whether this establishment [Crewe] could not be advantageously used, not only for the repair, but also for the construction, of engines. The plan was tried… and the cost was found to be much less than the price they had formerly paid.'[23]

For the next thirty years more and more railway workshops turned to the manufacture of their own locomotives (Fig. 2.12).[24] The gradual loss of the main-line locomotive market to these workshops had major repercussions for the independent industry, which went beyond the simple contraction of market potential. The initiative gained by the railway workshops included technological and design progress, which largely left the independent industry to undertake residual locomotive orders on a contract-only basis with a proliferation of designs specified by the railways themselves. The increasing loss of the British main-line locomotive market to railway-owned workshops from the late 1840s therefore diverted much of the manufacturers' attention towards the growing overseas and industrial locomotive markets.

However, the long delivery times and high prices, accentuated by the high value of the pound and import tariffs, also stimulated the establishment and growth of domestic locomotive industries in the German states, France, Belgium and Austria, further suppressing future potential for the British industry.[25] As railway networks expanded, continental manufacturers copied and adapted British and American locomotive technology, expanding quickly to fulfil the requirements of their own countries and become formidable competitors to the British industry.

In the autumn of 1847, the British economy again went into recession; the financial problems brought about by the 'tangled skein of credit disorders' adding to the strain already imposed by the corn and potato famines.[26] Well-established finance houses failed, including five Bank of England firms, and British investors sold some of their foreign government security investments. In 1848, the several political crises, in France and central Europe, also led to a considerable slowing of British investment in continental railways. With interest in railway schemes ending abruptly, locomotive orders dried up, considerably easing the pressure on the manufacturers and allowing them to catch up on their backlog of orders, which had built up to between two and three years.

The backlog was regained by 1850, but with a hesitant British market, now diluted by the railways' own workshop output, and an equally hesitant European market in the post-1848 political climate, locomotive production by the independent sector declined to 200 in the year. Manufacturers once again had surplus capacity which caused them to return to other markets. Some manufacturers withdrew from the locomotive business altogether to concentrate on their other markets, whilst other, quite large firms, such as Bury, Curtis & Kennedy in Liverpool, failed to secure profitable alternative markets and went out of business. The drop in the market was so severe that the number of firms which continued in locomotive manufacturing dropped to twenty in the early 1850s.

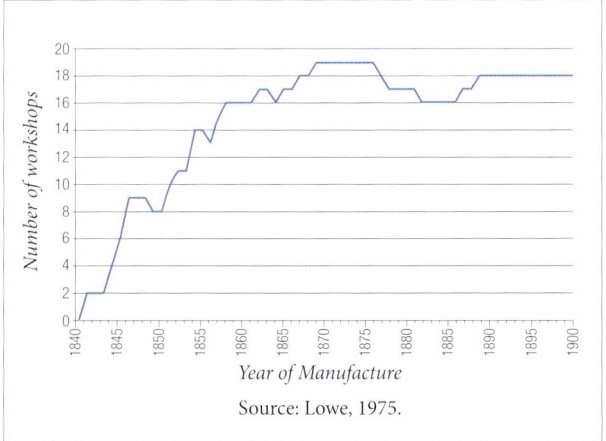

Fig. 2.12 Number of railway-owned workshops 1840-1900.

Market Evolution 1851-1870

The characteristics of the locomotive market changed radically in the twenty years after 1850. Not only had the industry to witness the increasing loss of the domestic market to the railway workshops, it also had to accept that Britain's decreasing influence in the main economic centres of Europe would favour the continental manufacturers. The early provision of portfolio capital for railway construction and equipment from the London market was subsequently emulated by other European capital markets. The nationalism that followed these changes, the political uprisings in Europe, the growth of national locomotive industries in major economic regions, and the increasing effects of import tariffs all decreased the potential for the British industry.

From the 1840s, French, Belgian, German and Austrian manufacturers took an increasing share of their respective home markets away from the British industry, and by the middle 1850s much of their national market requirements were met by their home industries. The withdrawal of the British locomotive industry from much of the western European market was thus due to the expansion of the capital markets of the western European nations, the differing evolution of designs brought about by changing operating requirements, and the tariffs imposed by countries to protect domestic manufacturing industries.

It has been claimed that it was highly improbable that tariffs were a major factor in Britain's trade losses against American and German competition,[27] but high levels of tariffs were imposed by both the United States and Germany to protect their domestic industries.[28] The fluctuating tariff levels in the United States remained high by international standards through to the First World War, although the German rates were much lower and closer to those of Britain. The ratio of overall duties to total imports, however, masks selectively high tariffs. Those for machinery and locomotives, for example, were fixed at 30% by the end of the century, which made Germany a 'closed country' for the British industry.[29]

Diverted from its dominant role in France, Belgium and the German states therefore, the London capital market re-directed its portfolio investments to new railway systems in Scandinavia, southern and eastern European countries, notably some Italian states, Iberia, Switzerland, Austria and Russia, in competition with continental financial syndicates. Much of the investment was arranged through the strong British contracting industry, dominated by Thomas Brassey, Samuel M. Peto and William Mackenzie.[30] With several railways in Holland also being financed through direct investment from London, their locomotive requirements were largely met from Britain throughout the century.[31] Although there was no direct linkage between portfolio investment and locomotive supply, it would seem that by the mid-1850s British contractors were encouraging orders to British locomotive firms, in common practice with their continental counterparts.[32]

British manufacturers had to adapt their design and product strategies not only to continental European requirements but also to the large potential of the early railways in the British Empire, notably in India and Australia, as well as to the growing industrial locomotive sector. The greatest influence in the second half of the century was the movement of the London capital market into foreign direct investments for railway schemes in the Empire and other areas that encouraged development capital, notably South America.

From the 1850s more British railway companies established their own locomotive workshops; about 18 were in being from the 1860s, outgrowing the activities of the independent firms. Some railways, such as the London & North Western with its Crewe workshops, and the Great Western with its Swindon and Wolverhampton workshops, became fully self-reliant for locomotive provision. Other companies, such as the Midland Railway, with its Derby workshops, constrained their new-build capacity to meet a relatively constant demand, with peak requirements only being met from the independent sector. Smaller railways, however, such as the Cambrian Railways, remained fully reliant on the independent manufacturers for the provision of their locomotives.[33]

Fig. 2.13 Locomotives for the smaller main line railway systems. Sharp Stewart & Co. 1863-built 0-6-0 (SS 1344/63) for the Cambrian Railways (No. 27, *CAMBRIA*). *Peter Wardle collection.*

The culture that developed, with each of the larger main line railways establishing an identity through the design and construction of its locomotives, was peculiar to Britain. Although a small number of railways in France and elsewhere constructed their own locomotives, the large majority of the world's railways preferred the negotiating strength of independent supply. There were merits and disadvantages in self-design and manufacture, and a cultural divide grew up between the two industries. Once a railway had invested in manufacturing facilities, it retained and enlarged them, in spite of fluctuating demand that could result in costs rising above prices from the independent sector.

The railway workshops and their locomotives became symbols of the railways themselves, and particularly of the locomotive engineers, some of whom have been described as being 'technical virtuosos with autocratic personalities'.[34] 'Lacking the economic motive for product differentiation in order to gain oligopolistic control of the market', the locomotive superintendents 'were able to establish 'private empires' in which idiosyncrasies of design and the continuing proliferation of locomotive types could flourish'.

The incentive for the railway directors in sanctioning the continued operation of in-house locomotive production is partly explained by the prevailing accounting conventions.[35] New locomotives, which were authorised as direct replacements for old stock, were chargeable against revenue account, thus avoiding the need to increase the capital account, with its implications for shareholder approval and stock market evaluation. The same incentive also led to a strong 'second-hand' market, which offered a cheaper source of motive power, particularly during economic recessions. With minimal expenditure, old main-line locomotives, replaced by faster or more powerful types, were re-built in the railway workshops and 'cascaded' on to less demanding operations. Alternatively, they could be sold on the open market to other railways, industrial concerns and contractors, further reducing the home market potential for the independent industry.[36]

Foreign direct investment increased during the 1850s, particularly for railways in Russia (post-Crimean war), Australia, India and South America, encouraged by 'guaranteed' government dividends.[37] Railway expansion in Europe, and the several new world markets, saw Britain's overseas locomotive production briefly exceed the home market in 1856/57.

The development of Britain's railways was slow to recover after the recession, both in terms of new routes and increased traffic, and development in the mid-1850s was at a rate considerably below that of the 1840s.[38] Although overall locomotive production remained at about 400 a year between 1854 and 1857, production dropped by 40% in 1858, due both to a reduction in European demand, and a one third reduction in locomotives for the home market. The demand for main-line locomotives was correspondingly low, whilst an increasing proportion was manufactured in railway-owned workshops, and although the demand for industrial locomotives had begun to grow, overall production for the home market by the independent manufacturers fell to 130 in 1858.

Manufacturers that survived the depression years were cautious in their interpretation of these market developments, and opinions were divided. Some firms anticipated further severe fluctuations, and sought to maintain their broad, capital machinery market base and craft-based production methods. Other firms, some new to the industry, such as Beyer Peacock & Co., saw good opportunities in the developing market that warranted greater specialisation in locomotive manufacture. As the new markets developed, these 'progressive' companies invested in new workshops and equipment, taking the number of firms back up to thirty, which helped to double the industry's production capacity to 800 locomotives a year by the late 1860s.

With the increasing loss of much of the main-line market, the home market might have reduced considerably but for the co-incident growth in the demand for industrial locomotives. The growth of British industry in the 19th century was largely dependent upon rail transport, for which large industrial sites required motive power for their internal rail systems. The market for small industrial locomotives expanded to meet the requirements of contracting, extractive, manufacturing, process, shipping

Fig. 2.14 Expansion through foreign direct investment. Beyer Peacock & Co. 1860-built 2-4-0 (BP 155/60) for the 5ft 6in gauge Great Southern of India Railway (No. 4). *The Locomotive Magazine, Vol. 17 (1911), p.134.*

and military operations in Britain and overseas. The market was manufacturer-led, rather than customer-led, allowing them to retain control of the market through standard designs. Although all manufacturers were engaged in this market to some extent, several, particularly the Leeds manufacturers, Hunslet Engine Co., Hudswell & Clarke and Manning Wardle & Co., together with the Avonside Engine Co. of Bristol, Andrew Barclay in Kilmarnock and Fletcher Jennings of Whitehaven took the strategic decision to specialise in industrial types.[39]

Fig. 2.15 An 1860s 0-6-0WT industrial tank locomotive. Fletcher Jennings & Co. (FJ 62/65) for the standard gauge Ebbw Vale Co. Ltd., *WILL O' THE WISP*. *Author's Collection.*

With improving balance of payments, British investors sought to resume their interest in western European railways. However, in spite of nearly £40 million of investment, the birth of the French 'Second Empire' in 1852, meant British investment was eroded by a resurgent French capital market, in particular the Crédit Mobilier favouring French engineering industry, including locomotive manufacturers.[40] Many of the substantial European railway-building programmes in the 1850s were railway concessions, financed and built through international syndicates formed by finance houses and railway-building contractors.[41] However, although French capital dominated railway development in Italy and Spain, as well as in France itself,[42] the collapse of the Crédit Mobilier in 1867, and the Franco-Prussian war of 1870 interrupted further French expansion.[43]

Locomotive orders from other major European economic centres also declined, however, as Belgian, German and Austrian capital, with the encouragement of their national governments, gave preference for their national industries, including the locomotive manufacturers. It was thus 'a surprise for Great Britain' to be unable to command any market open to foreign competition which she chose to supply.[44]

Those countries apart however, the emphasis of the British capital market shifted towards direct investments for railway projects in the regions of the world which had little or no financial or administrative experience with public works. Manufacturers were able to take particular advantage of the substantial growth in locomotive demand arising from the development of these railways, whose head offices were largely based in the City of London. Marketing and selling were made easy by the close community of overseas railway head offices, consulting engineers and the manufacturers' own offices or agencies. Foreign direct investment was very important, and these 'free-standing companies' were more deliberately managed, and their management concentrated in fewer hands.[45] The typical City head office for a free-standing company was small, normally comprising a corporate secretary and a board of directors, and with a brass nameplate.[46]

Fig. 2.16 In spite of Crédit Mobilier… William Fairbairn & Sons 1855-built 2-2-2 for the standard gauge Chemin de Fer du Nord (French Northern Railway) No. 164, *EUGENIE*. *The Locomotive Magazine, Vol. 36 (1930), p.157.*

Fig. 2.17 Residual order from a British main line railway company in the 1860s. Kitson & Co.1867-built 0-6-0 (K 1434/67) for the Midland Railway (No. 594).
Kitson collection, The Stephenson Locomotive Society.

There was no such link with Canadian railways, however, in spite of large portfolio investments. British manufacturers thus competed with the United States manufacturers, which, with some exceptions, dominated the North American market.

The fluctuations in the locomotive market further widened the views of the manufacturers between those which remained cautious about its progress, and those which anticipated its long-term development. This divergence resulted from strategic decisions that the locomotive firms took from the 1850s on investment for increased locomotive production, employment and the extent of product diversification. The more cautious firms reduced their dependence on locomotive work and maintained a big involvement in marine, colliery, process and manufacturing steam engines and other machinery, for which their 'craft' skills were well suited for small batch production. The 'progressive' firms invested in increased production capacity and further labour-saving capital equipment to reduce their dependence on craft skills, and allowing them to employ unskilled labour in numbers appropriate to the buoyancy of the market.

From the 1860s, between twenty-five and thirty large or medium-sized firms were engaged in locomotive manufacture at any one time, only a small proportion of which sought to specialise in the market. Whilst several manufacturers specialised in the industrial locomotive market, other firms in other manufacturing sectors also made small numbers of industrial locomotives, both for their own purposes and for supply to neighbouring companies when the market was buoyant and prices were high.[47]

There then followed a period of sustained growth in output for the locomotive manufacturers through to 1866, that took their annual production to over 800. This included a four-fold increase in locomotives for Britain's main line and industrial railways, reflecting the country's increasing industrial production and expanding economy. The main-line requirements went well beyond the capacity of the railway workshops, indicating that, after twenty years experience, some of the larger railways understood the economic benefits of stabilising their workshop capacity at an economic and sustainable level of locomotive production. Their peak requirements were ordered from the independent firms, which thus bore the brunt of the demand variation for the home market. The exceptions, however, were the London & North Western Railway and, from 1864, the Great Western Railway, whose Crewe and Swindon workshops, respectively, were large enough to accommodate their peak requirements.

The strong British market coincided with a more sustained growth in demand for locomotives for countries beyond Western Europe. Railway companies were formed to build railways in Russia, Scandinavia, South America, Canada, India and Australia with the encouragement of the host governments, to entice the large reserves of the London financial market, often with dividend guarantees, as a means of developing the economies of the countries concerned.[48] Following the Companies Act of 1862, a series of finance companies was formed, based on the French Crédit Mobilier system, to pursue the new investment opportunities.

British investment in India, particularly for railways, developed at an extraordinary rate in the wake of new government policies after the attempted revolution of 1857. It became so high in the early 1860s that it caused alarm as the British interest rate began to rise. The annual investment in Indian railways for the decade,

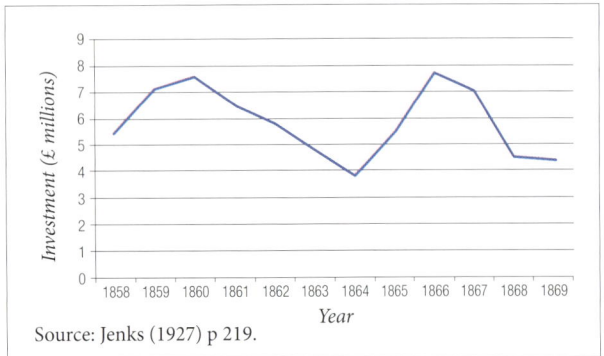

Fig. 2.20 Investment in Indian Railways 1858-1869.

mostly encouraged by the 'guarantee' system, is shown in Fig. 2.18. As the extensive railway routes in India were constructed and opened, the demand for locomotives quickly developed into the second largest market, after the home market, for the British locomotive industry. This new market was again cyclical, however, generally following the railway investment pattern, but a year or two behind as the new routes were opened and traffic expanded. 200 locomotives were constructed in 1867, for example, following the high investment year of 1866, but had dropped to only fifty locomotives in 1870 as the new building programme slowed down.

The influence of the consulting engineering profession from the mid-century gave rise to a strong market focus for the manufacturers, but with the penalty of decreasing discretion in locomotive specification and design. Engineers, such as Sir Charles Fox, Sir John Hawkshaw and Sir Alexander Rendel, and contractors, such as Thomas Brassey and Samuel Peto, were engaged to design and build the railways, and represent their clients in equipment procurement, including locomotives. The British civil and mechanical engineering professions played a major part in the development of railways in the Empire and other areas of the world, particularly South America. The engineers who worked for the railways took with them a strong expertise and culture which favoured the British manufacturing industry, and directly benefited the locomotive industry.[49]

London thus became the world centre for the locomotive market, in which manufacturers maintained close contacts with railways and consultants through their own London offices or through agents. The market was a loose 'cartel' which effectively blocked the participation of the continental and American locomotive industries. There thus developed a de facto link between foreign direct investment and market opportunities for British industry.

Between 1860 and 1876, some £50 million was loaned to the governments of the Australian colonies, chiefly for railway construction and other public works.[50] The sparseness of the population and the low service levels, however, meant that during these early years locomotive requirements, all supplied from Britain, were generally low to begin with, but rising as mileage increased.

The growth of the British economy was badly interrupted by the failure of the Overend Gurney & Co. discount bank in 1866, with debts of £10 million.[51] The effects on the railway contractors (Peto & Betts failed with debts of £4 million) and on the domestic railway building prog-

Fig. 2.19 Investment in the Indian railways. Kitson & Co. 0-8-0ST (K 1274/67) for the 5ft 6in gauge Great Indian Peninsular Railway's Bhore Ghat incline. *Kitson collection, The Stephenson Locomotive Society.*

CHAPTER 2 – DEVELOPMENT OF THE LOCOMOTIVE MARKET

Fig. 2.20 An early locomotive for Australia. Neilson & Co.1865-built 0-4-2 (N 1208/65) for the 3ft 6in gauge Queensland Government Railways No.1. *Author's collection.*

ramme were felt for some three or four years afterwards. The collapse revealed that much of the railway building boom in the 1860s had only been possible by an expanding credit system that could not sustain itself. Contractors had been willing to take shares in the railway companies for whom they were building, or to raise loans, which the new finance companies had proved eager to provide.[52] The collapse put an end to the raising of capital by contractors for railway construction purposes. The resulting loss of confidence, and its effects on the country's economy, led to a substantial falling off of locomotive orders. Annual production for the home market dropped by two-thirds to less than 200 locomotives between 1866 and 1869.

Following the Crimean War, the Russian economy was stimulated by capital from several western European financial centres, including London, which was partly used to build government-owned and concession railways.[53] Some railways, such as the Dunaberg-Vitebsk, were British-owned.[54] Other companies were funded by French, German and Austrian interests, and locomotive orders were generally, but not always, placed in the country with the prevailing financial interest. Several large orders were received by the British industry, over 150 being manufactured in 1870.

Fig. 2.21 The Russian market in the 1870s. Sharp Stewart & Co. Ltd. 1870-built 2-4-0 (SS 2062/70) for the 5ft gauge Rybinsko Bologovskaya Zh.D (Rybinsk-Bologoye Railway) No. A3. *Author's collection.*

Market Development 1871-1900

Fig. 2.22 The growth of the Indian market. Yorkshire Engine Co. Ltd. 1885-built 0-4-4T (YE 399/85) for the 5ft 6in gauge South Indian Railway (No.123).
Author's collection.

From the 1870s, the manufacturers' domestic market continued to be made up of residual orders from the main line companies and industry's growing requirement for internal motive power. This emphasised the importance of the export market, which was made up of several distinct geographical and political sectors, each influenced to varying degrees by British capital exports. From the late 1850s, these investments had increased significantly in the widening spheres of British influence around the world. It has been calculated that 41%, and possibly as much as 44%, of all Britain's overseas investments between 1865 and 1914 were for railway projects,[55] whilst the amount of British money invested in overseas railways by 1870 was very nearly equal to that which had been invested in the British railway system itself.[56] However, the export of capital and all kinds of capital goods from Britain, showed no direct relationship between the destination of exported capital goods and the apparent field of activity of British investment.[57] British capital exports fluctuated during the century, with railway investments following similar cyclical swings, and the correlation between capital export trends and the demand for locomotives explains these market fluctuations.

As the market developed in the last three decades of the century, the manufacturers experienced an overall growth in the demand for locomotives which was interrupted by three periods of low demand that severely tested their ability to remain profitable. The period also saw increasing competition from American and European manufacturers.

The growth in the size of railways and in the traffic they handled led to an increase in batch sizes which favoured the progressive manufacturers, whose continued investment in capital equipment allowed cost reductions to be made to the disadvantage of the 'craft' firms.

From 1870, the major expansion of London-based foreign direct investment in railways was in both the Empire, including South Africa and later other African nations, together with other regions of the world, particularly South America. Several railways in Europe, including those in Scandinavia, Holland and Russia, were either British-owned, with their head offices in London, or retained strong proprietorial contact with British manufacturers.

Fig. 2.23 Further example of the Russian export market. Kitson & Co.-built 2-4-0 (K1817/71) for the 5ft gauge Odessko-Baltskaya ZH.D (Odessa-Balta Railway) (No.176).
Kitson collection, The Stephenson Locomotive Society.

The London market for locomotives gave every opportunity for manufacturers to monitor the progress of demand in each country, although the economic and political influences on world railway development continued to create extraordinary fluctuations in demand that continued to be difficult to predict. Close contact was also maintained with the London-based representatives of colonial governments, or their agents, (particularly the Crown Agents for the less developed colonies towards the end of the century),[58] which sought to build railways to stimulate the development of under-populated areas, and for which guaranteed annual dividend payments were often made.

The industry's diverse overseas market opportunities, which had to be understood and acted on accordingly by the manufacturers, may be summarised in ten categories:

- railways funded through foreign direct investment, whose head offices in Britain (usually London) had responsibility for the acquisition of materials, including locomotives, usually with the assistance of consulting engineers,

- railways funded through foreign direct or portfolio investment, with head offices in the country of operation, whose specification of materials rested with British consulting engineers,

- railways funded through foreign direct or portfolio investment, for which the acquisition of materials was undertaken directly by a British-trained Chief Engineer who maintained contacts with British suppliers, including locomotive manufacturers,

- railways funded through foreign direct or portfolio investment, the acquisition of materials for which was undertaken by the Chief Engineer who pursued an open tendering policy,

- railways built for Colonial or other national governments, or the Crown Agents, financed by British capital (usually with guaranteed dividends), for which acquisition of materials and equipment was undertaken by British consulting engineers,

- railways built by Colonial or other national governments, financed by British capital, for which the acquisition of materials was undertaken by a British-trained Chief Engineer who maintained contacts with British suppliers, including locomotive manufacturers,

- railways built by Colonial or other national governments, financed by British capital, for which the acquisition of materials was undertaken by a Chief Engineer who pursued an open tendering policy,

- railways with minority British investment, but who engaged a British Chief Engineer or British consulting engineers, who favoured British suppliers, including locomotive manufacturers,

- railways with minority or no British investment, which pursued an open tendering policy,

- railways with no British investment, and for which the acquisition of materials and equipment was biased towards domestic industries by tariffs.

Britain's economy recovered strongly in the early-mid 1870s, in spite of major increases in raw material prices following the Franco-Prussian War, and the 1873 financial crash in Vienna.[59] Locomotive output from the independent industries achieved new highs of more than 900 a year in 1874 and 1876. In the latter year, a record 600 were made for the home market, a third being for industrial use, reflecting the high level of Britain's industrial activity. There was also a short-term boost arising from the extraordinary needs from France and Germany in the aftermath of the Franco-Prussian war, and nearly 500 locomotives were made for Western Europe in the mid-1870s.[60]

Fig. 2.24 After the Franco-Prussian war. Sharp Stewart & Co. Ltd. 1878-built 2-4-0 (SS 2744/78) for the standard gauge Chemin de Fer Paris Orleans (Paris-Orleans Railway, No.125).
Peter Wardle Collection.

From 1873, the upward trend in Britain's export income was reversed, and capital values fell in the so-called 'Great Depression'.[61] The country's surplus capital that had been re-invested abroad, including railway schemes, all but disappeared by 1876, and for the next generation, the country's overseas investments were limited to the 'secondary export of capital'. Britain's 1878 liquidity crisis was amongst the worst in the nineteenth century.[62] The resulting short-term adverse trade swing slowed manufacturing output, and locomotive production dropped by a third to below 600 in 1879, of which the domestic market accounted for only forty per cent.

The effects were worse in certain markets. Following the boom market for Russian locomotives up to 1876, for example, British investment in Russian railways, leading to locomotive orders, fell sharply. The vacuum was filled by continental investors, particularly those in Germany. In 1884, following a 'rapprochement', Germany made available a £15 million loan to Russia, which was seen as being 'essentially a railway loan' for which there would be 'some large orders for rails and railway appurtenances of all kinds.'[63] With Russia's locomotive industry unable to keep up with demand, the loan led to some large orders for

Fig. 2.25 The strong Indian market. Yorkshire Engine Co. Ltd. (YE 293/76) for the metre gauge Indian State Railways (No. F37). *Author's collection.*

the restored German industry, to the exclusion of British manufacturers.

The British industry's overseas markets were, however, broadly based from the 1870s. British investment in the railways of Argentina was extensive and, by the end of the decade, twenty-seven of the country's thirty-four railway enterprises were British controlled.[64] The locomotive market for South America as a whole reached between thirty and seventy per year by 1880. The strong Indian market in the mid-late 1870s did most to offset the declining domestic market, with a record of over 250 locomotives being exported there in 1880. Further markets arose from expanding railway projects in Australia, New Zealand and the Middle East, and new railways in southern Africa and the Far East.

British manufacturers experienced their first concerted competition outside Europe from the American locomotive industry, during the 1870s. In the wake of the 1873 American economic panic, and a reduction in its domestic market, the American industry competed for export orders. Even after the recovery of the American market in the mid-late 1870s, modest inroads continued to be made in markets previously considered to be 'British'. These were notably railways in Australia, New Zealand and Latin America, where the nature of certain routes, constructed to the more demanding, low expenditure standards of American railroads, favoured American designs over British ones.[65] The export of American locomotives between 1875 and 1879 was equal to 20% of the British export total.[66]

In the early 1880s, Britain's capital exports and trade recovered strongly. Australia benefited particularly, about half of all overseas investment being directed there in the late 1870s, and a quarter in the 1880s, mainly to expand the 'pastoral economy' and for railway construction.[67]

Fig. 2.26 Growth of railway networks in Australia. R. Stephenson & Co. 1871-built 4-4-0T (RS 1995/71) for the 5ft 3in gauge Melbourne & Hobsons Bay Railway in Victoria. *The Locomotive Magazine* (1941) Vol. 47, p.53.

Fig. 2.27
An increase in the South American market. Yorkshire Engine Co. Ltd. 1884-built 0-6-0T (YE 382/84) for the British-owned 5ft 6in gauge Buenos Aires Great Southern Railway (No. 6) in Argentina.
Author's Collection.

There was a large increase in locomotive demand, particularly for India, South America and Australia, and more than 100 were also made for railways in France, which its domestic industry could not fulfil at that time.[68] Production increased each year, to reach over a thousand locomotives for the first time in 1885. British manufacturers again faced capacity limitations and quoted longer delivery times and higher prices.

Taking advantage of the strong world demand, and with reduced home demand, American manufacturers expanded their locomotive exports, which increased to 30% of British-made exports in the early 1880s.[69] After 1885, however, the American industry switched production back to a rejuvenated domestic market, exports falling to 15% of the British total.

In 1886, there was another downturn in Britain's economy which depressed the locomotive market, the industry experiencing a 40% drop in production from the previous year, to 600 locomotives. Whilst the home market fell, also by 40%, the fall in overseas demand is more difficult to explain given the rapidly expanding volume of foreign investment. It has however been suggested that there was a short-term drop in the volume of capital exports in 1885/6, a pause in the 'long swing' rise in overseas investments, particularly to the Empire and to Argentina, through to 1890.[70]

A five-year trade recovery began in 1887, stimulated by Britain's expanded foreign investment programme which reached £120 million in 1890, the highest figure in the century. A major beneficiary was South America in which, by 1890, the nominal capital in a hundred railway enterprises was approximately £164 million.[71] A large network of rail routes spread throughout the sub-continent, particularly Argentina, for which the British industry manufactured nearly 350 locomotives in 1890.

Another expanding market was in the Far East, particularly Japan and China, in spite of competition from American and German manufacturers. The Australia and New Zealand market however was reduced by competition from America, and particularly by Australia's own growing locomotive manufacturing industry.

The British domestic main-line and industrial markets also strengthened in the late 1880s, and reached 400 locomotives annually in 1890/91, the highest for a decade. However, in the early 1890s, confidence in foreign securities was shaken by the Barings Bank crisis, the financial panic on Wall Street and other international banking-related problems.[72] Once again, Britain's economy and overseas investments went into a sharp downturn, depressing railway operations and locomotive demand. Production fell to just over 500 in 1894, half that of four years earlier, and the lowest since 1869. The home market fell by 40% from the 1891 figure, the majority being constructed were industrial locomotives. Other markets were also weak, although production for the Indian, South American and Japanese railways held up over the two years to carry the industry through the lean period.

Fig. 2.28 Thriving British industrial locomotive market. Hunslet Engine Co. 1895-built 0-4-0ST (HE 622/95), *LOMAS* for Lighthouse Colliery, Wakefield. *Industrial Locomotive Society 30752.*

The steep drop in British investment overseas was accompanied by more aggressive competition for capital goods, particularly from the United States and Germany, in some of Britain's areas of high investment. Both countries increased their share of overall imports into Argentina, for example, from about 8% in 1885/86 to over 13% in 1901-03, whilst the British share fell from 36.7% to 33%.[73] The American locomotive industry competed aggressively in Latin America, Far East, New Zealand and parts of Australia, and between 1890 and 1892 its export total rose to nearly 40% of the British export figure.[74]

The British economy strengthened from 1895 with the ending of the 'Great Depression', and returned to a boom period by the end of the century, with overseas investment regaining some of its lost ground. Foreign direct investment in South American railways alone exceeded £300 million by the Edwardian decade.[75] Britain's high industrial output generated unprecedented rail activity and, with railway expansion programmes resuming in all world markets, demand for locomotives rose substantially. In 1899, the industry's output exceeded 1,100 for the first time, equally divided between the home and export markets, including 300 industrial locomotives, reflecting the extraordinary growth of Britain's industrial sector.

The railway-owned workshops were also fully occupied, and at the turn of the 20th century their annual gross output, of £7.9 million, was 75% greater than the independent sector, although much of their work was related to maintenance and rebuilding.[76]

With both the independent sector and the railway workshops working to capacity, the industry as a whole was unable to meet the demand, leading to rising prices and lengthening delivery times. Orders were lost to American and European manufacturers. An order for eighteen locomotives in 1896 from Japan, for example, was transferred to a French firm, with a consequent loss of the 10% deposit, when the English manufacturer's delivery time slipped back unacceptably.[77] For the first time since the earliest railway era, three British main-line railways, the Midland, Great Northern and Great Central, actually acquired locomotives from American manufacturers, not forgetting a sole 2-4-2T locomotive built for the Lynton & Barnstaple Railway.[78]

Although the market remained buoyant into the Edwardian era, and the manufacturers continued to retain full order books, the loss of these orders had long-term effects for the British industry, which lost its dominance in the world market from this time. A contemporary American report calculated that the value of British locomotive exports dropped from $9 million in 1890 to $7.3 million in 1900, whilst in the same decade the value of American exports had risen from $1.3 million to $5.6 million.[79] For some countries, such as Japan, American companies manufactured almost as many locomotives as British firms.

The progressive manufacturers were generally successful in their pursuit of strategies to provide sufficient capacity to meet the growth in locomotive demand, and to improve their production facilities to provide the economies of batch production. The extraordinary surge in production at the end of the century was, however, partly created by a backlog arising from the protracted 'lock-out' dispute of 1897, which coincided with the high demand from the world markets. By the end of the century, the progressive firms were able to accommodate the sharp fluctuations in demand by a policy of employing unskilled labour according to demand, which minimised the risk of losing their skilled labour. The craft firms, however, without this high proportion of unskilled labour, were placed in severe difficulties when demand was low, since they could ill afford to employ men for whom they had no work.

The failure of the early export drive by US manufacturers did not deter later initiatives. From the mid-1870s, the American locomotive industry, which experienced a similar volatility in its large domestic market as the British industry, sought to use spare capacity when its domestic

Fig. 2.29 Strong market for locomotives for Argentina. Kitson & Co. 1893-built 4-6-0 (K 3589/93) for the British owned 5ft 6in gauge Ferrocarril Central Argentino (Central Argentine Railway) (No.3351) *Kitson collection, The Stephenson Locomotive Society.*

order books were low, by moving aggressively into exports. The industry, particularly the Baldwin firm that became the world's largest locomotive manufacturer, significantly increased its sales efforts in markets hitherto dominated by the British industry, including Latin America, Far East and New Zealand.[80]

The American locomotive industry had a relatively poor decade in the 1890s and, following a period of co-operation, eight of the smaller manufacturers amalgamated in 1901 to form the American Locomotive Company.[81] The new company claimed 46% of total American production against the 39% of the Baldwin Company. This marked the starting point for major investment programmes by both companies, which in turn prompted the amalgamation in 1903 of the three Glasgow locomotive builders, Neilson Reid & Co., Dübs & Co. and Sharp Stewart & Co. Ltd., to form the North British Locomotive Co. Ltd.[82] The defensive motivation for the merger was confirmed by a director of the new company 'because it appeared to us that we could economise in production and expenses generally'.[83]

Competition also arose from the German locomotive industry which, expanding quickly after the 1871 unification to meet the needs of the rapidly growing domestic railway systems, helped the country's economy to grow initially at an average annual rate of 4.6 percent.[84] The larger manufacturers, such as Borsig, standardised their products to effect scale economies.[85] From the late 1880s, the manufacturers turned to export markets, using liberal credit-loan arrangements to compete with growing domestic locomotive industries in Russia and other eastern European countries, as well as British and American export industries in several other world markets.[86] The German manufacturers formed a cartel, for which the domestic railways made a practice of apportioning orders in accordance with the export trade achieved by individual firms in the previous year.[87] From the 1870s, the Swedish locomotive industry had become sufficiently large to meet most of its main-line requirements, and the Russian industry was, similarly, able to meet most of its domestic needs from the 1890s.[88]

Conclusion

The strategic decisions taken by the locomotive industry in the 19th century were influenced directly by perceptions of its changing markets. Whilst the manufacturers controlled the industry's development in its earliest years, their subsequent decisions on expansion and capital investment, technological and design development, and employment were all conditioned by a market which, after 1850, it no longer controlled. The domestic market, in particular, for which the industry had previously provided all requirements to its own designs, gradually became a residual market consisting of designs undertaken by the railways themselves.

It was the inability to predict the extraordinary fluctuations in demand throughout the century that made strategic decision-making so difficult for the manufacturers. The risk to firms through over-commitment of investment, producing funding problems when demand was low, underlay the manufacturers' caution. It led directly to the industry's divergence between firms dependent upon craft skills, avoiding major capital expenditure, and the progressive firms which were stimulated to develop new capital equipment to reduce craft dependency. Employment policies for these firms were premised on the increasing use of un-skilled labour, which could be laid off and recruited according to the strength of the market.

These diverging policies, and the determination to survive the periods of low demand, led the manufacturers to base their corporate decisions, both of company structure and product diversification, on a combination of long-term growth in the locomotive market and a cautious interpretation of shorter-term demands. The raising of capital, through continued partnership status or through conversion to limited liability companies, was based upon this cautious interpretation. All firms relied on diversification to varying degrees, in their bid to survive through the periods of low locomotive demand. Several diversified away from the industry permanently, whilst others went out of business altogether through failure to provide alternative markets.

Fig. 2.30
High demand from Britain's main line railways. Dübs & Co. 1899-built 4-4-0 (D 3752/99) for the London & South Western Railway (No. 708). *Locomotives and Railways, Vol.1 (1900), p 72.*

Although the industry's dominance in the markets of America and the major European economic centres was lost to their home industries through tariff imposition and the increasing influence of European capital, the strong British capital market developed new locomotive markets through the major foreign direct investment programmes. The London market became the dominant influence for the locomotive industry, not only through direct contacts with the many overseas railways, but also with their consulting engineers. The increasing influence of the London-based engineering offices included their greater involvement in locomotive design, which further reduced the technological and design discretion for the industry.

It was true that in countries, such as the United States, Germany and France, which protected their home industries with tariffs, British portfolio investment was not able to generate benefits for British manufacturing industries. However, the relationship between foreign direct investment and the locomotive industry in other parts of the world, particularly in the Empire and South American countries, illustrates a close liaison between manufacturers and the railway companies, focused on London.

The manufacturers thus based their strategic decisions on their perception of the evolving locomotive market, and with some anticipation of the consequences of its volatility. They sought to reduce the uncertainties of locomotive demand, and the attendant risks to their business, through marketing and selling policies. These policies developed through the century as their major opportunities passed from domestic and European demand to London-based overseas markets.

Notes – Chapter 2

1. Wardley (1991), p 278.
2. Locomotive production figures have been analysed using 'Windows' Excel software adopting the most reliable information that is available. Published works lists have been used, and the author is pleased to acknowledge, with gratitude, the researches of the authors of these works. Other, non-published, lists have been assembled over many years by members of The Stephenson Locomotive Society, and the author is again pleased to acknowledge their research work with gratitude. These lists are lodged in the Society's library.
3. Cottrell (1975), Chapter 3 'The Growth of the portfolio, 1855-1914', pp 27-40.
4. White Jr. (1979), p 21; Also, Brown (1995), p 45.
5. Jenks (1927), pp 73-88; also Cottrell (1975), Chapter 2, 'Beginnings, 1815-1855', pp 17-25.
6. Jenks (1927), p 75.
7. White (1968), p 12.
8. Reed (1971).
9. White (1968), p 27.
10. Periodic listing of foreign railways from 1882 contained in *The Railway Engineer*, Vol.III onwards.
11. Analysis of production, *op cit* (2).
12. Lewin (n.d., but 1925), Chapter V, '1836-The Foundations of the British Railway System', pp 41-48.
13. Jenks (1927), p 84.
14. Cottrell (1975), Chapter I, 'The Volume of Capital Exports, 1815-1914', pp 11-16; also Jenks (1927), Chapter V, 'The Railway Revolution', pp 126-157.
15. Lewin (n.d. but 1925), Chapter IX, '1840-The Opening of Many Lines' &c., pp 77-96.
16. Jenks (1927), p 128.
17. Lewin (1936), *passim*.
18. *ibid*, p 146.
19. *ibid*, p 151.
20. Cottrell (1975), p 24.
21. Analysis of production from the Robert Stephenson & Co. collection, 'Order Books' (ROB/2/1) and 'Engines Delivered Books' (ROB/2/3). Also letter, Edward Pease to E.J. Cooke (for R. Stephenson & Co.), Darlington, 11 mo [November] 29. [18]44, Institution of Mechanical Engineers Library, Crow collection.
22. Copy letter R. Creed (for the London & Birmingham Railway) to R. Stephenson & Co., Pease-Stephenson collection, D/PS/2/64.
23. *Proceedings of the Institution of Civil Engineers*, Vol. XI, 1851/2, London, pp 466/7.
24. Figure is produced from a Windows Excel analysis of opening and closing dates of railway workshops shown in Lowe (1975), *passim*.
25. Article, 'Sketches of German Railways', *The Railway Chronicle*, October 11th 1845.
26. Jenks (1927), p 153.
27. Aldcroft, 'Introduction: British Industry and Foreign Competition, 1875-1914' in Aldcroft (1968), p 22.
28. Broadberry (1997), pp 138-142.
29. Evidence of J.F. Robinson (A Director of the North British Locomotive Co. Ltd.) before the Tariff Commission, reported in 'British Locomotive Building', *The Railway Gazette*, February 26th, 1909, p 281.
30. Walker (1969), pp 69-134 *passim*, also Jenks (1927), p 253.
31. Milward and Saul (1977), p 210.
32. Analysis of locomotive works lists, *op cit* (2).
33. Analyses of production, *op cit* (2).
34. Kirby (1988), p 288.
35. *ibid*, p 291.

36. For example, Bennett (1927), *passim*.
37. Jenks (1927), pp 193-198.
38. Lewin (n.d. but 1925), p 473.
39. For example Lane (1980), pp 146-154.
40. Cottrell (1975), p 23, also Jenks (1927), Chapter VI, 'Cosmopolitan Enterprise', and Chapter VIII, 'From Bill-Broker to Finance Company', pp 158-192 & pp 233-262.
41. Jenks (1927), pp 176/7.
42. Jenks (1927), pp 244-247.
43. Jenks (1927), p 172 and Milward and Saul (1977) p 21; also Stefano Fenoaltea, Chapter 3, 'Italy', in O'Brien (Ed.) (1983), pp 49-120.
44. Jenks (1927), p 192.
45. Jones (1997), p 23; Also, Corley (1994), pp 71-88.
46. Wilkins (1988), pp 259-282.
47. Lowe (1975), *passim*.
48. Jenks (1927), p 195, and Chapter VIII, op cit (43).
49. Burton (1994), *passim*.
50. Jenks (1927), p 231.
51. Jenks (1927), sub-chapter, 'The Failure of Overend Gurney & Co.', pp 259-262.
52. Cottrell, *The Journal of Transport History* (1975), pp 20-38.
53. Article, 'Russian Railways', *Engineering*, Vol. IX, May 13 1870, pp 331/2.
54. *The Railway Engineer*, op cit (17), *passim*.
55. Cottrell (1975), Fig. 1, p 14
56. Edelstein, *ibid*, Table 3.1, p 48; Also van-Helten and Cassis (Eds) (1990), Table 5.4, p 104.
57. Jenks (1927 re-published 1963), p 175.
58. Crown Agents' collection, British Empire and Commonwealth Papers, Bristol City Archives, Ref. 1999/221.
59. Cottrell (1975), p 36.
60. Ahrons (1914), p 40.
61. Jenks (1927), Chapter XI, 'At the End of the Surplus'; also Musson (1959-60), pp 199-228.
62. Collins (1990), pp 198-224.
63. Article, 'The Recent Russian Loan and Railway Business', *The Railway Engineer*, Vol. 5, No. 8, August 1884, pp 205/6.
64. Rippy (1959), p 33.
65. Brown (1995), p 45.
66. Analysis of production, *op cit* (2), compared with editorial, *The Railway Engineer*, Vol. XIV, March 1893, p 66.
67. Cottrell (1975), Chapter 3, 'The Growth of the Portfolio, 1855-1914', p 37.
68. Ahrons (1914).
69. Analysis of production, *op cit*, (2).
70. Cottrell (1975), p 35.
71. Rippy (1959), pp 38/39; also Cottrell, (1975), p 38.
72. Cottrell (1975) p 39.
73. Saul (1960), Table X, p 39.
74. Analysis of production, *op cit*, (2).
75. Corley (1994), pp 71-88.
76. First Census of Production 1907, quoted in Saul (1968), Table 1, p 192.
77. Consular Report No.427, Presented to Parliament June 1897, reported in *The Railway Engineer*, September 1897, Vol. XVIII, pp 276-279.
78. Ahrons (1914), p 307.
79. 'British and American Exports of Locomotives', *Railroad Gazette*, Vol. 33, Feb.1st 1901, quoted in Brown (1995), Note 63, p 265.
80. Brown (1995), p. xxv and p 46.
81. Brown (1995), pp 53/4.
82. Kirby (1991).
83. Evidence of J.F. Robinson, *op cit* (29), p 283.
84. Milward and Saul (1977), p 21.
85. Rainer Fremdling, Chapter 4, 'Germany', in O'Brien (Ed) (1983), pp 121-143.
86. Aldcroft (1968), p 20.
87. Saul (1968), pp 202/3.
88. Ahrons (1914), p 40.

Chapter 3
Marketing and Sales

Introduction

In the 1830s, the rapid growth of the locomotive market and the number of firms diversifying into locomotive manufacture introduced the need for marketing and selling practices as competition between manufacturers intensified. In contrast to the proprietorial or agency contact, which had been the normal selling practice for the heavy manufacturing industry, the proprietors were faced with the tactical decisions of marketing themselves to their potential railway customers in order to be included on the lists of favoured firms which would be asked to tender for orders. The introduction of 'transparent' tendering and contracting was required by the new main-line railways, which were joint stock companies or government-owned networks.

The manufacturers had to learn and adapt existing marketing and selling procedures, based on reputation for product quality, specification and design, as well as acceptable price, delivery terms and payment arrangements. They were required to learn new marketing techniques, aimed at raising and maintaining awareness about their products among railway customers or their representatives, to prompt enquiries of the manufacturer, or allow him to obtain a hearing on product sales. Although marketing and selling to home railways was often undertaken by the proprietors themselves, that to the railways of Europe and further afield was largely undertaken through commission agents, upon whom the manufacturers were dependent.

The competitiveness of the market depended upon the relationship between fluctuating demand and the capacity available to meet that demand. In adapting to market movements, the manufacturers sought to attract orders by offering better delivery quotations through parallel production scheduling, as well as through price and payment arrangements. When the market was strong, prices would rise and delivery dates would be spread to provide the best medium-term production schedule and cash-flow projection. When the market was weak, not only did prices fall and delivery dates become more critical, but the manufacturers made comparison with other machinery markets which could offer better returns on the use of men and equipment. The decisions on complete or partial diversification played an essential role in the prosperity, and even survivability of firms.

Fig. 3.1 Vulcan Foundry Co. Ltd. 1888-built 4-4-2T (VF 1291/88) for the Taff Vale Railway (No. 170). *Peter Wardle collection.*

The manufacturers had to adapt to the changing market through the century. The industry's success in its first twenty years had been due to its competitive marketing and selling, as well as its progressive design and manufacturing developments. Although tendering procedures for the reducing British railways market remained largely the same, they had to accommodate the growing influence of the locomotive superintendents.

The loss of markets in America and the major European economic centres, however, and the rise in the importance of the London market, shifted the marketing emphasis towards direct contacts in the capital with railway and government representatives and consulting engineers. For this purpose, new offices were obtained in London or representative agents appointed.

The loss of much of the manufacturers' technological and design discretion from the 1850s, brought about a change from a production-led to a market-led business. The manufacturers had very limited opportunities to offer technical progress as a marketing incentive, and relied instead on production quality and delivery reliability to keep their places on tender-lists. The economic benefits of their strategic decisions in favour of greater batch production were reflected in tender prices, but the proliferation of designs for both the home and overseas markets reduced opportunities to capitalise on them.

Proprietorial Contact

Marketing, which had begun as a novel concept in the 1830s, was successfully developed by the manufacturers and their agents. The evolution of marketing and selling responsibilities during the century, from predominantly proprietorial control to its delegation to senior managers and commission agents, was a prime example of the adaptability of 'partnership' enterprises. Such was the growth of the market, that partners quickly limited themselves to marketing contacts, usually at senior levels, whilst delegating negotiations and tender preparations to their senior aides. The proprietors' experience in developing contacts with railways, and the factors that led them towards delegation, therefore provide evidence of the evolution of management capabilities in the heavy manufacturing industry.

Marketing and selling of capital goods had been undertaken since the 18th century. Matthew Boulton, for example, prior to his partnership with James Watt, had developed an agency network throughout Europe to generate sales of his hardware articles since the 1770s.[1] Although the agents undertook some 'market research' amongst potential customers, including governments, it paved the way for Boulton and Watt to undertake 'sales' tours for their pumping and factory engines, negotiating price, installation and after-sales commitments.[2]

Prior to 1830, manufacturers, customers and agents were usually sole proprietors or small partnerships, contact between whom often developed close personal relationships. The proprietors of the Butterley Company: 'took a personal hand in selling their products', spending a good deal of time in London,[3] and when William Fairbairn visited Switzerland and France in 1823 he obtained extensive orders for water-wheels, mill-gearing and other manufactured items.[4] With the limitations on proprietors' time, however, representative agents were appointed to sell capital goods. By 1807, a sales agency was acting on the Butterley Company's behalf, and from 1810 Boulton and Watt received most of its overseas orders through London, Liverpool or Bristol merchants.[5]

The manufacturers of capital equipment who first diversified into locomotive manufacture, continued to pursue proprietorial sales contact with their customers. Following its success at the Rainhill Trials of 1829, Robert Stephenson & Co. was invited to manufacture the first locomotive fleets for the Liverpool & Manchester and other early main-line railways. With the Stephenson Company unprotected by patent cover, however, other manufacturers saw immediate market potential. Several offered locomotives speculatively to the railways, the most tenacious being Edward Bury who offered his first locomotive to the Liverpool & Manchester Railway as early as March 1830.[6]

Fig. 3.2 The first batch order for locomotives. Robert Stephenson & Co. 1830-built 2-2-0 (RS 10/30 [2nd series]) for the Liverpool & Manchester Railway (No. 9, *PLANET*). *I. Shaw, 1831, Plate VIII.*

The railway directors acquired some locomotives from other experienced manufacturers on the grounds of on-going trial of a new technology, but George and Robert Stephenson, who had responsibility to their railway clients for design and specification, sought to direct orders to their own Newcastle and Newton-le-Willows factories. Mounting pressure from other manufacturers, particularly Bury, brought about extraordinary episodes on the Liverpool & Manchester, London & Birmingham and Grand Junction Railways, in which the directors divided between those who favoured direct ordering of locomotives from the Stephensons and those who perceived a conflict of interest and sought to introduce an open tendering system.[7]

Acrimonious boardroom battles, which continued until 1836, degenerated into a trial of strength between the Stephensons and Bury and their supporters. Robert Stephenson was distressed by the severity of the personal attacks upon him. He wrote:[8]

> 'Our enemies viz Rathbone and Cropper [London & Birmingham Railway Directors] are raising a hue and cry about our having an Engine to build at Newcastle – they say another article will be brought out by [Dr. Dionysius] Lardner on the subject – They half intimate that I shall withdraw either from the Railway or the Engine Building – The revenge of these people is quite insatiable – This distresses me very much.'

Although the Stephensons lost influence on all three railways, and Bury was contracted by the London & Birmingham Railway to provide its motive power, the resolution of the struggle was the introduction of a formal and open specification, tender and contract system. This marked the beginning of 'transparent' competition for all manufacturers, which had to market themselves to achieve and maintain a place on the list of favoured firms kept by each railway or consulting engineer, to ensure that invitations to tender would be received. Obtaining a contract was then dependent upon the tender price, delivery times and payment details.

By the mid-century, manufacturers and their agents had gained sufficient experience to understand that such effort as was necessary should be directed principally towards maintaining a place on the tender lists of railway companies, government agencies and consulting engineers.

The ending of the Stephensons' domination of locomotive supply resulted in a rapid expansion of proprietorial approaches from the growing number of manufacturers promoting their firms and seeking invitations to submit tenders. As early as 1837, Robert Stephenson involved himself in the selling activities of Robert Stephenson & Co. He requested a briefing from his Head Clerk: 'of the present state of your orders and future prospects of completion – for my guidance in making further negotiations.'[9] Stephenson's high reputation as a consulting engineer, however, meant that he was too preoccupied with those duties to remain fully responsible for marketing and sales. With none of his fellow partners acting in an executive capacity, responsibility for marketing and sales rested with the Head Clerk.

For other firms, however, proprietorial contacts with the larger railways were a dominant feature of the locomotive industry for twenty or thirty years. High-profile proprietors, such as Charles Beyer (of Beyer Peacock), Edward B. Wilson, John Sharp (of Sharp Brothers) and Walter Neilson (Neilson & Co.) could command respectful audiences from senior railway managers, to ensure their firms were included on tender lists. Personal contacts with smaller railways often encouraged direct sales without formal tendering procedures.

Proprietors of the smaller manufacturers, and those seeking to build reputations, also worked hard to gain access to tender lists and to sell directly when the opportunity arose. In their case reputations were bolstered through references. A single proprietor, such as Timothy Hackworth would promote his business through references of existing customers and personal visits to locomotive superintendents.

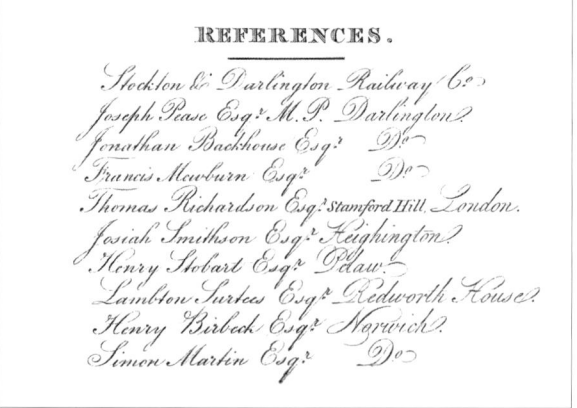

Fig. 3.3 Obverse and reverse side of Timothy Hackworth's business card from c.1840, showing list of references that could be provided. *Author's collection – monochrome, digitally enhanced.*

Fig. 3.4
A Neilson-built locomotive for the Caledonian Railway. 0-6-0 built in 1858 (N 460/58), running number 188.
The Locomotive Magazine, Vol. 47 (1941), p.123.

As late as 1865, John Fowler negotiated directly with William Martley, locomotive engineer of the London Chatham & Dover Railway to obtain an order for his new Leeds factory and secure a footing in the home market.[10]

In the 1850s it was not unusual for twelve or fifteen companies to be invited to tender, but by the 1890s ten companies was more usual.[11] Some regrettable statements were made to gain access to a tender list. In 1846, Arthur Jones of Jones & Potts wrote to his partner: 'Mr. Slaughter [of Slaughter, Grüning & Co.] has been soliciting work from Mr. Stephenson and recommends his engine as very vastly superior to ours... I said I thought Mr. Slaughter might have adopted a different way of obtaining orders than by pulling other people's engines to pieces – this he agreed to.'[12]

In some cases, a close supplier/customer relationship developed where a locomotive superintendent had previously been engaged by a manufacturer. Benjamin Connor, the Works Manager for Neilson & Co.'s Glasgow works, became the Caledonian Railway's Locomotive Superintendent in 1857. Whilst the railway regularly built locomotives at its own works, further new-build requirements were met from the nearby Neilson factory, no doubt influenced by Connor's relationships with the personnel there.[13]

Up to the 1860s, overseas visits by proprietors were important initiatives in developing and maintaining goodwill with customers, including government departments. William Fairbairn made several visits between the 1830s and 1850s, to Sweden, Russia and Turkey, the resulting sales including locomotives for his Manchester factory.[14] With Robert Stephenson's civil engineering work allowing him limited opportunity to visit European railways, he engaged, in 1840, the services of a representative agent, Edward F. Starbuck, to undertake the considerable marketing and selling work required on the continent. Their successful working relationship brought about several follow-up visits by Stephenson to railways on the continent and to Egypt, in which he combined consultancy advice with promotion of his Newcastle factory.

Following Starbuck's death in 1855, he was replaced by Charles Manby, the Executive Secretary of the Institution of Civil Engineers, whose Westminster address was adjacent to the Stephenson office. Manby's seniority and international reputation would, itself, have been a signal to customers of the importance placed on them by the Stephenson Company.[15] Manby, who had the status of a proprietor rather than an agent, subsequently moved his office into the Stephenson property, whilst continuing as Honorary Secretary to the Institution.

Return visits to Newcastle were also encouraged as part of the marketing effort. In 1841, Robert Stephenson & Co. received a visit from the Hamburg Railway engineer, which was 'for the purpose of seeing my new Engine, ... in order that he may be enabled to give an opinion to his Directors as to the propriety of having their Engines constructed on our new plan. – You will please to give him every information that he may require...'[16]

The close relationships with some of its customers developed by Beyer Peacock & Co. arose out of personal visits by Charles Beyer. As early as 1857, in a letter referring to the potential for further orders, G. H. Stieler, representing the Swedish railways, signed off: 'Many compliments to your associates from Your faithful friend.'[17] In 1863, Beyer undertook a continental tour, and met Brouwer von Hagendorp who was responsible for ordering locomotives on behalf of the Dutch railways. He laid the foundations for a successful supplier/customer relationship against increasing competition from the German Borsig company.[18] Von Hagendorp paid a return visit to the Gorton Foundry, from which orders were secured.

The personal connections of locomotive engineers, whose career moves promoted them to positions with manufacturers, consulting engineers and overseas railways, played an influential part with the overseas market. In a letter to their English agents, the Sydney & Goulburn Railway passed on the opinion of its engineer, James Wallace, who had just taken up office on arrival from England, that the first locomotives: 'should be procured from, if possible, either Messrs. R. Stephenson, Messrs. Fairbairn or Messrs. Sharp Roberts and Co.'[19] Similarly, Robert Bennett, formerly Beyer Peacock & Co.'s Works Manager, who was appointed Chief Mechanical Engineer to the New South Wales Government Railways, maintained a close relationship with Beyer Peacock, no doubt aided by his familiarity with personnel and procedures at Gorton.[20]

From the 1860s, with the passing of the first generation of proprietors and the significant increase of locomotive sales, the manufacturers' marketing efforts were developed principally by delegating responsibility for contacts to managers and agents working in the close-knit London market. These British firms vied with each other for inclusion on tender lists without regard to overseas competition. The market became the exclusive preserve of the British manufacturers and was an effective cartel which, until the end of the century, kept out European and American competition. Through their many contacts with the anglophile railway companies and engineers, the manufacturers could maintain their competitive position with little need for promotional material and exhibitions. The experience of the few manufacturers that had taken part in the early exhibitions had persuaded them that promotional marketing gave only limited returns.

Throughout the century proprietors of industrial locomotive firms retained close contacts with their British customers, for whom they made much other equipment, such as colliery machinery. The extraordinary growth of the industrial locomotive market in the second half of the century called for its own, different marketing and selling techniques. It was driven by the demand for inexpensive motive units with common specifications and economies of scale from batch production, with design initiative in the hands of the manufacturer. Industrial locomotives were more akin to the component supply market than the capital goods market, and were sold with demanding price and delivery terms. As the main-line and industrial markets diverged, therefore, manufacturers had to adapt their marketing and selling techniques accordingly.

The British endeavours were in marked contrast to the American industry which was dominated by commission agents until 1872.[21] Not until this practice was swept away by the manufacturers acting in concert, were they able to develop proprietorial representation, backed up by salesmen who travelled across North America on sales tours. The partners themselves then negotiated financial matters with the railway Presidents.

Agents

The use of representative agents by the manufacturers was, until the 1850s, an important feature of the industry's overseas marketing and selling activities which did much to ensure their success. The employment of such agents was not as widespread as for other capital goods, such as textile machinery, for which both home-based and foreign agents exclusively represented the manufacturers in particular geographic areas.[22] Representative agents undertook marketing work, canvassed orders, negotiated contracts, arranged shipments, insurance, payment arrangements and document translations, and represented their principals on other matters such as patent royalty payments. They represented only one locomotive manufacturer, but often represented other firms for other forms of goods, rails for example, which did not conflict with their locomotive interests. Commission payments could be quite high in buoyant years; E. B. Wilson & Co.,

Fig. 3.5
The demanding industrial locomotive market. Hunslet Engine Co. 1876-built 0-6-0ST (HE 171/76), *EMPRESS*, for Caphouse Colliery, Yorkshire. *Industrial Locomotive Society*: 30750.

for example, paid out £11,500 in commissions in 1857.[23]

An early example of the engagement of a representative agent was from 1837 when Edward F. Starbuck represented R.B. Longridge & Co., for whom he obtained some notable European orders. Starbuck was known to the Stephensons who first had contact with him in 1833. In 1840, however, Stephenson reported to Joseph Pease:

> You have probably heard that Longridge and Starbuck no longer carry on business together, the latter intends commencing a commission business in his own account and has applied to me to allow him to act for RS & Co more particularly on the Continent where he has already been instrumental in establishing a connection for Longridge & Co in the Locomotive department.[24]

As a London-based agent, Starbuck knew the locomotive market well, and the right price and contractual terms with which to win orders. In 1844, for example, he wrote: 'I am inclined to think we may be very firm in price & time of delivery, & where any increased dimensions are asked need not dread adding £50 or £75.'[25] He also kept close attention to the activities of competing manufacturers and their agents. He wrote, also in 1844: 'I believe Hawthorn has either gone or is just starting for Silesia,'[26] and later observed: 'I hear Samuelson [Sharp Bros.' agent] has not made himself very popular with many of the Continental R.W. Co.'[27]

The extraordinary success of Robert Stephenson & Co. in the 1840s and 1850s was largely due to Starbuck. He represented the company for all its products, including marine engines and bridges, as well as locomotives, and negotiated many continental contracts and some British ones. He earned respect from Stephenson over a contract for ten locomotives for the South Eastern Railway in 1850, although his commission was greater than the profit at a time of recession and low prices. 'If he had not gained access to the Chairman [of the South Eastern Railway] after a very great deal of trouble, we should have been shut out from tendering, and that at a time when we were almost standing still for want of an order.'[28]

Starbuck was paid 5% commission on all locomotives sold in Europe and re-imbursement of travel and accommodation expenses. There were occasional disagreements over commission entitlements, and in 1846 clarification was necessary on orders obtained from British railways. Stephenson considered that he was: 'Entitled to half Commission on those Home Contracts negotiated through him… but where the orders were negotiated by the Factory, without his intervention… no Commission should be allowed.'[29] This was further clarified the following year when Stephenson agreed his entitlement to the commission: 'on all transactions that may spring from his having obtained the first order or be the continuation of the original introduction.'[30]

Although Stephenson was by far his largest principal, Starbuck also represented about six other organisations, and his annual income was £4,000 in the busy years.[31] He was highly regarded by Stephenson, whom he represented personally on matters relating to royalty payments, Starbuck receiving 15% commission for this work.[32] He had proposed in 1842 that he should become a partner of the Stephenson company, but this was declined by Stephenson's partners. He pre-deceased Stephenson, who left a large sum of money to Starbuck's children in his will.[33]

The development of certain overseas railways, with head offices in their home countries, saw the recruitment of representative agents covering large geographic areas. Some of the larger manufacturers already employed agents to represent their engineering business, and those agents took on additional work as railway schemes got under way. In Australasia, diplomacy as well as commercial flair was a pre-requisite quality, as responsibility for locomotive orders rested with the colonial governments. William Fairbairn & Sons' agent, based in Melbourne, Victoria, by 1855 was described as being 'a gentleman of much enterprise and on the best terms with both the Government and Corporate Authorities'.[34]

From 1878, Beyer Peacock & Co. engaged a representative in Australia with a remit to cover every state, including Tasmania. His replacement in 1894 received an early letter from the company stating: 'If we decide to continue our Australian agency', indicating uncertainty at a particularly poor time for the locomotive industry.[35] The tenacity shown by Robert Fairlie in marketing his patent locomotive design also led him to engage agents, who represented him in Australasia and South America.[36] Agents were supplied with albums of photographs and folios of tracings to assist them in their marketing and selling endeavours.

Fig. 3.6 Example of a Fairlie locomotive built for use in South America. Avonside Engine Co. Ltd. 1872-built 0-6-0+0-6-0 Double Fairlie for the Anglo-Chilean Nitrate and Railway Co. of Chile (No.10). *The Locomotive Magazine*, Vol. 38 (1932), p.86.

Competition from American manufacturers increased from the 1870s, with the Baldwin Locomotive company engaging representative agents in Havana, Rio de Janeiro, Melbourne and Yokohama. The firm's partners also travelled abroad to secure business. In 1885, Baldwin's sales efforts included the circulation of 'broadsides' in Spanish announcing price cuts for both railway and sugar plantation locomotives.[37]

The overseas market was initially the preserve of commission agents, who were extensively used by British industry as the link to its foreign customers.[38] Furthermore, agents were the response to the enormous risks associated with exporting, particularly on matters of credit.[39] The employment of agents also brought about a reliance on intermediaries which prevented industry from developing a close relationship with its customers.[40]

However, the growth of British foreign direct investments in the second half of the century, and the dominance of London in the affairs of many overseas railways, led to the growth of a strong London market for locomotives with the opportunity for direct marketing and selling by the manufacturers. The ways in which the manufacturers responded to these changing requirements, establishing new marketing practices, agency networks, and pricing and selling techniques, are important indicators of their responsiveness to the new order and their capacity to take decisions necessary to maintain or improve market share.

Other agents, representing industrial concerns, railways and even governments, obtained locomotives on the best terms for their principals. These 'commission' agents, based both in Britain and overseas, were engaged to obtain a wide range of railway, mining and other industrial equipment, and would visit manufacturers with 'shopping lists' of requirements. Commission agents were similarly used in the United States in the 1850s, soliciting bids for locomotives for railroads with long credit terms or through partial payment in stocks.[41]

Many European railways were represented by commission agents in the 1830s and 1840s, and manufacturers sought to satisfy themselves about the advisability of doing business with some of them. Edward Starbuck wrote in 1844: 'I entirely agree with you that before we come to close quarters with parties such as Mons. Hirsch or Mons. Vogts a reference is requisite & such I shall certainly require before any operations are entered on.'[42] Occasional difficulties over commissions were encountered where representative agents obtained orders through commission agents, and, in the 1840s, the Stephenson Company sometimes found itself paying two commissions.[43]

Commission agents were adept at ascertaining competing prices from manufacturers. In 1851, William Bird & Co. advised its principal, Hambros, that: 'Stevensons' [sic] price at present in consequence of his having received some orders wd be £2,000 pr Locomotive & tender… but there are other makers of precisely similar Engines at full £400 or £450 less in price.'[44]

The different characteristics of the industrial locomotive market throughout the century meant that proprietors and senior managers were required to negotiate for sales with commission agents representing railways and industrial concerns. They negotiated discounts on list prices, payment, delivery and other contractual terms. Many of the agents were London-based, but others were in Glasgow, Liverpool, Manchester and elsewhere in the country and overseas. Maintaining contact with these agents was an important marketing effort for the industrial locomotive manufacturers.

Some of the London-based agents specialised in certain countries or parts of the world, and were, no doubt, themselves in competition with each other to represent their principals. Some names recur in the manufacturers' order books; Messrs. Fry, Meirs & Co., for example, appear frequently as representing a number of smaller South American railway and industrial concerns from the 1850s. Messrs. Mathieson & Co. represented several Spanish-speaking industrial organisations in the 1880s, before diversifying into the Far East market, where they represented railways in Japan and China. Black Hawthorn & Co. supplied tank engines to about forty agencies, several with repeat orders, in the last thirty years of the century.[45]

Fig. 3.7 Agent's plate (for Julius G. Neville & Co. fixed to the side of a 7ft gauge 0-4-0ST built in 1888 by The Falcon Engine and Car Works Ltd. (Fal.165/88) for the Junta Administrativa das Obras do Porto Artificial de Ponta Delgada (No. 3) on São Miguel in the Azores. The locomotive is one of a pair of British-built locomotives now preserved in Ponta Delgada in the Azores. *Courtesy R.F. Hartley.*

Fig. 3.8 Locomotives for Australia. Kitson & Co. 1888-built 2-4-2T (K 3088/88), *TASMANIA*, for the Victorian Railways.
Kitson collection, The Stephenson Locomotive Society.

The London Locomotive Market

With the locomotive market moving away from America and the several European economic regions in the 1830s and 1840s, and the shift in emphasis towards the London market, overseas agency representation declined in favour of London managers or, for the smaller firms, London agencies. In addition to several domestic railways, the establishment of overseas railway head offices and their consulting engineers in London from the 1850s, arising from the foreign direct investment programmes, focused the manufacturers' attention on that city. Some established offices in the capital to develop and encourage contact with decision-makers in the railway and government organisations, engineering consultancies and agents. This close contact did much to ensure that only British firms were approached to tender for locomotive contracts for such a large railway and industrial market. The dependence on London managers, working closely with their senior workshop colleagues, was a further demonstration of the devolution of decision-making within the industry.

By the 1860s, London had become the world centre of main-line and industrial locomotive tendering and contracting for both the home and overseas markets. Many railway companies operating in Africa, Latin America, India and other Asian countries had their head offices there, as did their consulting engineers, on whose recommendation tender invitations were made. Government organisations, with remits to acquire capital goods, including locomotives, were also based in London. These included the India Office, and, from the 1870s, the Crown Agents and Agents-General for the Australian colonies.

Robert Stephenson & Co. had led the way in demonstrating the advantages of a London office as early as 1840. In addition to Edward Starbuck's London office, Stephenson played an important proprietorial role for the firm in tandem with his Westminster civil engineering business. When part of the Stephenson premises was sold to allow expansion of the adjacent Institution of Civil Engineers in 1868, one-third of the site's asset value was recorded in favour of Robert Stephenson & Co.[46]

Many of the larger manufacturers had a London office, whilst others opted for representative agents. Companies as small as Dick, Kerr & Co. and Thomas McCulloch & Sons, both of Kilmarnock and specialising in industrial locomotives and other industrial equipment, had a London office and a London representative respectively.[47]

Manufacturers' order books show the complexity and diversity of the overseas markets achieved through London, with orders from railway companies and industrial users, obtained both directly and through commission agents, and others from national and provincial government representatives. The latter would be shown as being for: 'The Secretary of State for India', 'The Colonial Government of Cape of Good Hope', or 'The Crown Agents for the Colonies', whilst in 1870 the Agent-General for Victoria procured fourteen locomotives for the colony's North Eastern Railway.[48]

American and European manufacturers did not seek London representation to tap into the potential market until the end of the century, and it is likely that the London market was seen as a cartel for British manufacturers into which it would have been difficult to break. The concentration of their effort was on the markets not represented in London, namely the railways and industrial concerns of North America, Germany, Austria, Belgium and France, and railways in other countries financed by them.

Signs that London's dominance was to change came in 1895. Japan's railways had developed in the 1880s and 1890s, predominantly on the main island of Honshu, with British expertise and financial involvement.[49] Their equipment had been acquired by the Japanese Government through the London market, to the corresponding benefit of the British locomotive manufacturers. However, in 1895, perhaps under pressure from manufacturers in other countries, the Japanese took over direct responsibility for the acquisition of railway materials. A British Consular Report of 1897 stated:[50]

> British manufacturers have hitherto practically had a monopoly in furnishing rails, locomotives, rolling stock &c., but it would require renewed exertions on their part to continue to be the purveyors in this line… Up to the end of 1895, Government requisitions and indents were sent en bloc to London, and the materials required were purchased by Government agents under very rigid inspection and supervision. Since the end of 1895 all requisitions have been issued from the head railway office in Shinbashi, and this will probably continue to be the system adopted.

The prolonged industrial dispute in 1897, and the consequential back-log of locomotive orders in the buoyant economy of the late 1890s, opened the door for American and European manufacturers to break into traditional British markets. Burnham, Williams & Co., the then proprietors of the Baldwin Locomotive Works opened a London office, and was rewarded with orders for several overseas railways and six British railways. The Schenectady Locomotive Works engaged London agents, Messrs. Sanders & Co. to represent them, and they too were rewarded with an order.[51]

Marketing Methods

Although demand patterns of each country were difficult to determine, the long-term characteristics of the locomotive market were more readily perceived by the manufacturing proprietors. Although a significant volume of domestic business remained at the end of the century, the overseas market gave way in the 1850s to a market more dependent upon direct contacts within the London commercial area.

The start-up and growth of the industrial locomotive market from the 1850s gave rise to a third market category. The emphasis of all three categories was for the manufacturers to market themselves to ensure inclusion on tender lists, to provide themselves with the opportunity for quotations on time and price. Such fundamental commercial practices developed through the century, and it was incumbent upon the industry to adapt its marketing strategies accordingly.

Heavy manufacturing industry in the pre-railway era had been usually family-owned or partnership firms requiring only informal, personal and often local contacts for customers in the ship-building, colliery, iron, textile and other industries. When the locomotive market began in the 1830s, however, the new railway customers were joint-stock companies or government-owned railways, for whom 'transparency' was important. The introduction of tendering and contracting was therefore an important practice the industry was required to meet. As the market developed tendering procedures by railways and consulting engineers became more demanding, and the industry therefore implemented promotional marketing in addition to selling.

Promotion through marketing was a new discipline, which had to be learned and improved upon. The methods adopted supplemented the personal contacts that the proprietors, and their senior managers and agents were pursuing. Published and other printed material both informed and ensured a commonality of technical detail put forward by the managerial and agent teams, this being an important requisite in the delegation process. Whilst the success of the personal marketing contacts led to a perception that participation at international exhibitions would have only limited impact, trade fairs, in Britain and overseas, became an important outlet for the promotion of industrial locomotives.

With the introduction of formal tendering and contracting in the 1830s, manufacturers sought to raise or increase awareness of their developing capabilities. The Stephensons responded to their defeat by Edward Bury from 1837 with orders for the London & Birmingham Railway by writing a detailed published description of their *Patentee* locomotive, accompanied by good-quality arrangement drawings. Written by William Prime Marshall (1818-1906), one of Robert Stephenson's engineering team, the 67-page description was published as part of Thomas Tredgold's steam engine treatise published in 1838.[52] Understanding the marketing potential, the work was 'off-printed' as a separate volume, to promote awareness of the Stephenson Company's lead in locomotive development, although there is no evidence that copies were specifically forwarded to railway companies.[53]

This was followed in the 1840s and 1850s by several other books, in which competing manufacturers emulated the Stephenson company by having their own designs featured in them. Templeton's 1841 book, for example, included a description and engravings of a locomotive designed by

Fig. 3.9 Elevation drawing of a 2-2-2 locomotive designed by Kirtley & Co. in 1841.
William Templeton, The Engineer's Common-Place Book of Practical Reference &c, London, 1841 – Digitally enhanced.

Kirtley & Co.[54] A number of the books were re-published in Europe and North America, promoting awareness in overseas markets.

The Stephensons also benefited from the technical assessment of their locomotives by the French engineer, Guyonneau De Pambour, the two editions of whose publication were translated into English.[55] R. & W. Hawthorn responded by promoting its own 71-page publication on experiments carried out on two of its locomotives on the Newcastle & Carlisle Railway.[56] By 1850, books provided examples of best practice from several companies. Tredgold's work of that year included examples of locomotives built by Sharp Bros., Bury, Curtis & Kennedy, R. & W. Hawthorn and R. Stephenson & Co., and another of Thomas Crampton's patent design.

In 1862 Neilson & Co. produced a booklet illustrating its new locomotive works in Springburn, Glasgow, with photographs of the works and its machine tools to broadcast its move and its new manufacturing opportunities.[57]

Fig. 3.10 Neilson & Co. promotional booklet of their new Springburn workshops, 1862.   *Author's collection.*

It is likely that manufacturers of main-line locomotives were initially reluctant to produce catalogues of their products before the 1870s due to the proliferation of main-line locomotive designs required by railway companies and consulting engineers which prevented easy classification in a product catalogue. The tendering system generally did away with such a need anyway. In 1886 however Sharp Stewart & Co. Ltd. offered a standard range of locomotives for the export market and produced two catalogues of examples of their locomotive products, one for 'Broad Gauge Railways' and the other for 'Narrow Gauge Railways'.[58]

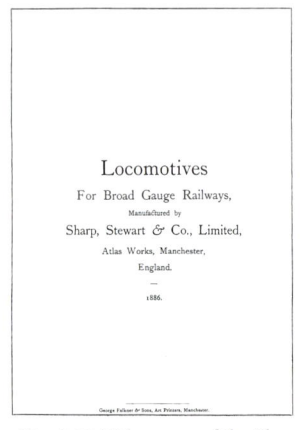

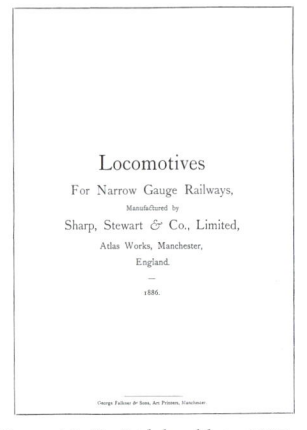

Fig. 3.11 Title pages of the Sharp Stewart & Co. Ltd. booklets, 1886. *Author's collection.*

The British position contrasted with the greater dependency on more regularised designs in the United States. The Baldwin Locomotive Works began issuing illustrated catalogues in 1872, providing: 'a series of descriptions of the various classes of engines now made, each type being illustrated by a photograph and tables being given of the principal dimensions of the various classes (forty in number) and of the duties which they were capable of performing'. The book was: 'got up in a style very unusual in works of this kind, and in excellent taste.'[59]

Industrial locomotive manufacturers, however, with their range of standard designs, began to produce catalogues of their products from the 1870s, an early example was a seventeen-page 'Illustrated Description of Some of the Tank Locomotives etc.', constructed by Fletcher, Jennings & Co., published in London on the company's behalf about 1870.[60] Another early example was published by Fox Walker & Co. of Bristol, also in the 1870s.

In 1892, the Atlas Engine Works, now with a change of ownership as Peckett & Sons produced: 'an elegant little catalogue' containing: 'illustrations and short descriptions of 25 different types of main line and tank engines'. Pecketts were said to keep several locomotives: 'in stock or progress', the catalogue thus being: 'useful for customers requiring locomotives at short notice'.[61]

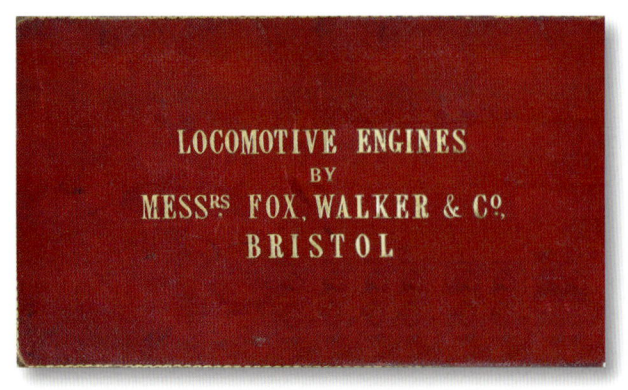

Fig. 3.12 Fox Walker & Co. product catalogue. *Author.*

Component manufacturers, however, regularly produced illustrated catalogues from at least the 1850s. The Salter Company, for example, which made industrial spring balances, including those for locomotive safety valves, was issuing illustrated catalogues by 1870.[62] William Fairbairn & Sons and Nasmyths Gaskell & Co. both produced catalogues of their standard machined castings and machine tools, but not of locomotives or other steam engines.[63]

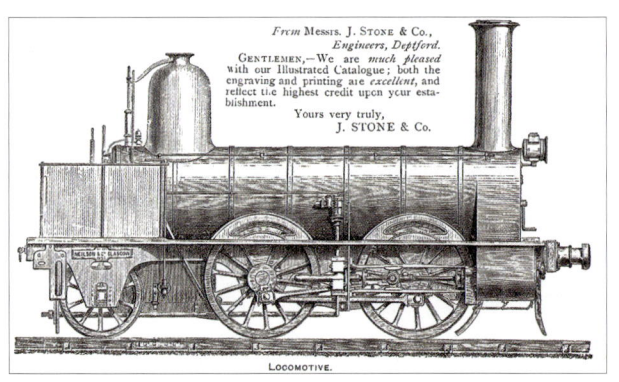

Fig. 3.13 Advertisement placed by J. Stone & Co. of Dartford showing a Neilson & Co. 0-4-2 locomotive built in that year, fitted with a cross-head water feed-pump probably made by the Stone company. *Illustrated London News, 1855.*

Photographs were used as a marketing aide by several manufacturers. Beyer Peacock & Co. began taking photographs in 1856, from when most types were photographed on completion at Gorton Works.[64] The company and its agents maintained albums of prints to show to prospective customers.[65] Baldwins had begun circulating lithographs of its engines from the early 1850s, but changed to photographs after 1860.[66]

In the second half of the 19th century it was common practice for British manufacturers to produce 'data' cards with photographic prints of locomotives accompanied by their basic dimensions on the reverse. These were useful to show to potential customers examples of locomotives that might be similar to ones that they required.

Fig. 3.14 Photographed for publicity purposes. Beyer Peacock-built 2-2-2 locomotive (thought to be BP 105/58) for the Dublin & Drogheda Railway, photographed in the manufacturer's works area. *Beyer Peacock collection.*

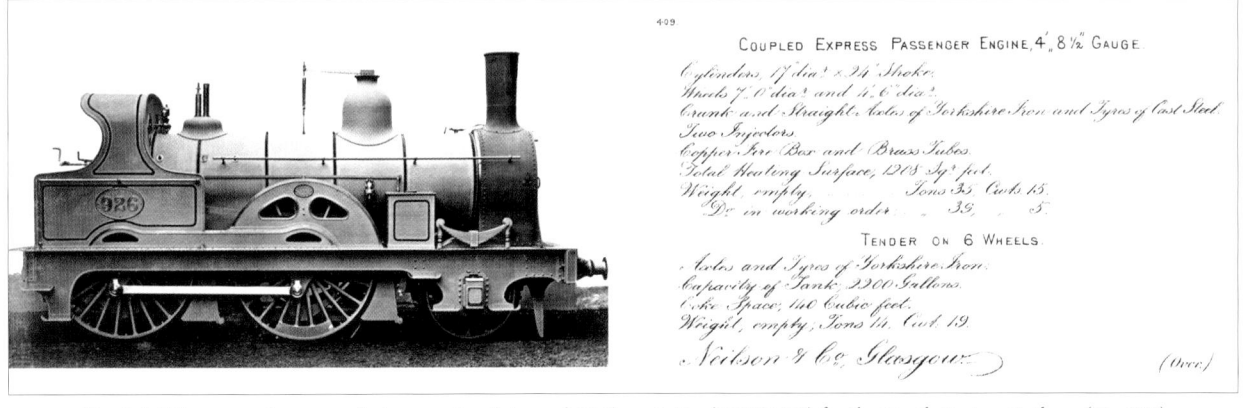

Fig. 3.15 Obverse and reverse of a locomotive data card. Neilson & Co. (N 1801/73) for the North Eastern Railway (No. 926). *Author's collection.*

Main-line locomotive manufacturers did not advertise, either in Britain or overseas. This again contrasted with the approach of the American industry, whose firms regularly advertised in the railroad trade press and other specialist periodicals. The Baldwin Company, at least from the 1880s, was also advertising in foreign railway publications, and subscribed to newspaper clipping services in London, Paris and St. Petersburg.[67]

Industrial locomotive manufacturers, however, regularly advertised in trade periodicals, this being an important part of their marketing efforts for their whole range of products. Advertisements appeared both in general periodicals, such as *Engineering*, and in specialist periodicals such as *Iron* which catered for the iron and steel industry. Advertising probably began in the 1860s, and by the 1880s, manufacturers such as Dick, Kerr & Co., Grant Ritchie & Co. and Barr Morrison & Co. were regularly advertising their locomotive and other general engineering products.[68]

Fig. 3.16 Advertisement for Barr, Morrison, & Co. tank locomotives in the journal, *Iron*, in 1883.

Fig. 3.17 Robert Stephenson Co.'s new 'Fitting Shop'. *Illustrated London News*, 15th October 1864, pp.392-394.

Published articles about locomotive factories may also have been regarded as a form of publicity, although how much marketing benefit was derived is questionable. Robert Stephenson & Co. of Newcastle upon Tyne played host to *Illustrated London News* after completion of its new machine and fitting shops in 1864, which were described and elaborately illustrated.[69]

In 1887, *The Railway Engineer* carried an illustrated article about Peckett & Sons' Atlas Works in Bristol. It may well have been company inspired from some of the phraseology used, including a welcome for: 'any visitors, be they buyers or not…'[70]

Although trade exhibitions developed in the late 18th and early 19th centuries, the first major opportunity for locomotive manufacturers to market their products was London's Great Exhibition of 1851. This was a turning point for two reasons. It was the first exhibition to promote international trade as, hitherto, although free trade had been seen as an economic ideal, manufacturing countries had been nervous to expose their home industries to foreign competition. Secondly, previous exhibitions had focused on technological advance rather than marketing. Although British industry considered that it led the world, it recognised the growth of industry in Europe and North America and, by inviting nations to take part in the friendly competition, sought to out-sell it.[71]

The Commissioners of the Great Exhibition had little or no experience of promotional events, and for a time there was a lack of response to their invitations. Robert Stephenson, one of the Commissioners, wrote at the beginning of 1851:

> I promised Col Reid. when there was a probability of there being a lack of exhibitors to send an Engine or two and I had in my mind the notion of sending the old Engine with what the Stockton & Darlington was opened – the Rocket and one of our last improvements… Being a Commissioner I did not after some reflection think it right to force any thing upon the commission of substance…[72]

Stephenson's words reflected little understanding of marketing benefits that could be derived from the Exhibition and, in the event, he did not send any locomotive, old or new. Indeed, only five manufacturers exhibited locomotives in their own name, namely: R.&W. Hawthorn, Kitson Thompson & Hewitson, E.B. Wilson & Co., William Fairbairn & Sons and George England & Co. Only Hawthorn appears to have seen the marketing potential, its: 'first-class patent passenger locomotive', with 'several novelties', being accompanied by a twelve-page descriptive booklet.[73] The other manufacturers exhibited tank locomotives, none representing outstanding examples of design.[74]

Thomas Crampton, however, renowned for his patent locomotive designs, exhibited one of his express locomotives, *FOLKSTONE* [sic], built by Robert Stephenson & Co. for the South Eastern Railway. He saw opportunity to promote his designs amongst interested railway and locomotive manufacturers, and the Stephenson Company may have indirectly benefited from his initiative. With different motivation, two railways exhibited locomotives, including the London & North Western Railway with another of Crampton's patent locomotives, *LIVERPOOL*. This had been built by Bury Curtis & Kennedy, which, however, went into liquidation during the year.

Fig. 3.18
Exhibited at the Great Exhibition, London, 1851 by Thomas Crampton. Robert Stephenson & Co. 1851-built 4-2-0 (RS 787/51) for the South Eastern Railway (No. 136, *FOLKSTONE*). *Author's Collection: Original Photograph by H. Fox Talbot.*

The Great Exhibition sparked considerable international interest, and in the second half of the century the numbers, frequency and scale of international exhibitions grew significantly. In 1855 a bigger event was held in Paris to begin a cycle of exhibitions in Paris, London, Vienna and America.[75] Although the railway audience, being representatives of administrations from around the world, was global rather than European, the British locomotive manufacturers participated in only a limited way. Some major firms, such as Beyer Peacock & Co. and the Vulcan Foundry Co. Ltd., never exhibited locomotives at any of the international exhibitions.

Foreign manufacturers, on the other hand, particularly those in France, Belgium, Germany and Austria, were constantly striving to develop their European markets and, for them, the exhibitions provided an important marketing medium. Two locomotives from Belgium and one from France were shown at the Great Exhibition in 1851, including one from John Cockerill & Co. which put on a major exhibition and was awarded a 'Council Medal'.[76]

For the continental manufacturers, the exhibitions circumvented the London market, as opportunities developed in the Far East and South America. At the 1867 *Exposition Universelle* in Paris, the Würtemberg manufacturer, Kessler, exhibited one of twenty locomotives it had built for the East India Railway.[77] This rare example of an Indian order not being undertaken by a British manufacturer sent a clear message to the British industry. Indeed, it was noted at the time, in referring to both the Indian locomotive and another built by Schneider & Co. for the Great Eastern Railway in Britain, that: 'These two engines afford incontrovertible proof of the possibility of getting English designs carried out abroad quite as well as at home, and at a cheaper rate…'[78]

The British railway supply industry, as a whole, had only a minimal involvement with the exhibitions, which, in 1867 in Paris, was cause for official comment:[79]

> In the English section, although in individual cases the 'exhibits' are unsurpassed, if not unequalled, the Exhibition, as a whole, does not come up to the standard of what might have been expected, either in numbers or in importance of the objects exhibited. In it there are neither goods engines, railway carriages, vans nor goods trucks; nor, with the exception of some steel springs, are there any specimens of the various locomotive and carriage fittings which are exhibited in such numbers by other nations.

By the Vienna Exhibition of 1873, the lack of participation by British manufacturers was the subject of critical public comment:[80]

Fig. 3.19 A rare British exhibit at the Exposition Universelle in Paris. Kitson & Co. 1867-built 2-4-0 (K 1423/67). *Kitson collection, The Stephenson Locomotive Society.*

It is a matter for regret that in such a collection of locomotives as that now to be found at the Vienna Exhibition, the makers of Great Britain, France and the United States are not fairly represented… Germany and Austria are both large exhibitors, and Belgium is fairly represented; but we miss the names of those makers who have gained England its reputation for locomotive engineering…

Later in the year, *Engineering* was moved to issue a warning to the manufacturers:[81]

…The Exhibition at Vienna, more strikingly than any other, has shown the British manufacturer how great the producing, and how much greater are the imitating powers of the leading Continental makers, and it has taught him the salutary lesson that maintaining the lead is no longer the comparatively easy matter that it was, even at the period of the Paris Exhibition in 1867. Never before have the Continental nations put out their strength as at this Exhibition, and never has so grand an occasion arisen for the study and consideration of the actual position England holds among manufacturing countries. It has shown us that, harder pressed in the great race than ever before, she still, in the main, holds her own, and must continue to do so…

The manufacturers themselves would perhaps have regarded these comments as naive, disregarding both the commercial realities of the European market and the significant deviations in scale and design of the continental locomotives.

By 1900, the international exhibitions had become very large affairs dominated by continental interests. Some 68 locomotives were shown in Paris that year, of which just five were from Britain. Four of those were railway designed and built, the attendant publicity benefiting the railways as carriers, but with no opportunities for generating manufacturing interest. As at the previous Paris Exhibition of 1889, Neilson Reid & Co. was the only British company to attend, exhibiting an express locomotive for Holland.[82]

The marketing policies of the British main-line locomotive manufacturers were, therefore, quite different from those of their continental competitors and from the industrial market. Their limited reliance on printed material and exhibition promotion reflected their dependence on personal contacts within the London market, in which their primary objective was inclusion on tender lists.

As overseas competition developed in the 1880s and 1890s the British industry was shown to lack promotional marketing experience, in contrast to the continental and American industries. As manufacturers sought to pursue new customers in the developing open market, notably in Latin America and the Far East, through financial centres other than London, they were faced with a need to develop new marketing strategies.

Manufacturers were faced with direct competition, from American and German manufacturers in particular, as the influence of the London market, outside of its Empire interests, began to decline. Their lack of experience of exhibitions, trade fairs and advertisements thus put them at a disadvantage. Their lack of promotional material, a consequence of the proliferation of main-line locomotive designs, was also a problem. Experience had told them that, unlike consumer products, their efforts would have no effect on the volume of demand. This point was made about the American locomotive industry, noting that Baldwins sent no fewer than sixteen locomotives to the 1893 Chicago Exposition, but that the marketing effort 'came to naught.'[83]

Industrial locomotive firms, that manufactured mining, colliery and other industrial equipment, and other types of steam engines, regularly exhibited at trade exhibitions. They initially attended the international exhibitions, but gained most benefit from the growing number of trade fairs in Britain and overseas from the 1870s.

Their first endeavours were at the London and Paris International Exhibitions in 1862 and 1867. The long-established Lilleshall engineering company exhibited its first locomotive in London and, after building only a few industrial locomotives, exhibited an 'express passenger' locomotive at the Paris Exhibition, for which it was awarded a silver medal. It derived no benefit, however, and re-built it for sale as a colliery tank engine.[84] Henry Hughes & Co., and Ruston Proctor & Co. both received 'honourable mentions' for their tank engines at the 1867 Paris Exhibition.[85] Hughes also exhibited a tank engine at the Vienna Exhibition of 1873, as did Fox Walker & Co., the only British manufacturers present.

An early trade fair was the first Russian industrial exhibition, held in Moscow in 1872. Nine locomotives were exhibited, of which five were of Russian manufacture, telling a sceptical world that Russian manufacturing had come of age. One locomotive was from Germany ('without doubt the best in the Exhibition'), and two from Britain, a narrow-gauge Fairlie locomotive built by Sharp Stewart & Co., already in service in Russia, and a three-year old crane locomotive built by Dübs & Co., also in Russian service. The ninth was a twenty-five year-old locomotive exhibited for historical purposes.[86]

The number of trade fairs, both in Britain and overseas, increased considerably from the 1880s. In 1883, two manufacturers, Fowler & Co. and Dick, Kerr

Fig. 3.20 The growing industrial sector in the late 19th century. Avonside Engine Co. 1898-built 0-4-0ST (AE 1387/98). *Author's collection.*

& Co., exhibited at the Engineering and Metal Trade's Exhibition in London.[87] The latter company also exhibited a locomotive at the Calcutta International Exhibition of 1883-84, where it gained five medals for the various railway items it exhibited.[88] R. & W. Hawthorn Leslie & Co. Ltd. exhibited one of its crane tanks at the Adelaide Exhibition in South Australia in 1887.

In contrast to the main-line market, the industrial locomotive sector successfully developed marketing and selling expertise in the British and London international market. The specialisation of several firms in industrial manufacture reflected the different proprietorial approach that was required. The industry was more akin to the component supply industry and marketed itself accordingly. Although perceiving little benefit from the international exhibitions, following its minimal involvement in Paris (1867) and Vienna (1873), the manufacturers did perceive opportunities from the specifically targeted trade fairs around the world. The very competitive nature of the market, and its close relationship with the supply of other industrial capital goods, led to vigorous marketing, both through trade catalogues and trade fairs, targeted at customers and commission agents alike.

Through their close association with their railway and industrial customers, the manufacturers were fully conversant with their developing expectations for improved locomotive performance and economy. The gradual withdrawal of technological and design discretion as Britain's larger railways and consulting engineers developed their own capabilities during the century was to convert the industry from a manufacturing-led to a largely customer-led business.

Sales Practices
Selling locomotives called for tactical decision-making in accordance with the prevailing market intelligence, raw material prices and anticipated production programmes. With the rapid growth in the locomotive market, the immediacy of the tendering process required increasing delegation of responsibilities for pricing, delivery quotation and methods of payment. Guidelines were laid down by the proprietors and agreed with senior managers and representative agents who became adept at reading the market.

After 1836/7, the tender/contract system for locomotive acquisition was widely used by Britain's main-line railways, although some smaller railways and early overseas lines continued to obtain their requirements without wide comparisons of price or availability. The larger, London-based overseas railways would also issue invitations, either directly, or through their consulting engineers, or through one of the government agencies. Other invitations from overseas railways were obtained by representative agents.

Some of the largest British main-line railways offered full 'transparency' by advertising in journals, such as *The Railway Times*, providing the opportunity for any manufacturer to tender.[89] It is likely, however, that tenders were only seriously considered from firms with whom they regularly contracted. Over time, railway companies responded to changes in the manufacturing sector by dropping under-performing firms and including firms which had impressed with locomotives for other railways, or which had been brought to their attention by their marketing efforts.

As tender invitations became more detailed, and as railway companies became more demanding with their design and material requirements, manufacturers were invited to see arrangement drawings and specifications at the railway offices. The invitations were accompanied by a deadline for receipt of tender. An early example was in 1839 when I.K. Brunel, Engineer to the Great Western Railway, wrote to manufacturers:[90]

> I am instructed to inform you that the Directors of the Great Western Railway are desirous of receiving offers for the immediate supply of a certain number of locomotive Engines and Tenders to be made according to drawings and specifications which are prepared for the inspection of yourself and of the other manufacturers to whom copies of this circular have been addressed… I enclose a printed copy of the specification to enable you to apply to the above, but the drawings can be seen at the Engineer's office at the Company's station at Paddington and copies will be furnished if your tender is accepted.

Some invitations to tender contained considerable detail and required much time and thought by the manufacturers. In 1844 Robert Stephenson & Co. received: '…what you most appropriately term the little Volume from the Rhenish R.W. Co. it is being translated… but even to a good Translator these documents require much care & time.'[91]

Some railways were unable to anticipate their locomotive requirements in good time and, finding themselves with an urgent need, appealed to manufacturers for early quotations and deliveries without going to tender. Often, this would be achieved by substituting other customers' locomotives. Charles Beyer, of Beyer Peacock & Co., wrote in 1857: 'They want five passenger Engines at Warsaw immediately and I offered them by telegraph the 5 of Talabot's [for the Lombardo Venetian Railway] we have standing here in the yard.'[92] In 1863, he received an urgent enquiry from the Inverness & Aberdeen Junction Railway: 'If the Inverness Rw. wants Goods Engines, they can have two first class one's [sic] at once. The two Egyptian engines are nearly finished and we can replace them in time for the Pasha.'[93]

Fig. 3.21 Vulnerable to short-term re-direction to another customer. Beyer Peacock & Co. 1857-built 2-2-2 (BP 56/57) for the Societé IR Privilégiée des Chemins de Fer Lombards-Vénitiens et de l'Italie Centrale of the Kingdom of Lombardy-Venetia.
The Locomotive Magazine, Vol.15 (1909) p.175.

Some smaller railways speeded up the tendering process by requesting prices and delivery times for small numbers of locomotives of existing designs. In 1857, Charles Beyer wrote: 'I have had a visit today from Mr. Needham Engineer of the Dundalk and Enniskillen Rw. and have to give them a tender for 2 Engines and Tenders… by tomorrow afternoon. They have [also] asked Fairbairns and Sharp…'[94]

Invited manufacturers submitted tenders which included specifications, indicating materials and relevant design characteristics, general arrangement tracings, numbers of locomotives offered towards the desired fleet requirement, estimated delivery dates and price per locomotive. Some railways provided standard tender forms to allow direct comparison between manufacturers. In other cases manufacturers had their own standard tender forms, which benefited them when frequent tenders were being prepared. The complex mix of tenders was considered by the Boards of directors, or their nominated committees, with the advice of their engineering superintendents or consulting firms. Negotiations over points of detail often preceded contract signing.

The start-up of locomotive manufacturing on the continent led to drastic moves to catch up on Britain's fast-moving technological progress. John Cockerill Company of Liege in Belgium resorted to copying British-built locomotives in order to understand how they were constructed.[95] Having learned how to make locomotives, the Cockerill Company made them for both Belgian use and abroad where they then competed with British manufacturers. Future imports to Belgium had a tariff imposed equal to some six to eight per cent.

A less formal tendering system was applied on some of the first continental railways in the 1830s and 1840s. Manufacturers submitted tenders, without formal

Fig. 3.22
Example of Robert Stephenson & Co.'s 'Long Boiler' design. 1844-built 0-6-0 (RS 470/44) for the Great North of England Railway (North Eastern Railway (No. 37).
National Railway Museum, R.E. Bleasdale collection.

invitation, when they learned about procurement intention. Prices, delivery and payment terms were negotiated by agents, and sales could be encouraged by reduced prices at times of low demand. Railways which developed close relations with manufacturers became adept at negotiating beneficial terms. Edward Starbuck, the Stephenson Company's agent, played a central role in such negotiations, and built up a significant expertise in selling to the continental railways, either directly or through commission agents. In August 1844, he wrote to his principal:[96]

> … you may have heard direct from the Cologne Minden Co. regarding 2 Engines they purpose taking, 2 of you – 2 of Sharp & Co – 2 of Borsig, Berlin – 2 of [Cockerill] Seraing – to test the qualities of each maker! Now on their writing I recommend your allowing me to reply… for I am quite au fait at all the maneuvres [sic] of this Company.

Some tenders were made directly through proprietorial contact and without competitive bids. In September 1846, for example, Gilkes, Wilson & Co., which was then seeking to re-start locomotive manufacturing encouraged by the buoyant market, wrote to the Stockton & Darlington Railway:[97]

> Referring to thy conversation with our E. Gilkes… respecting our building a Locomotive Engine for the Stockton & Darlington Railway Comp., we beg to say that in accordance with thy proposal we are prepared to commence with an Engine similar to Robt. Stephenson & Co's patent Engine on the following terms That when the Engine is completed she is to be taken by the Ry. Co at the current price, should they so incline, that should they not incline to take her we shall be then at liberty to sell her to any other party…

New characteristics or component features occasionally prompted letters of reference to assist with a tender. In 1867, Beyer Peacock & Co. obtained a detailed letter from the Norwegian Government Railway regarding the capabilities and performances of the narrow-gauge locomotives on their Drammen line, which they forwarded with their tender for similar locomotives for Adelaide, South Australia.[98]

A successful patent design was a strong selling point for a manufacturer. The Stephenson Company promoted its 'long-boiler' patent design in the 1840s: 'We enclose you Extract from the York & North Midland R.W. Co. on the working of Engines similar to the one now required by you, which as you are probably aware, is our patent plan & cannot be supplied by any other maker.'[99]

Manufacturers were usually willing to pursue sales opportunities, from wherever they arose. In 1863, for example, Charles Beyer wrote:[100]

> Yesterday I had a private letter from Mr. Wilson W[est] M[idland] Rw., saying he thought he could sell for us the 2 Llangollen Tank Engines if we had them still on hand. I wrote him we could make two in 3½ months and price was £2,050 and if he could sell them for us we shall be glad to pay him a commission for his trouble…

Although the international exhibitions were largely marketing opportunities, an unexpected exception occurred at the 1862 London Exhibition, where Neilson & Co. had exhibited a large passenger locomotive for the Caledonian Railway designed by their former Works Manager, Benjamin Connor. The Viceroy of Egypt, who visited the exhibition, was so taken by: 'Its striking appearance with its magnificent wheel' that he bought it and had it shipped to Alexandria. After successful trials, two more examples were ordered, a rare example of a British railway-designed locomotive export.[101]

56

Fig. 3.23
Acquisition at an exhibition. Neilson & Co. 1862-built 2-2-2 (N 850/62), designed by Benjamin Connor for the Caledonian Railway, exhibited at the 1862 London Exhibition, and later purchased and delivered to the Egyptian Railways. *Author's collection.*

Occasionally, manufacturers had locomotives on hand resulting from 'frustrated' orders when customers withdrew from contracts. They were offered to other potential customers, usually at a discounted rate. In 1847, for example, Robert Stephenson & Co. offered to the York, Newcastle & Berwick Railway: '3 Goods Engines & Tenders… which were ordered by a Railway Company who have requested us to substitute for them, engines of a different construction.'[102]

Some locomotives could be difficult to re-sell. In 1866, John Fowler & Co. had four locomotives on its hands when the Irish Midland Railway could not raise the purchase money. Built to the 5ft 3in Irish track gauge, they sought to offer them cheaply to another Irish line, and eventually sold them to the Waterford & Kilkenny Railway.[103]

The poor reliability of certain components up to the 1850s was acknowledged by manufacturers and customers alike. It was normal practice for tenders for batches of locomotives, particularly export orders, to include provision for 'duplicate' components such as cylinders and crank-axle wheel-sets. In the event of failure, they could be replaced, avoiding the locomotive being out of use for a long period. They were separately priced and generated significant additional revenue for the manufacturers. Improvements in material technology, and in the railways' own repair-facilities in the 1840s and 1850s, gradually reduced the requirement for duplicates. By the end of 1861, only one order received by Robert Stephenson & Co., required duplicates. This was for an Italian railway which had inadequate facilities for major repairs.[104]

Pricing
Most manufacturers kept detailed records of the direct costs of locomotive production. These formed the basis of their 'list prices' which were generally determined on the long-established cost plus percentage basis. In the 18th century, Boulton and Watt had calculated its steam engine prices by keeping close records of manufacturing costs and adding on approximately 30% profit margin for home orders and 100% for overseas orders.[105]

As early as 1834, Robert Stephenson & Co., having costed in 'Trade Expences' as a common 15% of manufacturing costs to accommodate its overheads, then added a standard 25% charge to determine its target price. The latter charge accommodated agency fees, where applicable, and the firm's profit.[106] In 1878, Beyer Peacock & Co. employed a similar method, applying a common 20% 'profit' margin to its cost figures to determine its list prices. This probably included an allowance for agency fees as, in 1863, the company had applied a 10% profit margin in a direct quotation to the Great Eastern Railway.[107]

List prices were based on standard fittings and recommended materials, but all manufacturers would vary quotes for specification changes. In the 1830s, Robert Stephenson & Co. and Charles Tayleur & Co. recommended copper fireboxes, which had longer operating lives than wrought iron ones, but had a higher first cost (£100-£120 compared to £20-£30). To compete with other manufacturers, they were obliged to quote also for wrought iron.[108] In 1878, Beyer Peacock's policy was that 'Slight deviations or modifications' to its specifications incurred no extra charge, whilst significant changes, such as 'crucible cast steel' crank axles instead of 'best selected Yorkshire scrap iron' should be charged for at cost price.[109]

Changes in raw material costs were promptly reflected in list price changes and, subject to the competitive situation, passed on to the customer. In its 1878 policy statement, Beyer Peacock & Co. instructed its Australian agent that variations in list prices due to raw material costs would be made in four stages of 2½%, to a maximum of 10%. The changes were communicated from Gorton by a simple telegraphic message; 'five up', for example, requiring a 5% increase on list price.

Whilst most of the communications between the proprietors, agents and factory managers were by correspondence, the urgency of many of the sales matters led to early use of telegraphic communication. The earliest recorded use of the telegraph was by Robert Stephenson &

Fig. 3.24 Batch order for India. Kitson & Co. 1879-built 4-4-0 (K 2259/79)for the Great Indian Peninsular Railway, one of a batch of ten locomotives. *Kitson collection, The Stephenson Locomotive Society.*

Co. in 1853, but most, and probably all manufacturers used the telegraph by the end of the century.[110]

Negotiated prices depended upon current market conditions. Manufacturers increased prices at times of high demand, but reduced them with low demand to stimulate orders. In 1854, for example, Thomas Fairbairn, on behalf of William Fairbairn & Sons, sought to increase an order for four locomotives for the Lancashire & Yorkshire Railway to as many as fifteen or twenty as: 'prices will rise during the next couple of years.'[111] In 1878, reductions to Beyer Peacock's list prices were permitted at the discretion of its Australian agent when negotiating for orders and in 'severe competition' with other manufacturers. These were, again, to be in four equal stages to a maximum of 10%. If a commission agent was involved in the sale, then a consequential increase on the list price would be imposed. Manufacturers of main-line locomotives closely guarded their list prices from competitors. Beyer Peacock & Co. instructed its agent: 'In no case must you show or give copies of the prices list we have sent… but keep it strictly private…'[112]

Where a close proprietorial relationship had developed, a gentlemanly negotiation over price would take place. Charles Beyer, for example, who became friendly with Carl Pihl of the Norwegian Government Railway, sought to justify a major increase in price for comparable locomotives between £1,200 quoted in 1858 and £1,400 in 1865, explaining: 'At that time we were in the wrong and would have lost money had we obtained your order and the price we now ask I expect will yield no more than an ordinary trade's profit.'[113]

List prices were reduced for large batch orders, for which progressive manufacturers well understood the economy of scale benefits through acquisition of materials and sub-components, and from manufacturing economy. Beyer Peacock & Co. encouraged orders of more than six locomotives by 'small' reductions on list prices. However, not until the 1870s did the larger railways, particularly in India, regularly take advantage of large batch production, making significant list price reductions possible.

There were substantial risks and benefits for manufacturers when arranging payment terms for their locomotive business. Adequacy of working capital was essential to avoid unnecessary interest payments or, at worst, insolvency. The proprietors, senior managers and agents developed an expertise with which to interpret the locomotive and financial markets, in order to take the necessary tactical decisions on payment terms to maximise their liquidity opportunities.

In the earliest years of locomotive contracts, payment was made by Bills of Exchange for which credit fell due after a stated time. This was usually three months, but could be fixed at any time from a week to twelve months and reflected the strengths and weaknesses of both the locomotive and money markets. Contracts usually specified that half the price should be paid when the locomotive was delivered and the balance on satisfactory conclusion of the proving mileage. For large locomotive orders, multiple payments would be made, phased during the delivery programme.

At times of peak demand when delivery times were extending, manufacturers not only increased prices but also required an advance payment to secure delivery within an acceptable time. In 1838, Robert Stephenson &

Co. resolved: '… that with all future contracts an advance of one third be stipulated.'[114] In 1844 the company wrote to the Sächsisch-Baierische Eisenbahn-Compagnie (Saxon Bavarian Railway Company): '…our practice as you are aware is to receive about one third of the amount of the order on its being given, & cash for each shipment, deducting the amount advanced from the last…'[115] The advance may have been increased later that year, as the company's Head Clerk received the advice: '… I cannot doubt but it will be your care to follow up the well known exhortation 'make Hay while the sun shines.'[116]

The advances greatly assisted the manufacturers' cash-flow, assisting them with bulk purchase of raw materials and components. In 1858, for example, Beyer Peacock & Co. received; '£9,090 on account from the Madras [Railway]. Money never so plentiful as at present.'[117] It also allowed the larger manufacturers with lengthening order books to reach agreements with other manufacturers to undertake subcontracted orders.

When funding for railway companies was difficult or when small railways were starting-up, they offered part-payment in shares. In 1857, for example, Sharp Stewart & Co. agreed to accept half-payment for two locomotives for the Dundalk & Enniskillen Railway in preference shares on delivery, and the other half with a twelve months bill.[118]

As was also the case for Baldwins in America, these 'credit' sales posed serious cash-flow risks to the manufacturers, who were obliged to balance a low order book with a shared risk with their customers.[119] In the early 1850s, Baldwins were obliged to accept total payment in the stock of some railroad companies. The practice does not seem to have been widely used in Britain, other than at times of low demand. In 1858, when manufacturing more than one locomotive per week, Robert Stephenson & Co. rejected a call from the London-based owners of the Turkish-based Smyrna-Aiden Railway for half payment in shares for six locomotives.[120]

Payment for locomotives for overseas railways whose offices were not based in London were, typically, 'Terms net cash payable in London on presentation of Bills of Lading' which normally safeguarded the manufacturers from the potential of bad debts. For 'Firms of undoubted stability' a manufacturer would forward the invoice and Bills of Lading through its bankers with a draft bill of, say, '30 days for the full amount including shipping charges, bank exchanges and insurances.'[121] Overseas railway companies would be expected to pay the shipping companies for the freight on discharge of the locomotives at the destination port.

Continental firms made payment available through an international banking house. Robert Stephenson & Co., for example, received an 'acceptance' to its 'draft for £1,070' which Edward Starbuck, would: 'probably negotiate… to Messrs. Rothschilds & Sons paying the Amount when received to Messrs. Glyn & Co. to your acco't as usual.'[122]

Industrial locomotives for larger customers in Britain were also obtained through the tendering system, but with the design initiative resting with the manufacturers, modified to meet the specific requirements of the customer. The large market for small industrial locomotives, however, was dominated by customers or commission agents applying for current prices and discount possibilities of standard locomotives.

The system in the United States appeared less regulated. Once a railroad had defined its requirements, Baldwins prepared a formal proposal accompanied by a detailed set of technical specifications, which served as a basic contractual agreement.[123] It did not go nearly as far with its advance payment requirements as in the 1830s it required half payment when a locomotive was half completed, and the remainder on delivery. During periods of low demand, Baldwins accepted half payment on completion, with the balance due six months after delivery.[124]

Fig. 3.25
Full payment required in cash. Robert Stephenson & Co. 1860-built 4-4-0 (RS 1206/60) for the Smyrna-Aiden Railway of Turkey.
The Locomotive Magazine Vol. 48 (1942), p.25.

Delivery and Commissioning

Attractive delivery dates were usually as important as price when competitive tenders were being considered. Manufacturers developed an expertise in anticipating locomotive demand, with which to balance production schedules with their perception of market price. When demand was high, production capacity was increased as far as possible through a combination of facility expansion and recruitment, which required a 'lead' time to implement. The judgement required of the proprietors and their senior managers was thus a combination of this estimated lead time, and of locomotive production scheduling for parallel deliveries to two or more customers.

Locomotive manufacturing times decreased only gradually during the century, improvements in manufacturing techniques and production practices being offset by larger and more complex designs. Six months for delivery of the first locomotive was normal at first, but this reduced to about five months by the 1870s-1880s.[125] At times of low demand, manufacturers would reduce anticipated delivery times in an effort to win orders. Beyer Peacock & Co. stated in 1863 that: 'We can deliver any kind of Engine, provided they are our scheme in 3½ months, that is as fast as an engine can be made.'[126] In the 1894 slump in orders, Beyer Peacock & Co. was quoting 3½ - 4 months for delivery.[127]

Competitive delivery dates could win or lose orders. Beyer Peacock & Co. lost an order in 1855 as: 'Kitsons of Leeds have got the order by promising an earlier delivery.'[128] Once railways and industrial customers had decided to proceed with an order, they usually wanted early delivery dates. The largest manufacturers anticipated this urgency by quoting relatively early delivery dates for the first two or three locomotives and spreading the remainder with deliveries over several months, or even years. This allowed them to build locomotives for up to, say, six customers simultaneously. It also had the advantage that urgent orders for similar locomotives could be substituted at a premium rate to the railway concerned. Main-line railways responded to the prospect of long delivery times by dividing their orders between two or three manufacturers. This practice was gradually discontinued for all but the largest orders, as manufacturers' production capacity increased.

Manufacturers were generally bad at maintaining their predicted delivery dates, and through the century railways sought to introduce penalty clauses in their contracts to encourage adherence to the agreed programme. In the 1840s, manufacturers were defensive about delivery dates in communications with their customers, a view undoubtedly influenced by the unpredictable changes in demand. With demand for its products and prices both increasing, the Stephenson Company accepted a large number of orders for which it had insufficient manufacturing capacity, and it became notoriously bad on delivery dates. Edward Starbuck wrote, however: '… it is not Messrs. Robert Stephenson & Co.'s custom to allow the Insertion of a clause (in a contract) giving a penalty when the Engines are not ready to the day named. We are exceedingly punctual in the performance of our Contracts to their spirit & à la lettre – but the penalty clause you will be good enough to resist in future, if in your power.'[129]

In 1883, the manufacturers came under intense pressure from the Secretary of State for India to accept a penalty clause in respect of the Indian State Railways, which led to a flurry of meetings, proposed penalty definitions and arguments between the members of the Locomotive Manufacturers Association. The manufacturers sought definitions that did not leave themselves open to penalties for reasons outside their control, due to changes in design or specification by the customer, late deliveries of components or raw materials, or problems with shipping arrangements. They eventually agreed that a member's tender for an Indian contract could include a penalty clause, but subject to a further compromise exclusion clause and the agreement of all the Association members on each occasion. The clause read:[130]

> If the Contractor shall have been delayed in the execution of any part of the work by alteration in design, or by any other cause which the Secretary of State in Council shall consider to have been beyond the Contractor's control, or may admit as reasonable cause for extra time, the Secretary of State in Council will allow such addition to the time for the delivery thereof as he may consider to have been required by the circumstances of the case.

In addition to price, payment terms and delivery times, locomotive contracts also specified place of delivery, guaranteed 'proving' mileage, delivery and commissioning. These were, again, important to win contracts, and the judgement for offering competitive terms fell on the proprietors and their senior managers. The willingness of manufacturers to improve upon standard terms of contract could assist them to win contracts over other firms with, for example, the commissioning and the training of overseas railway operating and maintenance personnel.

Early tenders specified either 'delivery' at the factory or at a port, with an additional sum to cover shipping costs,[131] but as the railway network developed, contracts usually specified rail delivery to a main centre of operations.[132] Overseas locomotives were occasionally quoted for delivery 'at the works', although the normal arrangement was 'free on board' (fob) a ship in a nominated port, for which customers would arrange shipment through an agent. When overseas railways requested delivery at a port near their operations, manufacturers would arrange 'carriage, insurance and freight' (cif), the cost being included in the contract price.[133]

From the earliest contracts in the 1830s, provision was included for the first 1,000 miles of satisfactory operation before final payment was made. This 'proving mileage' became a regular contractual requirement, but was increased to, typically, 2,000 or even 3,000 miles by the end of the century.[134] Although expertise in locomotive technology quickly devolved throughout Britain, overseas railway expertise took longer to develop. Locomotive delivery became a major concern for manufacturers who had to send responsible superintendents to undertake discharge, transport and erection of the locomotives, train the railways' personnel, occasionally set up a maintenance workshop and remain for the proving mileage.

Manufacturers' costs in having experienced superintendents away from the factory for months at a time would have been quite high, but there is no evidence to confirm that this cost was added to the contract price. Thomas Wardropper, one of Robert Stephenson & Co.'s senior foremen, accompanied the company's first locomotive, PROVORNY, to the Tsarskosel'skaya ZH.D (Tsarskoe-Selo Railway) in Russia in September 1836.[135] He worked continuously in St. Petersburgh until at least the following February when he wrote to Newcastle: 'There has not been anything said about a further agreement yet but… if I stop another winter here…'

Fig. 3.26 Start-up of railways in Russia. Quarter-scale model of PROVORNY, the first 2-2-2 locomotive built by Robert Stephenson & Co. in 1836 (RS 163/36). Displayed at the Central Museum of Railway Transport in St. Petersburg. *Courtesy, Peter Davidson.*

Robert Weatherburn spent several years overseas as a roving representative engineer for Kitson & Co., discharging, delivering and commissioning locomotives in Russia, Austria, Denmark, Germany and France. He recorded his pioneering experiences with locomotive deliveries and setting up maintenance facilities, which served to emphasise the importance of such personnel in the export market of the locomotive manufacturers.[137]

The manufacturers showed themselves to be astute in selling locomotives in the competitive market. In spite of their reducing opportunity for product initiative, and the proliferation of customer-led designs, their tactical decisions on prices, methods of payment, production and delivery scheduling showed an awareness of the market opportunities as they arose. Although at first many proprietors and their agents were personally involved in preparing tender documents or in direct selling, the expansion of the market, and its concentration in London, saw these responsibilities delegated to London managers in consultation with their workshop colleagues. This necessary delegation, to ensure up to date knowledge of sales opportunities, raw material prices and production scheduling, well illustrates the evolution of managerial responsibilities within 'partnership' and 'limited company' enterprises.

The Locomotive Manufacturers Association (LMA)

Until the early 1870s there had been no moves by the manufacturers towards any form of trade association. It is likely that this was partly through lack of perceived need and partly because most manufacturers continued to see locomotive manufacturing as part of the wider heavy engineering industry. In the USA in the 1850s and 1860s there had been three attempts to form an association of locomotive builders, with a price-fixing motivation, but they had all failed.[138] The use of commission agents by American railroads had been the cause of their concern, and the formation of the Locomotive Builders Association (LBA) in 1872 succeeded in forcing railroads to order locomotives directly from manufacturers. An attempt to collude on prices again failed however, and the LBA was wound up the following year.

The price depression experienced by British industry from 1873 led to the formation of trade associations, particularly in the iron and steel, textile, chemical and manufacturing industries, whose aims were to fix prices, allocate market quotas and liaise on technical matters.[139] The Locomotive Manufacturers Association (LMA) was formed in this period, and a consideration of the ways in which the industry adapted to the market changes, particularly through this trade association, demonstrated the adaptability of the industry.

The motivation that led to the formation of the LMA in Great Britain was the protection of the home market from railway workshop competition. The British market had expanded after 1870, leading to lengthening delivery times and rising prices. The Lancashire & Yorkshire Railway (LYR) found this unacceptable and, at a time when consideration was being given to amalgamation with the London & North Western Railway (LNWR), it contracted with the latter for six locomotives to be made at Crewe

Works. These were completed in 1871, a rare example of inter-railway co-operation for locomotive manufacture.

No opposition was raised by the independent manufacturers, probably because of their lengthening order books. A further thirty-seven locomotives in several orders, built for the LYR at the LNWR's Crewe Works, were delivered up to 1874, again without comment from the manufacturers. In March 1875, however, the LYR sent out an invitation for fifty locomotives, towards which the LNWR tendered to manufacture twenty-five.[140] The independent manufacturers, whose orders were 100 fewer in 1875 than in 1874, thus shortening delivery times, were alerted to the danger of a lost order by the LNWR tender.[141] In the words of the LMA's opening memorandum:[142]

> Some time in the month of April 1875, information was received that the London & North Western Railway Company had entered into competition with the locomotive manufacturers of the country, and had undertaken to construct a number of engines for the Lancashire & Yorkshire Railway Company. In consequence of this information, Mr. E. Sacré, of the Yorkshire Engine Company, entered into communication with the various firms in England and Scotland, with a view to gathering their opinion upon the legality of the action…

The manufacturers' proprietors met in April and resolved: 'It is the opinion of this meeting that it is now necessary to take steps for the protection of engineers and others against the competition of railway companies as manufacturers for sale.'[143] Under the chairmanship of John Robinson of Sharp Stewart & Co., the manufacturers collectively obtained the services of Counsel, and on the 4th June the LMA was formally established.[144] Application was made, in the name of the Attorney-General, to restrain the LNWR in the High Court of Justice (Chancery Division), and judgement was made in December 1875 in the LMA's favour through a 'perpetual injunction' against the LNWR.[145]

The LMA members met again in March 1876 and confirmed their intention to continue the Association: '… with a view to any action it may be necessary to take for the mutual protection of the interests of its members.'[146] As early as January 1877, the LMA's Parliamentary advisers let it know of the actions of the Great Eastern Railway (GER) which proposed to build locomotives for the London, Tilbury & Southend Railway (LTSR). The GER fought long and vigorously in the High Court and Appeal Court to be allowed to fulfil its agreement with the LTSR, and not until May 1880 did the House of Lords confirm the injunction in the LMA's favour.[147]

The legal battle served to strengthen the LMA, which then diversified its activities into a wider trade association representing its members on several issues of common concern. By 1889, it described its 'Object' as being 'To overcome the evils resulting from excessive competition, by means of a friendly combination of the principal firms in the trade.'[148] Not all manufacturers were members. There were fourteen firms in its early years, but only ten firms from the early 1880s to the end of the century, namely:

Beyer Peacock & Co.	Nasmyth Wilson & Co.
Dübs & Co.	Neilson & Co.
Hunslet Engine Co.	Sharp Stewart & Co. Ltd.
Kitson & Co.	Robert Stephenson & Co.
Manning Wardle & Co.	Vulcan Foundry Ltd.

The LMA campaigned on both legislative and commercial issues. Early in its existence it dealt with the imposition of tonnage dues for deck shipments which would increase the price of exported locomotives,[149] and dissuaded the Great Western Railway from re-gauging and exporting twenty-six broad gauge locomotives to Russia in competition with the new-build potential for LMA members.[150] It also considered such diverse issues as penalty clauses in contracts for late deliveries, the undertaking of metal tests to satisfy customers and insurers over safety requirements, and the inspection of locomotives on delivery overseas.[151]

The LMA produced and printed a 'Private Memorandum and Agreement entered into by sundry firms in the Locomotive Building Trade &c', and which was several times re-issued as 'The Amended Rules of the Locomotive Makers' [sic] Association'.[152] It set out lengthy rules that bound the members to a cartel, requiring each member to advise the value of all tenders it proposed to make to the LMA Secretary, who would determine average tender quotations. The difference between the lowest and average price for each order was to be added to the actual tenders of each member, as a means of weighting the quotations to provide support for the smaller companies.

The cartel was strengthened from 1894 following the sharp downturn in the home and overseas markets. Members were then required to include an allowance of 2½% in their tender quotations. The allowance was to be passed by the successful company to the LMA Secretary, who would then distribute the sum amongst the unsuccessful member companies in agreed proportions, to offset their tendering costs. Further modifications were made in the mid-1890s, including an increase in the allowance to 5%, as the LMA sought to prop up its weaker members. In spite of the rapid expansion in demand from 1898, and the resulting rise in prices, it failed to secure the future of Robert Stephenson & Co. Ltd., which went into liquidation the following year.

Although the LMA had arisen out of a threat to the industry's domestic market, in the 1880s and 1890s it increasingly provided a forum for the protection of its members in a wider market context, thus developing into a cartel. This cartel served to accommodate the inefficiencies of the smaller, craft-based firms, and the consequent raising of prices at a time of increasing international competition, particularly from the American and German industries, was indicative of an industry that had got out of touch with wider railway developments after years of dependency on the London market.

Notes – Chapter 3

1. Tann (1978), p 364.
2. Tann (1978), p 372.
3. Riden (1973), p 43.
4. Byrom (2017), pp 74/5.
5. Tann (1978), p 364.
6. Lecount (1839), p 377.
7. Minutes of the Liverpool & Manchester Railway Board, National Archives, Rail 371; London & Birmingham Railway, London and Birmingham Board Committees, Rail 384; Grand Junction Railway Board, Rail 220, *passim* 1830-1836.
8. Letter, Robert Stephenson to Michael Longridge, London, 26th January 1835, Pease-Stephenson collection, D/PS/2/67.
9. Post-script to letter, Robert Stephenson to E.J. Cook, 16th June 1837, Library of the Institution of Mechanical Engineers, Crow Collection.
10. Lane (1980), p 138.
11. For example, Minutes of the Midland Railway's Locomotive Committee, National Archives, Rail 491/168-192, *passim*.
12. Letter, Arthur Potts to John Jones, London, March 31st 1846, Robert Stephenson & Co. collection, (ROB/5, Folder 19).
13. Thomas (1964), pp 85-96.
14. Byrom (2017), Chapters 3 and 4, *passim*.
15. For example, letter, Charles Manby (for R. Stephenson & Co.) to Carl Pihl (for the Norwegian State Railways), 30 December 1858, Beyer Peacock & Co. collection, MS0001/258.
16. Letter, Robert Stephenson to Edward Cook, London, 2 Oct 1841, Deutsches Museum Library, Munich, 1981-19.
17. Letter, G.H. Stieler to Beyer Peacock & Co., Gothenberg, 16th March 1857, Beyer-Peacock collection, MS0001/255.
18. Hills and Patrick (1982), p 45.
19. C.A. Cardew, 'Copy of Centenary Notice of the Opening of the Sydney and Parramatta Railway', Robert Stephenson & Co collection, (ROB/5, Folder No. 7).
20. Hills and Patrick (1982), pp 56/65.
21. Brown (1995), pp 42/3 & 50/51.
22. Farnie (1990), p 151.
23. Letter, Charles Manby to R. Stephenson & Co., Westminster, June 8 1858, Crow Collection, *op cit* (9).
24. Letter Robert Stephenson to Joseph Pease, London, 24 Oct 1840, Pease-Stephenson Collection, D/PS/2/56.
25. Letter, E.F. Starbuck to E.J. Cook (for R. Stephenson & Co.), London, 8 August 1844, copy letter book, Starbuck collection.
26. Letter, E.F. Starbuck to E.J. Cook (for R. Stephenson & Co.), London, 4th April 1844, copy letter book., Starbuck collection.
27. Letter, E.F. Starbuck to E.J. Cook (for R. Stephenson & Co.), London, 17th April 1844, copy letter book., Starbuck collection.
28. Letter, Robert Stephenson to W.H. Budden (for R. Stephenson & Co.), Westminster, 18 Sep 1851, Starbuck collection, 131/45.
29. Letter, J.E. Sanderson (for Robert Stephenson) to W.H. Budden (for R. Stephenson & Co.), Westminster, 6 February 1846, Starbuck collection, 131/42.
30. Letter, J.E. Sanderson (for Robert Stephenson) to W.H. Budden (for R. Stephenson & Co.), Westminster, 22 Jany 1846, Starbuck collection, 131/44.
31. E.F. Starbuck commission statements, 1845-55, Starbuck collection, 131/53 and 131/54.
32. Letter, Robert Stephenson to W.H. Budden, London, 18 Sep 51, Starbuck collection, 131/45.
33. Probate Document for Robert Stephenson, Science Museum Library, Stephenson papers, MS 2033/95, Schedule 16, Schedule of Pecuniary Legacies.
34. Letter, Wm. Fairbairn & Son to Messrs. C.J. Hambro' & Son, Manchester, 20th July 1855, Hambro's Bank Ltd. loan papers archives, Guildhall Library, London, Hambros papers, MS 19158.
35. Copy correspondence, Beyer Peacock collection, MS 0001/546.
36. *Engineering*, Vol. XIV, p 15, July 5th 1872.
37. Brown (1995), pp 45/6.
38. Payne (1988), p 41.
39. Chapman (1992), pp 129-166.
40. Payne (1988), pp 524/5.
41. Payne (1988), p 19.
42. Letter, E.F. Starbuck to E.J. Cook (for R. Stephenson & Co.), London, 15th June 1844, copy letter book, Starbuck collection.
43. Letter, J.E. Sanderson (for Robert Stephenson) to W.H. Budden (for R. Stephenson & Co.), also, letter, E.F. Starbuck to E.J. Cook (for R. Stephenson &

44 Letter, Wm. Bird & Co. to C.J. Hambro & Son, London, 12 January 1851, Hambros papers, *op cit* (34), MS 19129.
45 Analysis of manufacturers' Works lists prepared by the Industrial Locomotive Society.
46 *Report from the Council of the Institution of Civil Engineers Upon the Building Question* &c., Presented at a Special General Meeting of Members & Associates on the 7th April 1868, Institution of Civil Engineers archives, register No 213.
47 *Engineering*, 25th May 1883 and 9th July 1886, shown in Wear (1977), pp 346/369.
48 *Engineering*, Vol. IX, April 22nd 1870, p 270.
49 Ericson (1996), pp 32-36.
50 Consular Report No. 427, Presented to Parliament, June 1897, reported in *The Railway Engineer*, Vol. XVIII, September 1897, pp 276-279.
51 Brown (1995), p 208. Also Minutes of the Locomotive Committee of the Midland Railway, 1895-1899, National Archives, Rail 491/182.
52 Tredgold (1838).
53 Marshall (Ed.) (1838).
54 Templeton (1841).
55 De Pambour (1836 and 1840).
56 Hawthorn (1840).
57 *Neilson & Co., Locomotive Engine Makers, Glasgow.* Booklet retained in the National Railway Museum Archives, York.
58 *Locomotives For Broad Gauge Railways, Manufactured by Sharp Stewart & Co. Limited,* Atlas Works, Manchester, England, 1886, and *Locomotives For Narrow Gauge Railways, Manufactured by Sharp, Stewart & Co. Limited,* Atlas Works, Manchester, England. 1886.
59 Brown (1995), p 34. Also, *Engineering*, Vol. XIV, November 15th 1872, pp 335/6.
60 Ottley (1983), entry 2963.
61 *The Railway Engineer*, Vol. XIII, November 1892, p 315.
62 Brochure of the Salter Co. (n.d. but *c.*1870), Staffordshire Record Office, CXD 4721/J/1/1.
63 Examples of Nasmyth Gaskell & Co.'s machine tool catalogues from 1839 and 1849 have survived, in Musson (1957-58), p 125; also Hayward (1971), p 2.10.
64 Hills and Patrick (1982), p 26 and *passim*.
65 Beyer Peacock collection, photographic collection, and MS0001/546.
66 Brown (1995), p 34.
67 Brown (1995), pp 34/43-45.
68 Wear (1977), *passim*.
69 'Stephenson's Locomotive Manufactory at Newcastle-on-Tyne', *The Illustrated London News*, October 15th 1864, pp 392-394.
70 *The Railway Engineer*, Vol. VIII, No 10, October 1887, pp 304-309.
71 Greenhalgh (1988), p 10.
72 Letter, Robert Stephenson to Edward Starbuck, Suez, 1 Jany 1851, Robert Stephenson & Co collection, (ROB/5, Folder 18).
73 Hawthorn (1851).
74 *Official Catalogue of the Great Exhibition of the Works of Industry of All Nations, 1851*, p 34, Section II, Machinery, Class 5, 'Machines for Direct Use, Including Carriages, Railway and Marine Mechanism'.
75 Greenhalgh (1988), pp 3-24.
76 Official Catalogue, *op cit* (74).
77 Sir. D. Campbell Bart., *Reports on the Paris Universal Exhibition 1867*, Presented to both Houses of Parliament, London, 1868, Vol. IV, 'Containing Reports on… Railway Apparatus', p 512.
78 *ibid*, p 487.
79 *ibid*, p 516.
80 'Locomotives at the Vienna Exhibition No.1', *Engineering*, Vol. XV, June 6th 1873, p 404.
81 'The Vienna Exhibition', *Engineering*, Vol. XVI, November 7th 1873, p 381.
82 *British Official Catalogue Paris – Exposition Universelle de 1900*, published by the Royal Commission; Group VI Civil Engineering and Transportation, Class 32, 'Railway and Tramway Plant', p 171, 'Locomotives and Rolling Stock'.
83 Scranton and Licht (1986), quoted in Brown (1995), p 262, note 31.
84 Lowe (1975), p 380.
85 Dunod (Ed), and Taillard (Text) (1867), p 171, XXXI and XXXII.
86 *Engineering*, Vol. XIV, September 1872, p 172.
87 *The Railway Engineer*, Vol. IV, September 1883, pp 231-238.
88 *Engineering*, 9th July 1886, shown in Wear (1977), p 346.
89 In a random year, 1875, *The Railway Times* carried four series of advertisements for three railway companies, viz. Midland Railway, Great Northern Railway and South Eastern Railway, being invitations to tender for batches of locomotives.
90 Multiple letter, I. K. Brunel, Engineer, Great Western Railway, to locomotive manufacturers, London, 4th March 1839, Robert Stephenson & Co. collection, (ROB/5, Folder 15).
91 Letter, E.F. Starbuck to E.J. Cook (for R. Stephenson & Co.), London, 17th April 1844, copy letter book, Starbuck collection.
92 Letter, Charles Beyer to H. Robertson, Manchester, Sept: 8/1857, Beyer Peacock collection, MS 0001/255.

93 Letter, Charles Beyer to H. Robertson, Manchester, February 22nd 1863, Beyer Peacock collection, MS 0001/256.

94 Letter, Charles Beyer to H. Robertson, Manchester, Sept: 8/1857, Beyer Peacock collection.

95 Select Committee Report on Operation of Laws Affecting Exportation of Machinery, First Report, 1841, p.46, Evidence of Grenville G. Withers.

96 Letter, E.F. Starbuck to E.J. Cook (for R. Stephenson & Co.), London, 8 August 1844, letter book, Starbuck collection.

97 Letter, Gilkes Wilson & Co. to John Pease (for Stockton & Darlington Railway), Middlesbrough, 9th month 2nd 1846, Stockton & Darlington Railway papers, National Archives, Rail 667/773.

98 Letter, C. Pihl, Jernbane-Direktøren (Norway) to Beyer Peacock & Co., Kristiania, 30th Novr. 1867, Beyer Peacock collection, MS 0001/261.

99 Letter, E.F. Starbuck to Eastern Counties Railway, London, 27 May 1844, copy letter book, Starbuck collection.

100 Letter, Charles Beyer to H. Robertson, Manchester, February 22nd 1863, Beyer Peacock collection, MS 0001/256.

101 Thomas (1964), pp 95/96.

102 Letter, W.H. Budden, Ppro R. Stephenson & Co., to Edw. Fletcher Esq. (York, Newcastle & Berwick Railway), Newcastle upon Tyne, 5 Octr. 1847, York, Newcastle & Berwick Railway archive, National Archives, Rail 772/96.

103 Lane (1980), p 139.

104 List, 'Orders on Hand', prepared by R. Stephenson & Co., Dec.27th 1861, Pease-Stephenson collection, D/PS/2/75.

105 Tann (1978), p 384.

106 Cost & Profit Notebook of R. Stephenson & Co. (probably prepared by Wm. Hutchinson), Bidder collection, Arch:Bidd 27/8.

107 List Prices of Engines & Tenders, including 25% Profit, fob in this country (but calculated as 20% profit throughout), Beyer Peacock collection, MS 0001/546. Also letter, Charles Beyer to H. Robertson, Manchester, January 21st 1863, Beyer Peacock collection, MS.0001/256.

108 Bailey (1984), pp 300/1.

109 Beyer Peacock & Co. instructions to its Australian agent, W.S. Brewster, Feby 27 1878, Beyer Peacock collection, MS 0001/546, pp 228-232.

110 E.F. Starbuck Account Sheets, Starbuck collection, 131/66. The telegraphic address for Robert Stephenson & Co. was 'Rocket', Newcastle-on-Tyne, as set down on a brochure, *The 'Rocket' Oil Engine, R. Stephenson & Co. Ltd.*, n.d. but 1893, author's collection.

111 Letter, Thomas Fairbairn (for William Fairbairn & Co.), to George Wilson (Deputy Chairman of the Lancashire & Yorkshire Railway Company), Wilson Papers, Manchester Public Library, Local Studies Unit, cited in Byrom, (2017), p 316.

112 List Prices of Engines & Tenders, Beyer Peacock collection, p 228.

113 Draft letter, C.F. Beyer (for Beyer Peacock & Co.) to C. Pihl, Norwegian State Railways, Manchester, Aug 9/65, Beyer Peacock collection, MS 0001/259.

114 Partners' Minute Book, Robert Stephenson & Co. collection (ROB/1/1), entry for 10m [Oct] 20th 1838.

115 Letter Edwd. J. Cook, Ppro Rob't Stephenson & Co. to the Directors of the Saxon Bavarian Railway, Newcastle, 5th May 1844, copy letter book, Starbuck collection.

116 Letter, Edward Pease to E.J. Cooke (for R. Stephenson & Co.), Darlington, 11 mo [November] 29. [18]44, Crow collection, Institution of Mechanical Engineers.

117 Letter, Charles Beyer to H. Robertson, Manchester, July 13th 1858, Beyer Peacock collection, MS 0001/256.

118 Letter, Charles Beyer to H. Robertson, Manchester, Sept: 8/1857, Beyer Peacock collection, MS 0001/255.

119 Brown (1995), pp 12/13.

120 Letter, Charles Manby to R. Stephenson & Co., Westminster, June 8th 1858, Crow collection, Institution of Mechanical Engineers.

121 Bailey (1984), p 232.

122 Letter, E.F. Starbuck to R. Stephenson & Co., London, 23rd February 1844, Crow Collection, Institution of Mechanical Engineers.

123 Brown (1995), pp 42/3.

124 Brown (1995), pp 12/13.

125 Locomotive contracts retained in the archives of 19th century British railway companies, for example South Eastern Railway, National Archives, Rail 635/225-230.

126 Letter, Charles Beyer to H. Robertson, Manchester, February 22nd 1863, Beyer Peacock collection, MS 0001/256.

127 Beyer Peacock & Co., instructions to its Australian agent, W.J. Adams by Sept. 27/94, Beyer Peacock collection, MS 0001/546.

128 Letter, Charles Beyer to H. Robertson, Manchester, June 13th 1855, Beyer Peacock collection, MS 0001/256.

129 Letter W. Winfield (for E.F. Starbuck) to Mons. F. Kunitz (agent in Hamburg), London, 30 April 1844, copy letter book, Starbuck collection.

130 *First Minute Book of the Locomotive Manufacturers Association 1875-1900* (LMA Minute Book), Railway Industry Association Collection, National Railway

Museum, entry for December 12th 1883.
131 For example, tenders for Locomotive Engines for the London & Birmingham Railway, National Archives, Rail 384/265-269.
132 B.S. Stafford, Memorandum, 'Engines pr Railway', Locomotive Department, London & Birmingham Railway, 15/1/39, recording detail of locomotives being delivered to the London & Southampton, London & Brighton and London & Croydon Railways, Dendy-Marshall Collection, the Newcomen Society.
133 General Specification of Locomotives built by Messrs. Beyer Peacock & Co. Gorton Foundry, Manchester, Beyer Peacock collection, MS 0001/546, p 232.
134 For example, locomotive contracts, *op cit* (125).
135 Diary of Thomas Wardropper, 28th September 1836 to 28th January 1837, Tyne & Wear County Record Office.
136 Letter, Thomas Wardropper to William Hutchinson (Head Foreman for R. Stephenson & Co.), Trotsky Bridge [near St. Petersburgh] 16th February 1837, in private ownership.
137 Robert Weatherburn, 'Leaves from the Log of a Locomotive Engineer', *The Railway Magazine,* Part I, Vol XXXI, July-December 1912, p 289, to Part XXX, Vol XXXVI, January-June 1915, p 240, *passim*.
138 Brown (1995), pp 50/51.
139 Wilson (1995), p 99.
140 Draft letter, Messrs Hargrove, Fowler & Blunt (solicitors for the locomotive manufacturers) to the Secretary of the London & North Western Railway, n.d. but 4th May 1875, in LMA Minute Book, *op cit* (130).
141 Analysis of manufacturers' records, Chapter 2, Reference 1.
142 Opening entry, April 1875, in LMA Minute Book, *op cit* (130).
143 Minutes for 29th April 1875, LMA Minute Book, *ibid*.
144 Minutes for June 4th 1875, LMA Minute Book, *ibid*.
145 Minutes for December 16th 1875, LMA Minute Book, *ibid*.
146 Minutes for March 31st 1876, LMA Minute Book, *ibid*.
147 LMA Minute Book, *ibid, passim*.
148 Private Memorandum and Agreement of the Locomotive Manufacturers' Association, adopted at a meeting in London, April 11th 1889, *op cit* (130).
149 Minutes for March 31st 1876, LMA Minute Book, *ibid*.
150 Minutes for November 23rd 1877, LMA Minute Book, *ibid*.
151 LMA Minute Book, *ibid, passim*.
152 Printed Memoranda and Rules (of the Locomotive Manufacturers Association) 1889-1898, Railway Industry Association Collection, *op cit* (130).

Chapter 4
Technology and Design

Introduction

Britain's heavy manufacturing industry had developed from the 18th century through its ability to innovate with new technologies and designs. The industry's partnerships had formed through combinations of technical and business expertise to develop machines and structures that fulfilled customers' requirements. Such partnerships diversified their activities to bring about the emergent locomotive sector, which rapidly developed thermodynamic, mechanical and material technology, and made available improving designs for the nascent railway industry. Until the 1850s, the effectiveness of the manufacturers' marketing and selling efforts were sustained by offering products that met the developing requirements of their railway customers.

To provide for the long-term success of their businesses, it was incumbent upon the manufacturers to pursue long-term strategies through the development of locomotive technology and design. They largely lost the initiative from the 1850s, however, as their changing British main-line railway market led them away from research and development to become a mostly contract-only industry, largely manufacturing to the designs of their railway customers. Only the industrial locomotive sector, with its limited opportunities for technical advancement, remained as an industry-led activity.

In discussing innovation in the locomotive industry, previous studies of locomotive technology and design have noted that there were two 'intensive bursts' in the years 1829 to 1841, and 1896 to 1911, with incremental progress in between.[1] However, with the growing international nature of the locomotive market through the century, the motivation for, and means of achieving, technological progress in locomotives needed to address global advances in technology and it is in this context that the contribution of the British independent industry needs to be assessed.

The study of the history of technological innovation now considers the social construction of technological thought, as well as the evolution of technological change.[2] The historical context for considering technical change is both

Fig. 4.1 Kitson & Co. 1864-built 2-2-2 passenger locomotive (K 1213/64) for the Waterford & Limerick Railway.
Kitson collection, The Stephenson Locomotive Society.

local and specific; innovations are either new products (or variations on old ones), or artefacts or processes designed to raise the quality of commodities, or techniques that lower production costs.[3]

The long-term development of all manufacturing firms depended on continued exploitation of new technologies and materials, and design progression to fulfil and stimulate market requirements. Innovative firms could license as well as exploit their inventions. Many firms in the 19th century transformed an invention into a working design but were unable to pursue its large-scale commercial exploitation. Often more likely was the eventual diffusion of the technique among a number of firms.[4] In the capital goods sector, this diffusion could have been promoted by the market through the tendering system, since manufacturers were obliged to meet the, often quite detailed, specifications laid down by would-be railway and industrial customers.

The manufacturers were therefore faced with significant strategic decisions regarding, firstly, the depth of what is now known as research and development to stimulate new locomotive technology; secondly the risks associated with implementing innovation based on their own work, balanced against the benefits that could be derived from issuing licences to others; and finally the level of design initiatives to improve or maintain market share. They were also faced with decisions regarding the balance between design-led business strategies, using their in-house design resources, and contract-led business strategies, manufacturing to the designs of railway customers and consulting engineers.

Main-line locomotive technology showed evidence of all three processes, as it developed from an optional form of motive power, dependent upon the engineer to demonstrate an economic case for its use, to become a form of demand-led technology with railways requiring ever more demanding engineering and economic standards. It is rewarding to follow the causal link between the need for technological advance of locomotives and the innovations which made it possible, particularly as they were achieved in a short period of time.[5] Distinction can be made between conceptual developments, relating to thermodynamics, materials and components, and manufacturers' design innovation which was generally developed in response to external influences and market requirements, deriving benefits through patent protection, licensing and royalty income.

The development of the first acceptable locomotive, upon which the start-up of the main-line railway era depended, was prompted by personal ambition and reputation, and dependent upon invention, but which succeeded through innovation. Once the locomotive had proved itself, the economic incentive for greater speed and power drove technological and design improvements through generations of mutations, re-combinations and hybrids.[6]

The independent locomotive industry was subject to three major market-based transitions in the century, which withdrew its discretion to innovate. The first arose from the emergence of a railway's right to specify the type and design of locomotive best suited to its requirements. The introduction and development of tendering and contracting in 1835/6, gave locomotive superintendents increasing influence in specification and design. By the early 1840s, the manufacturers were responding to the rapidly developing railway requirements whilst pursuing further innovation, and at the same time seeking component standardisation with which to contain manufacturing costs.

The second transition arose from the commencement of locomotive manufacture in railway-owned workshops. The increase in number and capability of these workshops in the 1840s and 1850s was accompanied by an expansion of both technological and design expertise. Railways' own designs and specifications became more detailed, and, over time, their developing expertise curtailed manufacturers' opportunities for technological and design development.

The third transition, from the 1850s, arose from the increasing involvement in locomotive design and specification of the London-based consulting engineering firms, representing the growing numbers of overseas railways. The introduction of railways into India, Australasia, the Middle East and South America, was under the direction of consultants, such as Sir Charles Fox (1810-1874), Sir John Hawkshaw (1811-1891) and Sir Alexander Rendel (1828-1918), whose growing expertise further curtailed manufacturers' freedom for design initiatives.

There were eight major technological developments during the century, of such economic benefit that they were quickly adopted internationally. Indeed, the locomotive was not a specific machine but a generic form, within which were considerable opportunities for thermodynamic, material, mechanical and design innovation. In addition, locomotive design was itself a dynamic process which accommodated developing requirements for size and performance capability, together with evolving specifications for different traffics, track standards, track and loading gauges, operating speeds and gradients. Locomotive design progressed through several significant innovations, which were scale and performance related, and built on the opportunities of technological advancement, as well as meeting market demands. As railway and consultancy design teams began to dominate locomotive progress from the 1850s, the resulting proliferation of designs became an overriding characteristic of the British locomotive industry, which substantially reduced its opportunities for standardisation and the manufacturing economies of large batch production.[7]

By the last quarter of the century, the locomotive industry's technological and design progress was judged increasingly against the designs of the overseas industries, notably of America and Germany. The very different economic and technical backgrounds of both industries were built on design practices which diverged from their British origins in the 1830s and 1850s respectively. As locomotive markets became more internationalised towards the end of the century, the comparative merits of the British, American and German design schools were brought sharply into focus. Comparison between these design schools provides a better measure of the technological and design progress of the British industry, than an insular consideration of locomotive technology for the London-based market.

Origins of the Main-Line Locomotive
The partnership structure of the heavy manufacturing industry readily allowed diversification into locomotive development through the combination of technical and material knowledge, craft experience and entrepreneurial drive. Such was the urgency to develop the first main-line locomotives that tactical decisions were taken to pursue innovatory features, whilst relying on existing skills and materials. However, the immediate design success was tempered by the inadequacy of the materials and railway requirements for further improvements. This quickly called for manufacturers to pursue systematic enhancement of locomotive technology, design and materials.

Robert Stephenson's research and development programme between 1828 and 1830 that led to the main-line locomotive was one of the most remarkable in the history of mechanical engineering.[8] Through his tenacity a unified programme was carried out, combining new technological principles, design features and material developments. It conducted systematic examinations of major components, and innovatory features were tried on a series of experimental locomotives. The motivation for Stephenson's programme was to provide motive power suitable for inter-urban operation on the Liverpool & Manchester Railway, stimulated by the strong claims made for locomotive operations by his father, George Stephenson, the railway's Chief Engineer.[9] It culminated, after just thirty-three months, in the prototype PLANET locomotive, the progenitor of the 'Stephenson'-type adopted by the world's railways for more than a century.

Fig. 4.3 Replica of a *PLANET*-class locomotive, the first main line locomotive class initially adopted for use on the Liverpool & Manchester Railway in 1830. *Author.*

Whereas locomotive design arrangements had previously been advanced empirically and on an experimental basis, with millwrights and fitters using sketch drawings to prepare and fit components, Stephenson's development programme included the services of a design draughtsman for the first time.[10] Arrangement drawings introduced design techniques to accommodate components within the space and weight constraints of the early locomotives. By the completion of the development programme in the early 1830s, the foundations of drawing office design work had been laid, which led on over the next decade or more to detailed sub-assembly and component drawings for manufacturing purposes.

Design, materials and construction methods were required to keep in step for successful innovation, and the extraordinary speed of Robert Stephenson's programme had outpaced the availability of suitable materials and satisfactory methods of construction. During the

Fig. 4.2 Replica of *ROCKET*, one of the prototype main line locomotives, manufactured to participate in the Rainhill locomotive trials of October 1829. *Author.*

1830s, therefore, the early manufacturers were under considerable pressure to develop more reliable materials and new manufacturing methods, as well as to pursue further design innovation. The most significant material improvements related to those adopted for crank-axles, wheels, fireboxes and fire-tubes, the frequent failures of which had led to considerable anxiety by the railways, not least when accidents occurred.

Fig. 4.4 Crank axle of the replica 1837 *NORTH STAR* 2-2-2 locomotive, built by Robert Stephenson & Co. for the Great Western Railway, displayed in the 'Steam Museum' in Swindon. *Author*.

Better quality wrought iron, a more robust design and better forging and machining methods were necessary to overcome crank-axle failures. By 1839, John Moss, the Chairman of the Grand Junction Railway, reported to a Parliamentary Select Committee that: 'we had many accidents in the first instance of axles breaking, but we have not had any for some time.'[11] Even after the crank-axle improvements, failures occasionally occurred until the adoption of inside and mixed frames in the 1840s. Similarly, it took nearly four years to improve design, materials and casting techniques to provide wheels free from stress fracture problems. Firebox and fire-tube failures were directly attributable to unsuitable materials, and when copper plate and brass sheet, respectively, were confirmed as preferable, new industries had to be established to produce sufficient quantities to meet the new demand.

The Stephenson development programme introduced several new technological and design features, none of which had patent protection, in spite of George Stephenson's early experience of patents, which were vested with the Stephenson Company on its formation in 1823.[12] There was, therefore, no deterrent to imitation and, as early as 1824, Stephenson's partner, Edward Pease, had felt nervous about protecting locomotive designs from the interests of potential competitors:[13]

> …if it be possible we must have GS to adopt some improvements for these Engines & <u>get a new patent</u>, I mean to write him in a day or two to enter a caveat in the patent office, for improvements, for I cannot doubt such is the enquiry about Railway & any but these engines will be a most important thing & <u>ought to leave us no small sum</u> for either making or Licences.

Fig. 4.5 0-4-0 locomotive built by Bury Curtis & Kennedy of Liverpool in 1846 for the Stockton & Darlington Railway No. 89, *HUDDERSFIELD* (later North Eastern Railway No. 1089). *National Railway Museum, R.H. Bleasdale Collection, 1875 Stockton & Darlington Railway Jubilee commemoration.*

Fig. 4.6 0-6-0 Hackworth-type mineral locomotive built by William & Alfred Kitching of Darlington in 1840 for the Stockton & Darlington Railway (No. 26, *PILOT*) shown at the Stockton & Darlington Railway Jubilee celebrations as No. 10 *AUCKLAND* (built by Timothy Hackworth in 1839). *National Railway Museum, R.H. Bleasdale collection, 1875 Stockton & Darlington Railway Jubilee commemoration.*

Stephenson's neglect in the 1820s seems prompted by an arrogant belief that, as the country's leading railway engineer, there was no need to take out patents. A contributory factor may have been the extraordinary pace of events, both Stephensons being so taken up with the supervision of surveying, route construction and mechanical research and development work, that they allowed themselves no time to brief a patent agent. Thus, by the completion of the programme, other manufacturers, notably Edward Bury & Co. of Liverpool, had begun to take an interest, and a period of imitation and further innovation took place.[14] Only then did the Stephensons appreciate that their opportunity for protection had been lost.

Three design 'schools' emerged in the early 1830s as the Stephensons' two main competitors took advantage of their neglect. In addition to the main Stephenson school, with its sandwich frames, rectangular fireboxes and, from 1833, three axles, Bury developed a variation with wrought iron bar frames, 'D'-form fireboxes and two axles. The third school, pursued by Timothy Hackworth, continued with the incremental development of the 1820s 'colliery'-type locomotive, better suited to mineral haulage than main-line operation.

In 1830, as a defensive response to the challenges of their competitors, rather than as a strategic policy, the Stephensons began to take out patents for their further, but relatively less important, inventions. In developing a new type of wagon axle, Robert Stephenson revealed a lack of patenting experience when he wrote to his father:[15]

> I hope you will think it well over, but as it is new and likely to answer, let us take a patent for it, the patent cannot cost much and if [it] does get introduced upon Railways a very small additional price on each carriage would produce a great deal of money –

Stephenson took out a patent the following summer,[16] followed a month later by his father's first solo patent, for a wrought iron spoked wheel.[17] Thereafter, novel design features were increasingly patented, firstly by the independent manufacturers, but, subsequently, also by railway-employed and independent locomotive engineers. Nineteen patents were taken out in the first decade, some relating to component improvements, such as wheels, others relating to whole locomotive schemes.[18]

The first significant patent, taken out by Robert Stephenson in 1833, sent a clear message that he had learned the lesson of his previous failure.[19] The patent introduced the features of a three-axled locomotive, which, to emphasise the point, was known as the *PATENTEE* type. It was adopted by several manufacturers, in addition to the Stephenson Company, and it is likely that Stephenson received significant royalty payments, although their extent is not known.

Fig. 4.7 *PATENTEE*-type 0-4-2 locomotive *LION* built by Todd, Kitson & Laird of Leeds in 1837 as preserved and restored. Shown at Liverpool Road station, Manchester – now displayed in the Liverpool Museum. *Author.*

Thus, by the introduction of the trunk railway building programme in the mid-1830s, proven locomotive designs were available to railways, but largely in the gift of the Stephenson and Bury companies. The influence of the two 'schools' was such that engineers of new railways specified one or other of the two types in their tender invitations, allowing other manufacturers to gain a relatively inexpensive way of entering locomotive manufacture.

Evolution of Locomotive Design
In the 1830s and 1840s the long-term development of manufacturing firms depended on continued exploitation of new technologies and materials, and design progression to fulfil and stimulate market requirements. Innovative firms could license as well as exploit their inventions. Many firms in the 19th century transformed an invention into a working design but were unable to pursue its large-scale commercial exploitation. Rather, the technique was eventually diffused among a number of firms.[20] In the capital goods sector, this diffusion could have been promoted by the market through the tendering system, since firms were obliged to meet the, often quite detailed, specifications laid down by would-be customers.

The manufacturers' initiative in promoting the early locomotive types set the bench-marks for design evolution until the 1840s. In accordance with the market growth and development, their design strategy was both to increase the power, speed and economy of main-line locomotives, and to develop new types to suit each new kind of railway and industrial operation, as determined by railway or consulting engineers. The strategy was sharpened by the increasingly competitive nature of the locomotive market, and several of the design and component innovations were patented.

By the time locomotive design was considered by Parliament's Select Committee on Railways in 1839, the evidence was a re-iteration of the arguments between the adherents of the Stephenson and Bury types.[21] This drew attention away from improvements in reliability that had taken place, through better materials and construction methods. Trains now kept better time: 'owing to the engines being themselves in much better order, very much improved, and owing to their getting a better supply of coke.' The engines had been much improved: 'Not in principle, but they are better made.'[22] Indeed, as the potential of technological and design development became evident, main-line railways sought to improve the speed, haulage capability and economy of their locomotive fleets, and they began to take their own initiatives for improvements based on the extensive operating experience they had gained.

During the 1840s, through increasing loss of initiative, manufacturers began to experience a shift from a product-led to a demand-led business. The introduction of railway-owned workshops in Britain was accompanied by their introduction of drawing offices and technical development teams which took over the initiative for improvement and design, and increasingly subordinated the manufacturers to largely contract manufacture.

Fig. 4.8 Neilson & Co. 1859-built 'Crewe'-type 2-4-0 locomotive (N 488/59) for the Caledonian Railway (No. 193) designed by Benjamin Conner, the railway's Locomotive Superintendent.
The Locomotive Magazine, Vol. 47 (1941), p.123.

The Bury type largely stagnated however because of his adherence to two axles, and, although in 1846 railway pressure forced him to introduce three-axle variants, his designs were found wanting, and he turned to contract construction of railway-designed locomotives. With the demise of Bury, Curtis & Kennedy in 1851 the manufacture of his design type was discontinued in Britain, but his bar-frame type had become firmly established in North America, being better suited to the poorer track. After 1850, the Stephenson type dominated locomotive design development for railways in Britain, and overseas railways of British influence.

The advent of British railway workshops from the 1840s witnessed the increased employment of talented mechanical engineers, and the companies sought to license their patented arrangements. Design initiative therefore gradually passed to them as their drawing offices and development teams experimented with new arrangements.

The growth of the consulting engineering firms from the 1850s, took further specification and design initiative away from the manufacturers in respect of locomotive types for the several new kinds of railway around the world. The choices confronting the manufacturers, therefore, were to seek to maintain their own design initiatives, offer licensing deals to independent patentees, or become passive designers limited to the detailing of customers' designs for manufacturing purposes. The extent to which the manufacturers were able to pursue a combination of these options, and the effects on their long-term ability to promote new products, can be judged from a review of design evolution through the century.

Fig. 4.9
A surviving three axle (2-2-2) locomotive manufactured by Bury, Curtis & Kennedy of Liverpool in 1848 for the Great Southern & Western Railway of Ireland, now preserved in Kent Station, Cork. *Author.*

Fig. 4.10
Consulting engineer-designed locomotive. Kitson & Hewitson 1859-built 4-4-0 locomotive ([one of batch] K 635-638/59) for the Copiapó & Caldera Railway of Chile (No. [one of batch] 9-13), probably to the specifications of Edward Woods (1814-1903), the railway's Consulting Engineer. *Kitson collection, The Stephenson Locomotive Society.*

The locomotive industry well understood these market challenges and responded to them with diverging strategic policies. The manufacturers decided either to retain their development and design capability, and retain their broad heavy engineering market base, or to move towards greater specialisation in the largely sub-contracted locomotive market, with a limited design capability. In responding to their subordinated role, two broad technical issues were addressed by the manufacturers. These were the motivation, influences and innovatory processes of technological and design evolution, and the design proliferation and consequent lack of manufacturing economy which resulted.

The economics of locomotive manufacture were affected by the large variety of designs, which were multiplied by the individual specifications of railway superintendents and consulting engineers. British capital goods generally were produced to customer specifications and even quite specialised enterprises were forced to maintain a wide product range.[23] With the Empire-building ambitions of British locomotive superintendents leading to the diversity of locomotive designs, the resulting proliferation led to a lack of standardisation, thus developing fragmented and relatively small market opportunities.[24] Proliferation was exacerbated by the continued existence of a large number of railways, mergers being suppressed by Parliament from

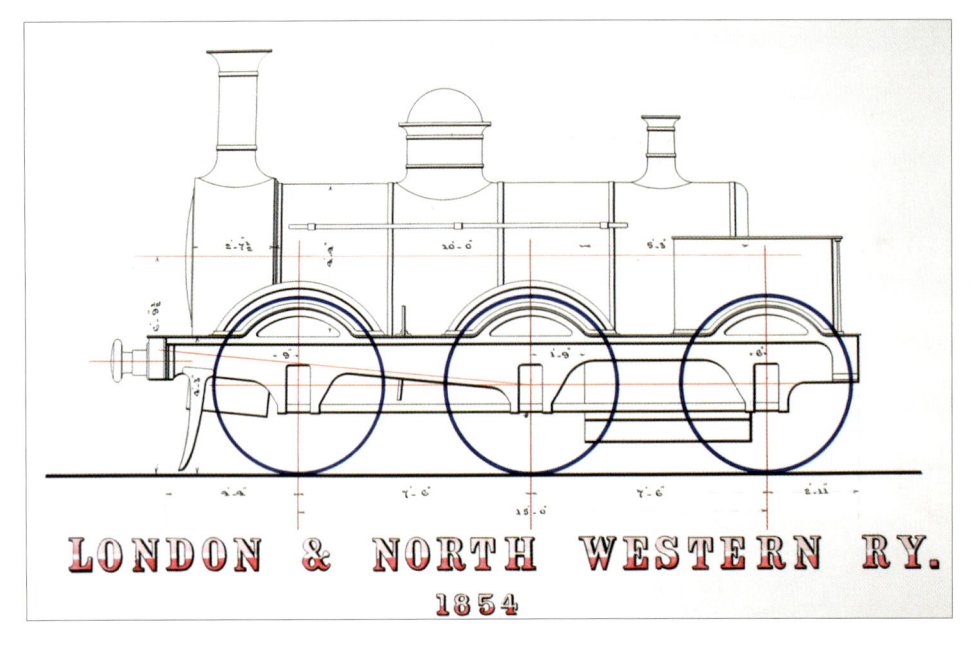

Fig. 4.11
Outline drawing of 0-6-0 locomotives designed by the London & North Western Railway and built in 1854 by Kitson, Thompson & Hewitson of Leeds ([one of batch] K 372-382/54).
Kitson collection, The Stephenson Locomotive Society.

the early 1870s, which thus removed opportunities for consolidation.[25] This, in turn, reduced the opportunity for larger locomotive batches.

When the larger railways set up their workshops from the mid-1840s locomotive superintendents supervised drawing offices which produced distinctive locomotive designs, manufacture of which was frequently shared between their workshops and the independent firms. An early example was James McConnell's passenger locomotive type for the London & North Western Railway in 1851, the class of 40 being shared between Sharp Brothers, Kitson & Co. and the railway's own Wolverton workshops.[26] Further 0-6-0 locomotives were built by Kitsons in 1854.

As the capability of railway drawing offices grew, tender invitations were accompanied by more detailed arrangement drawings. Whereas the earliest specifications of 1836 had been listed on two foolscap pages and were accompanied by two or three arrangement drawings,[27] by 1864 they took up, typically, fourteen double foolscap pages and were accompanied by more than fifty component and arrangement drawings.[28]

Locomotive weights and axle-loads increased as main line track improved, and greater tractive power was achieved through more wheels and higher boiler pressures. Design development was, therefore, directed towards increases of size and enhanced specification as well as through innovation. In a major paper in 1883/84, Edward Darbishire identified the several demand-led requirements which had influenced locomotive design in the previous twenty years:[29]

> First, undoubtedly, is the enormous development of traffic, demanding every day more and more powerful engines to conduct it.

> The passenger traffic… in carrying third class passengers by all trains, gave an impetus to travelling which has resulted in an immense increase in the weight of trains, which, moreover, are run at higher speeds than formerly. Local or suburban traffic has developed as facilities have been taxed to meet the demand for engines to suit the special conditions of this traffic. Goods traffic is worked at higher speeds than formerly, and, on many of our leading lines, through goods trains are run at a higher rate of speed for long distances without a stop. In fact, in all directions the tendency has been to require heavier and more powerful engines for all classes of traffic…

Design proliferation was not unique to Britain, there being a similar problem amongst American railroads. The railroad master mechanics had discretionary authority which not only gave them the freedom to dictate design specifications but also allowed them to shape the course of innovation.[30] As the railroads expanded in the last quarter of the century, their technical departments became larger, more formalised and more bureaucratised.[31] Responsibility for technical matters passed to staff offices, which sought to limit change in the pursuit of standardisation. From this grew the practice of restricting locomotive contracts to one manufacturer, which therefore obtained scale benefits of production denied to the British industry.

Other locomotive markets were also subject to an extraordinary proliferation of designs. The extensive use of British consulting engineers for overseas railways and the smaller domestic railways brought about this proliferation, further preventing the manufacturers from exploiting what would have been substantial economies of scale in production.[32] This restricted their competitive position against the American and German manufacturers towards the end of the century. With consulting engineers influencing production practices, this market feature played a key role in the manufacturers' attitudes to batch production.

Substantial overseas railway developments also increased locomotive performance requirements, for which specifications and detailed arrangement drawings were increasingly prepared by consulting engineers. For example, locomotives specified in 1854 for the Sydney and Goulborn Railway, the first main line railway in Australia, were to the London & North Western Railway designs of James McConnell, who was also the line's consultant.[33]

The consulting firms recruited locomotive designers and superintendents from both British railways and independent manufacturers. Edward Snowball, for example, Robert Stephenson & Co.'s talented Chief Draughtsman, was recruited as Superintendent of the Scinde Railway in India, although he was subsequently attracted to a career with Neilson & Co of Glasgow,[34] where he earned the high salary of £300 per annum.[35] From the 1850s, therefore, the manufacturers became increasingly restricted in their scope for design innovation, but, as they experienced improved arrangements and components from their larger customers, they were able to pass some of these on to their other, smaller railway customers, subject to any patent licensing arrangements.

The following summary of design innovations outlines the several changes in market requirements and highlights the transition of design initiative from the independent industry to the railway workshops and consulting engineers.

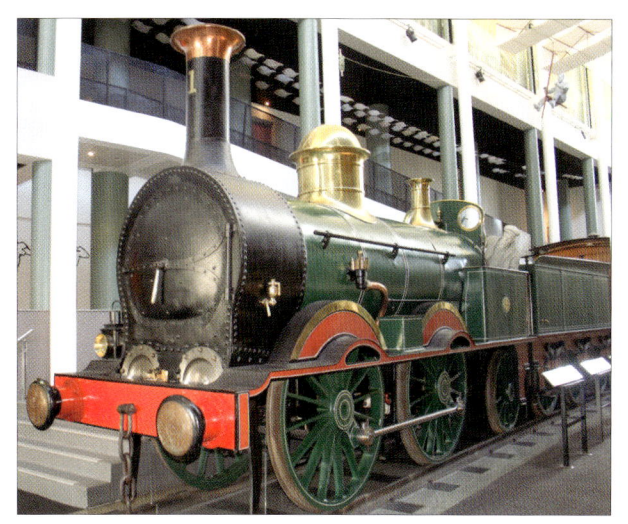

Fig. 4.12 0-4-2 tender locomotive built in 1854 by Robert Stephenson & Co. (RS 953/54) for the Sydney & Goulborn Railway in New South Wales to the designs of James McConnell. On display in the Powerhouse Museum, Sydney, New South Wales. *Author.*

The long-boiler design type, patented and introduced by Robert Stephenson & Co. in 1841, was motivated by fuel economy.[36] It provided an increased heating surface, whilst minimising: 'the escape of a large quantity of waste heat up the chimney.'[37] Its distinctive boiler/wheel layout, also introduced inside plate frames, vertical slide valves and boiler feed-pump eccentric drives.[38] Although constructed under licence by several manufacturers in Britain and on the Continent, it was more widely adopted by European railways, particularly in France and Germany, than in Britain.

Fig. 4.13 2-4-0 'long-boiler' locomotive built in 1846 by R. B. Longridge & Co of Bedlington for the Shrewsbury & Chester Railway (No.5). *Author's collection.*

Fig. 4.14
Example of an early 2-2-2WT tank locomotive built by Sharp Brothers of Manchester in 1850 (SB 650/50) for the Manchester, Sheffield & Lincolnshire Railway, (No. 73, *VENUS*)
The Locomotive Magazine Vol. 37 (1931), p.190.

The mixed frame locomotive, or 'Crewe Type', with inside bearings for the driving axle and outside bearings for carrying axles, was the first design initiative taken by a railway engineering team. It was developed by the Grand Junction Railway in the early 1840s, and was subsequently taken up by independent manufacturers, including Jones & Potts, E.B. Wilson and Kitson, Thompson & Hewitson.[39] It circumvented the Stephenson 'long-boiler' patent, and became popular amongst several railways in Britain, for whom the latter type was less suitable at higher speeds.

The 1840s also saw the introduction of entrepreneurial engineers, such as Thomas R. Crampton (1816-1888), whose inventiveness led to several lucrative patents, and who encouraged several railways to specify his designs leading to patents, licensing and royalty income.[40] Whilst employed by the Great Western Railway (GWR), Crampton pursued a radical departure in locomotive design when he patented a 'single-driver' locomotive, whose large, rear-mounted wheels were the particular characteristic. Although not taken up by the GWR, a few independent manufacturers in Britain and several European railways did manufacture the type.

In the 1840s, increased speeds were constrained by yawing motion, which led the Manchester-based Swiss engineer, John Bodmer (1786-1864), to pursue balancing the reciprocating mass of wheels and motion by counter-forcing the piston action. He worked closely with Sharp Roberts & Co., which built a small number of his balanced locomotives[41] until the simpler expedient of adding balancing weights to driving wheel rims, developed on the Eastern Counties Railway, became widespread.

Significant economy, through all-up weight reduction, was achieved for short-distance operations, by eliminating the tender, and introducing the tank locomotive, accommodating coke and water on the locomotive itself. Apart from a few experimental 'tank' locomotives, the first true tank design was introduced by Charles Tayleur & Co. in 1846 and taken up by Sharp Bros. and Jones & Potts in the following two years.[42] These were the fore-runners of many main-line tank locomotives for secondary duties.

The widely-adopted design type from the mid-19th century was the industrial tank locomotive for the expanding industrial market sector. The light-weight, un-complicated, short wheel-based locomotives, were competitively priced. The first standard examples, manufactured by Manning Wardle & Co. in 1858,[43] were followed by examples from several other manufacturers who retained full design control, allowing a high degree of standardisation whilst accommodating customer variations.

Fig. 4.15 0-4-0ST built by Manning Wardle & Co. in 1862 (MW 46/62) for the Marquis of Bute's Railway, Bute Docks, Cardiff (No.5).
The Locomotive Magazine, Vol. 30 (1924), p 204.

Fig. 4.16 An early narrow gauge locomotive. Kitson & Co. 2-6-0 built in 1869 (K 1596/69) for the 3 ft 6 in gauge Queensland Railway (No. 18). *Kitson collection, The Stephenson Locomotive Society.*

Narrow gauge railways, which were cheaper and more flexible than standard gauge for certain applications, opened new opportunities for the manufacturers. Robert Stephenson & Co. and Slaughter Grüning & Co. constructed the first 3ft 6in gauge locomotives for Norway in 1862, but an 1866 design by Beyer Peacock & Co. set the standard for narrow gauge main-line development.[44] Narrow gauge industrial lines in Britain and main line overseas railways developed extensively from the 1860s, further expanding opportunities for the independent industry, which adapted designs for several different gauges.

From the 1860s, steeply graded mountain railways, beyond reliable adhesion, called for a new type of locomotive. The consulting engineer, John Fell (1815-1902), developed and patented a 'centre-rail' adhesion system for application to the Mont Cenis Railway in Alpine Italy.[45] He licensed British and continental manufacturers to build the locomotives, which were subsequently employed on mountain railways around the world. In 1882, a new rack system using solid bars with vertical teeth machined into them was devised by Carl Roman Abt (1850-1933), a Swiss locomotive engineer. Two or three of these bars are mounted centrally between the rails, with the teeth offset.

A mountain railways rack system was also patented, in 1886, by Herman Lange (1837-1892), Chief Engineer and Joint Manager of Beyer Peacock & Co. and James Livesey (1831-1925), consulting engineer to several South American railways, who had himself been trained by Beyer Peacock.[46] Several British manufacturers were licensed to construct locomotives adopting one of these rack systems.

The introduction of the Metropolitan Railway's underground services in 1863 required a condensing

Fig. 4.17 An Abt system rack tank locomotive still in use. Dübs & Co. 0-4-2 RT built 1897 for the 3 ft 6 in Mount Lyell Railway & Mining Co. in Tasmania (Dübs 3730/97). In heritage service with the West Coast Wilderness Railway. *Author.*

Fig. 4.18
A condensing locomotive for the Midland Railway (No. 205). 4-4-0T built in 1864 by Beyer-Peacock & Co. (BP 776/64).
National Railway Museum, R.H. Bleasdale Collection.

apparatus to minimise smoke nuisance for passengers and train crew. The first locomotives, provided by the GWR with crude condensing equipment, were soon replaced by the Metropolitan Railway's own fleet, designed and built by Beyer Peacock & Co.[47] The simple condensing apparatus, which returned exhaust steam to the water tanks, was adopted by other railways, such as the Midland Railway, throughout the steam era.

In 1864 the consulting engineer, Robert Fairlie (1831-1885), patented his double-bogie, double-boiler articulated locomotives for use on secondary routes with sharp radii, particularly overseas.[48] After initial hesitation of the 'double-Fairlie's' novel design, by both railways and consulting engineers, he formed, in 1869, the Fairlie Engine & Steam Carriage Co., which however ceased within a year after only five locomotives had been built.[49] Thereafter, he energetically promoted the type, and licensed six British independent manufacturers and others in Europe and the United States to manufacture them. By 1875, Fairlie had received orders from forty-one railways around the world with track gauges ranging from 2ft to 5ft 3in.[50]

In 1868, a first articulated locomotive was built to the patented design of Jean Meyer (1804-1877), an Alsatian consulting engineer. The type became more popular in the 1890s when the Hartmann Works in Germany and Kitson & Co. in Leeds began to build variants for particular applications on steeply-graded mountain lines. A further articulated type, patented by Anatole Mallet in 1884, was first adopted for narrow-gauge use, but was later adapted for main lines, and in the twentieth century became the most widely adopted of all the articulated types.

Fig. 4.19 A metre gauge Double-Fairlie locomotive. An 0-6-0 + 0-6-0 built in 1900 by The Vulcan Foundry Co. Ltd. for the Burma Railway.
Author's collection.

CHAPTER 4 – TECHNOLOGY AND DESIGN

Fig. 4.20 A Kitson-Meyer locomotive for South America. An 0-6-0 + 0-6-0 locomotive built in 1895 by Kitson & Co. (K 3532/95) for the Anglo Chilean Nitrate & Railway Co. *Kitson collection, The Stephenson Locomotive Society.*

Manufacturers increasingly lost the opportunity to design main-line locomotives as specifications produced by consulting engineers became increasingly detailed. In 1884, for example, Robert Stephenson & Co. manufactured some 2ft 6in gauge tank engines for the steeply-graded Antofagasta Railway, to the 'general design and specification' of its consulting engineer, Edward Woods.[51] They incorporated Webb's compound system, Joy's valve gear, Cleminson's patent radial axle-box, Adam's bogie, Friedmann's injector and Hancock's 'inspirator'. The Stephenson Company undertook only the detailing work, prior to manufacture.

However, manufacturers did retain the capability for preparing new arrangements where, for example, consulting engineers issued 'Heads of Specifications' only. In such cases, they would undertake sufficient arrangement work to enable them to tender, leaving full detailing work to be undertaken once a contract had been awarded. In 1880, for example, James Livesey invited tenders for eight tank locomotives on behalf of the Chilean Taltal Railway.[52] The invitation explained that: 'As time is of the utmost importance, an outline specification only is given, so that manufacturers can adopt such patterns, dies, and templates as they may have. The tender to be accompanied with an outline tracing, and a detailed specification.'

It is thus apparent that manufacturers' design strategies for main-line locomotives were conditioned by their developing markets, and that they were obliged to accept an increasingly passive role in design innovation from the 1850s. Their drawing offices retained the facility to undertake arrangement work for their smaller customers in Britain and overseas, from whom they received broad specifications only. In this respect, the 'craft' firms, which retained a broad market base, manufacturing small batches of machinery and engines, were better placed than the 'progressive' firms which specialised in larger batch production of pre-specified locomotive designs. Although the larger British railway drawing offices provided manufacturing drawings, all other contracts, particularly those for overseas, required detailing work, which was undertaken by the manufacturers' own drawing offices.

Industrial locomotive manufacturers retained full control over their design work, albeit forming a few standard types, but which often involved additional detailing work to accommodate varying track gauges and other customer requirements.

Fig. 4.21 An 1880s standard industrial tank locomotive. Hudswell Clarke & Co. 1881-built 0-4-0ST (HC 216/81) *AIRE*, built for Atterbury & Shaw of Northampton. *Author's collection.*

Development of Thermodynamic Technology

A fundamental strategy for locomotive manufacturers was the search for improving thermodynamic technology, which would allow for increased efficiency through reduced fuel cost. Stephenson's *PLANET* class locomotives in the 1830s have been shown to have had a draw-bar thermal efficiency of about 2%.[53] Development work was undertaken by both manufacturers and railway workshop teams, and by the end of the century, a four-fold increase in efficiency had been achieved. Cheaper fuel, through the substitution of coal for coke, in addition to the efficiency increase, gave a larger reduction in fuel costs, although it is not possible to quantify this reduction. The developmental work was conducted empirically, as the scientific principles of thermodynamics were not fully understood. Only towards the turn of the century were scientific trials carried out, that led to the introduction of superheating with its further increase in thermal efficiency.[54]

As the new industry got under way in the 1830s, the potential for reduction in the volume of steam used through variable expansion, was sought for its most economic use with speed, load and gradient. The principle had been understood since 1828,[55] but, although the manufacturers improved the early, cumbersome valve gears, the search for a variable cut-off mechanism required considerable technological evaluation and design work.

The first improvement, in 1840, arose from experimentation in railway service, when the Liverpool & Manchester Railway's superintendent, John Dewrance (c.1803-1861), modified locomotives to allow steam expansion, and increased the blast-pipe diameter resulting in a 'sweeter draught, which did not tear the fire to pieces'.[56] Dewrance's experiments, which reduced coke consumption by over 50%,[57] not only made a major contribution to locomotive technology, it also demonstrated that manufacturers, without day to day operational contact, could lose initiative in locomotive development. This message was reinforced when Dewrance was entrusted to design and build locomotives in the company's Liverpool workshops, the first railway-built examples.[58]

In 1842, the staff of Robert Stephenson & Co. derived the variable cut-off 'link' motion, which was demonstrated to Stephenson himself using a small wooden model.[59] Criticising certain parts of its action, he requested a full-sized model to confirm its potential: 'If it answers it will be worth a jew's eye and the contriver of it should be rewarded'.[60] The successful trials of the 'Stephenson' link motion, which was not, however, patented, quickly led to general use of variable cut-off valve gears, a number of different forms being introduced by other manufacturers, notably Daniel Gooch (1816-1889) on the Great Western Railway in 1843, Egide Walschaerts (1820-1901) on the Belgian State Railways in 1844, and David Joy (1825-1903) as an independent venture in 1879.[61]

Fig. 4.22 'Link'-motion fitted to the 1844-built 2-2-2 *L'AIGLE* locomotive built by Robert Stephenson & Co. (RS 401/44) for the Chemin der fer Marseille à Avignon. Locomotive displayed at the Cité du Train Museum, Mulhouse, France (see Fig. 4.27). *Author*.

The next major technological advance was the development, in the 1850s, of the coal-burning firebox, which fulfilled smoke emission bans.[62] The motivations for the switch from coke were the reduction in fuel and maintenance costs and line-side fires, and it is notable that all development work was carried out by railway development teams, rather than by the independent manufacturers. Comprehensive in-service trials, during the 1840s and 1850s, were undertaken by locomotive superintendents, the research incurring incremental design improvements of fireboxes manufactured in railway workshops. Joseph Beattie (1808-1871) of the London & South Western Railway, James McConnell (1815-1883) of the London & North Western Railway and James Cudworth (1817-1899) of the South Eastern Railway each developed large and complex coal burning fireboxes, which they patented in their own names.[63]

Without the in-service trial opportunities available to the railway teams, the independent manufacturers were unable to contribute to the programmes. In 1857, however, Beyer Peacock & Co. saw the commercial opportunities of the coal-burning firebox and reached agreement with Beattie to tender for locomotives incorporating his design.[64] The arrangement was short-lived, however, as the definitive coal-burning firebox, incorporating a brick-arch and fire-hole deflector plate, was derived by the Midland Railway during trials in 1859/60. The features were already known, although not previously arranged as a coherent design, and there was no opportunity for patent protection. The breakthrough was announced to the Institution of Mechanical Engineers in 1860,[65] and the design was adopted by all European manufacturers, ending the use of coke.

The potential for improved efficiency through the 'compound' use of exhaust steam in a further, low-pressure cylinder provided the next research effort, again by railway development teams, particularly in countries, such as France, that lacked cheap coal.[66] In 1876, a prototype compound locomotive was made by Anatole Mallet (1837-1919) for the Bayonne & Biarritz Railway, and shown at the 1878 Paris Exposition. The fuel-saving potential soon led to the introduction of compound locomotives elsewhere in France and Germany. Mallet read a paper to the Institution of Mechanical Engineers in 1879,[67] prompting several British railway locomotive superintendents to pursue compound development programmes, which took several forms, leading to a number of patents.[68]

Compound development programmes were carried out by independent manufacturers in France (Société Alsacienne de Constructions Mécaniques – SACM), and the United States (the Baldwin Locomotive Works' 'Vauclain' type), with the close co-operation of railway locomotive engineers.[69] By the 1890s, compounding was more widely used on European than on British railways who remained equivocal in its application.[70] The higher maintenance costs of compound locomotives were generally thought to be greater than the savings from reduced coal consumption, and were generally less acceptable because of the conditions under which they were required to work.[71]

Again prompted by high coal costs, oil-burning trials were undertaken in 1868 on the French Chemin de Fer de l'Est, but the cost of oil was greater than coal, and the project was abandoned.[72] Liquid fuel refining in the 1880s provided a residual fuel oil which was adopted on the Gryaze-Tsaritsynskaya Zh.D (Gryaze-Tsaritsin Railway) in Russia.[73] Although there was little economic incentive to pursue oil-fuel burning in the 19th century, an embarrassing surplus of waste oil from the Great Eastern Railway's gas-making plant (for carriage lighting) in the 1880s, led it to experiment with the fuel, which was adopted for use on several dozen express locomotives.[74]

The scientific principles of thermodynamics were first pursued in the 1890s, although superheating was not adopted until the early years of the twentieth century. The German engineer, Wilhelm Schmidt (1858-1924), experimented with stationary engines, before the first locomotives fitted with superheaters were built in 1898, by Henschel & Sohn and the Stettin Maschinenbau A.G. Vulcan. The major improvement in thermal efficiency with a low maintenance requirement quickly endeared it to the world's railways. Superheating has been recognised as: 'the greatest step forward in steam locomotive technology since the days of Stephenson', and was quickly and extensively fitted to British locomotives from the mid-Edwardian era.[75] Its introduction did much to displace the need for compounding, particularly in Britain.

The independent locomotive industry thus played little part in efficiency improvements after the introduction of the link motion in the early 1840s. As a direct consequence of the greater role of the British railway development teams and the subordination of the manufacturing industry to a largely 'sub-contracting' role, it carried out little further research and development work but encouraged railways to accept standardised types. This was in marked contrast to the European and American industries, which maintained their development roles, in close co-operation with their railway customers and with technical institutes who pursued scientific enquiry from the 1890s. Even if the manufacturers had wished to pursue their thermodynamic work through co-operative ventures with British railway development teams, they were effectively prevented by the aspirations of the railway superintendents building on the capabilities of their workshop teams.

Development of Material Technology

The development of materials to meet new component requirements was as important to design development as arrangement and component detailing. Keeping material technology in tandem with design progress had been a major problem for the manufacturers since the 1820s.[76] The manufacturers actually initiated material technology development whilst pursuing locomotive design progress. As designs evolved, their implementation was slowed until suitable materials were available, sometimes requiring new raw material and component supply industries. By its nature, material development depended on both the supply industries and the manufacturers' railway customers.

The rapid growth of the locomotive industry in the early 1830s placed unprecedented demands on its material suppliers, particularly the iron industry, in terms both of quality and quantity. The suppliers had to develop materials that could meet demanding specifications on strength, stress and weight. From 1839 for example, locomotives supplied to Austria had to undergo an hydraulic boiler test of three times their working pressure under the supervision of 'a learned Professor', which required strengthened copper firebox plates.[77]

The manufacturers therefore worked closely with suppliers to develop materials of requisite strength that

Fig. 4.23
A locomotive subjected to extreme pressure testing by a 'learned Professor'. 0-4-2 locomotive *AJAX* built by Jones, Turner & Evans of Newton-le-Willows in 1841 for the Kaiser Ferdinands-Nordbahn in Austria. On display in the Technisches Museum in Vienna. *Author.*

could demonstrate safety and economy to the railways. The failure of several components in the first decade of operations, however, was of such concern that railway locomotive superintendents began to specify materials for particular applications. These were especially high-quality iron grades, such as that from Low Moor (Bradford), and other grades from Yorkshire and Staffordshire. By 1850, the superintendents of the largest railways had developed personal contacts with suppliers and became quite specific in their material requirements for most locomotive components. One example in 1847 required: 'best Lowmoor plates' for the boiler, smokebox and front tube plate, 'best Staffordshire iron plates' for the ashpan, and 'best copper plate' for the firebox.[78]

For industrial countries with a relative scarcity of good quality iron ore, such as Germany, imports of iron products from Britain incurred heavy import tariffs. The German iron industry was therefore persuaded to undertake considerable development work to provide suitable grades of steel. Its more durable characteristics and higher tensile strength offered the prospect of longer life, reduced maintenance and replacement costs, and weight reduction. Each type of steel had different tensile characteristics, however, which had to be independently developed for industrial use, including locomotive components. These early forms of steel were beset with metallurgical problems, which required considerable development work for over a quarter of a century before becoming a safe and cost-effective alternative to wrought iron.

The Krupp company of Essen developed crucible cast steel for tyres in 1851.[79] Trials confirmed its hard-wearing properties, but considerable research was necessary to minimise fracture problems. Krupp set up an agency in London to promote sales of the tyres, the first imports being in 1856, whilst, in 1859, Naylor & Vickers of Sheffield began supplying them to the London & North Western Railway. Cast steel was then widely specified by railways in Britain, France and Germany as, in spite of its higher initial expense, costs were less over the longer life of the tyres. Bessemer steel, and an improved rolling mill developed by George & Co. of Rotherham in 1864, later provided cheaper rolled steel tyres. Although they were quickly brought into general use by the British steel industry, metallurgical problems persisted, and, as late as 1876, tyre fractures remained the subject of enquiry and research.[80]

The Krupp company also developed axles of crucible cast steel, the first examples being shown at the Great Exhibition of 1851. British railways were wary of them, because of a high failure rate, particularly of crank axles, the material

Fig. 4.24
Example of steel crank axle made by F. Krupp and retro-fitted to the 1841-built 0-4-2 tender locomotive *AJAX* (Fig. 4.23 above). Author.

being brittle and less able to withstand running forces. Axle steel was supplied for forging into crank and straight axles, the caution by British locomotive manufacturers arising out of lack of metallurgical understanding. Unlike the German firms, they were inexperienced in working the steel to ensure perfect homogeneity.[81]

From the 1850s, therefore, the British and German steel industries and, from the 1860s, the larger British railway workshops were in the forefront of development work for the several kinds of locomotive steel. The subordinated role of the independent industry from this time, however, meant that it had no opportunity to participate in this work. Its knowledge of the use and working of steel resulted from experience gained from orders which specified the use of each steel grade.

The potential uses of steel were considered early by the railway industry, in close collaboration with the developing British steel industry. The LNWR's Locomotive Superintendent, John Ramsbottom (1814-1897), introduced a Bessemer steel-making and rolling plant into Crewe Works in 1864.[82] Ramsbottom's assistant, Francis Webb (1836-1906) undertook research and development work into steel boilers and fireboxes that was keenly watched by other locomotive and consulting engineers. A further steel plant, for the Horwich locomotive works of the Lancashire & Yorkshire Railway, began production in 1886.

Motion parts, particularly piston rods, valve spindles, slide-bars and connecting rods were switched to steel with the same caution following the first use of Krupp steel in 1862. Forged Bessemer steel axles were introduced by Naylor & Vickers in 1866, but, although successful trials were carried out, there was no major cost benefit, Yorkshire wrought iron axles being only slowly displaced over the following twenty years.[83] As late as 1884, Beyer Peacock & Co. retained an option for the use of wrought iron:[84]

> ... in cases where the selection of materials for axles ... is left entirely to us, we shall use mild steel of good quality, instead of wrought iron ... but you need make no increase in your prices over those which you have been in the habit of quoting for iron. If however crucible steel made by Vickers, or Krupp, or any other special steel is specified, you must quote the additional price as at present arranged for steel ...

The introduction of Bessemer steel plate rolling mills in the 1860s, made possible rolled steel frames and boiler plate. From 1867, frame plates could be rolled in one piece, offering considerable cost savings over wrought iron sectional frames welded under steam hammers. Locomotive superintendents were particularly cautious about adopting steel plate for boilers, however, as the plates pitted more quickly than wrought iron. After considerable development work and trials by the LNWR's Crewe Works, confidence was gained, and steel boilers became generally specified from the late 1870s.[85] By 1886 Webb stated that: 'Since commencing to make steel plates, we have made in the [Crewe] works 2,752 locomotive boilers ... [and] 230 stationary and marine boilers, and of all the material used, not a single plate has ever failed.'[86] Their introduction permitted higher steam pressures, rising from typically 120 lbs to 175 lbs per square inch in fifteen years, providing a significant economic justification for their use.[87]

Firebox steel was first made in Sheffield in 1860, and trial fireboxes were made in 1862 by Daniel Adamson & Co. of Hyde. Trials with different kinds of plate, however, revealed that, when worn, they became too weak for the screwed stays, and frequently gave way: 'with a loud report when on the road.'[88] Although steel fireboxes were cheaper than the normal British copper type, overall costs were higher due to their short life.[89] After several trials, including a major series undertaken by the LNWR in 1872-73, British railways remained unconvinced about steel fireboxes. In spite of widespread use in North America, locomotive superintendents persisted in specifying copper fireboxes until well into the 20th century. As late as 1927 it was concluded that: 'The riddle of the failure of the steel firebox in British locomotive practice still remains unsolved.'[90]

The independent locomotive industry thus played no further part in material development after its pioneering work in the 1830s and 1840s, being a further example of responsibilities passing to the railway development teams and consulting engineers. When inviting tenders for locomotives, they specified both the steel grades and a short-list of suppliers to the manufacturers, who were then required to provide the necessary capital equipment and expertise to work the materials.

The European and American locomotive industries were also dependent upon steel supplies from specialist firms, such as Krupp and Pennsylvania Steel, with whom they co-operated on development work, although as early as 1873, the Baldwin company acquired its own steel works to reduce its dependency on external supplies.[91] In Britain, however, the lead in steel development for locomotives remained vested with railway companies, particularly at Crewe Works, working closely with the British steel industry.

Construction of copper fireboxes, in the early 1830s, created the need for an entirely new copper-plate industry.[92] Copper producers, such as Thomas Bolton & Sons of Birmingham, developed rolling mills to meet the new demand. With the predominant use of copper for fireboxes and tube-plates throughout the century, in spite of trials with steel plates, the copper-plate industry grew significantly with several firms participating, especially Boltons, which established large rolling mills in Widnes to supply the locomotive industry.[93]

The manufacturers carried out trials of several copper alloys to provide brass, bronze and gun-metals suitable for locomotive bearings and other wearing surfaces and fittings. There were considerable variations in bearing life between the manufacturers, as much due to surface quality as the varying composition of the metal.[94] The introduction of the harder wearing phosphor-bronze in the 1870s introduced a new chemistry. The Vulcan Foundry, for example, had derived six copper-alloy compositions for phosphor-bronze for different components by 1885.[95]

Fig. 4.25 Copper firebox crown-plate recovered from shipwreck site off Islay, Hebrides. Fitted to an 1857 Neilson-built 4-4-0 locomotive (N 386/57) for the Nova Scotia Railway in 1857 (see Fig. 4.33). *Author.*

Fig. 4.26 Cast bronze crosshead boiler feed-pump recovered from the 1857-built Neilson 4-4-0 locomotive (N386/57) for the Nova Scotia Railway in 1857, recovered from shipwreck site off Islay, Hebrides (see Fig. 4.33). *Author.*

Patents

The importance for manufacturers to adopt patent protection for novel locomotive design features had been well illustrated by the Stephensons' omissions in the late 1820s. Patent law was an important legal procedure for inventors, providing them with a period of exclusivity to protect themselves from imitators and pursue financial exploitation, through manufacture and sale, or through licensing royalties. It has been suggested that patents merely indicated the potential profits inventors perceived they could make from inventions.[96] However there was no theoretical debate concerning the nature of invention, and the patenting process was a largely anonymous activity. Following the Patent Law Amendment Act of 1852 the patenting process become the subject of considerable controversy and a platform for the popularisation of 'heroic' inventors.[97]

As with other industrial sectors, the number of railway, particularly locomotive, patents grew markedly, and by 1850 represented the second largest industrial category after textile machinery.[98] The 374 railway patents in the 1840s were two and half times more than in the 1830s, representing some eight per cent of the total. The majority of patents, however, could demonstrate neither technical nor financial benefit and were not taken up, and only a handful produced significant incomes for the patentees.

The most successful early patent was taken out by Robert Stephenson in 1841, for his company's 'long-boiler' locomotive type, with which he demonstrated a determination to maintain a design lead over Edward Bury and other aspiring manufacturers.[99] The partners of Robert Stephenson & Co. minuted:[100]

> Rob. Stephenson having stated that he has some improvements in Locomotives in view which he apprehends it would be to the advantage of this concern to take out a patent for, he is encouraged to effect the same in his own name on our behalf.

No discussion was recorded regarding sharing the royalties, and it would seem that the partners were content that their interests would benefit from an increase in orders for the Newcastle factory. Stephenson received all royalties personally over the patent's fifteen-year life.

The 'long-boiler' type was widely adopted in Britain and Europe, with manufacture initially being limited to the Stephenson Company for the British market, and to British and foreign firms for the European market, royalty negotiations for which were handled by Stephenson's agent, Edward Starbuck. When the market was low in 1843 and early 1844, Stephenson resisted all approaches from manufacturers seeking to make 'long-boilers' for the British market. Starbuck wrote curtly to Kitson, Thompson & Hewitson:[101]

> You are probably aware that he [Robert Stephenson] has already repeatedly stated to Parties both in England & on the continent his objection to this permission [to adopting his patent arrangement of Engine] & his interests have been so injuriously affected by the circumstance of other manufacturers having offered & undertake to make Engines on his patent arrange't that he is compelled to adhere to this decision…

Relationships with other manufacturers became strained over matters of infringement. When, without agreement, R.&W. Hawthorn, the Stephenson Co.'s neighbours in Newcastle, began making a long-boiler locomotive in the spring of 1844, the Stephensons' Head Clerk promptly wrote:[102]

> Gentlemen   Having been credibly informed that you are about Exporting one or more Locom've Engines, upon a construction which embraces our Patent, & which is a manifest infringement of it, we feel called upon for the protection of our interest, to protect against such use being made by yourselves or any other Engine Builders of our Patent right & to state that we cannot consent to your delivery of such Engines. The writer will be in Newcastle in the early part of next week, & will then be glad to receive any communication you may be disposed to make upon this subject.

Hawthorns at first denied any infringement, but, after a meeting with Stephenson, they conceded the point, agreeing to a £50 royalty per locomotive. Stephenson felt: '… <u>very strongly</u> the shameful manner he has been treated by H. & Co. in denying in the <u>first</u> instance… the infringement…'[103] Hawthorns later objected to the agreement and claimed that the positioning of their rear axle was sufficiently different from the patent to make it void. The Stephenson Company promptly sought legal advice, which found in its favour. Starbuck wrote:[104]

> … as a matter of <u>Equity</u> H & Co must know themselves wrong – as a matter of <u>Law</u> the high opinion of Webster [barrister] makes them so. The maintenance of our just Right in this question will have great weight on the Continent, such affairs give a prestige much in favour of the Inventor.

The poor relationship between the two companies was thus further strained, but the litigation confirmed to the industry as a whole the full strength of a patent. As the locomotive demand rose in the 'mania' years of 1845-47, however, the Stephenson Company could not keep up with the demand and allowed orders to pass to other manufacturers in return for royalty payments.

Different royalties were negotiated by Starbuck for each market. He offered French manufacturers an agreement to make any number of locomotives for £1,000, or £50 for each one made.[105] In Holland he obtained £25 for each locomotive, whilst noting that the Dutch patent laws lapsed if a patent was not exercised for two years.[106] In 1846, the royalty was: '£10… an Engine… including everything – Wheels &c', the latter being duplicate components.[107] In 1854, the Paris & Orleans Railway was charged £10 per locomotive for a fleet of twenty-six. Three smaller orders in the same year generated a royalty of £30 per locomotive,

Fig. 4.27
A long-boiler locomotive adopted in France. Built by Robert Stephenson & Co. in 1846 (RS 401/46) *L'AIGLE* for the Chemin der fer Marseille à Avignon. On display in the Cité du Train Museum, Mulhouse. *Author.*

suggesting a stepped rate of £30 each for less than ten, and £10 above that number.[108]

Keeping track of licensees' output, and of manufacturers who sought to infringe the patent, was a time-consuming activity for Starbuck, who checked all orders to claim Stephenson's royalties, of which he, in turn, received a 15% commission. Starbuck's commission accounts show that Stephenson earned several thousand pounds in royalties, particularly in the peak year of 1846 when he grossed over £12,000.[109] In the reduced market conditions of 1848, Stephenson looked back at his substantial income over the previous four years and paid Starbuck a gratuity of £500 over and above his commission.[110]

The Stephenson Company had won few friends with its rigid application of the patent rights. This helped to persuade other manufacturers, notably Sharp Brothers and E. B. Wilson & Co., to develop the alternative, 'mixed-frame' types that, without the royalty premium, could be more competitively priced.[111] These were judged to be significantly better than the 'long-boilers' which were judged to 'hunt' in motion under the higher speed operating conditions in Britain. The use of 'long-boilers' in Britain was, therefore, limited, but this contrasted with Continental railways, particularly France, Belgium, Holland and Germany, whose slower operating speeds made them popular long after the patent had expired.

Although the Stephenson Company's link motion of 1842 was such an important technological step, it was not patented.[112] William Howe, who helped develop the valve gear, gave evidence to a patent tribunal in 1851 stating: 'I do not think there was ever one [patent] applied for. There seemed to be a doubt whether it would act effectively or not at the time.'[113] The tribunal, which enquired into precedence of expansion valve gear, arose out of action by John Gray (1810-1854), patentee of the 'horse-leg' valve gear of 1839,[114] against the London & North Western Railway for its use of the Stephenson link motion. Gray's gear may have influenced Stephenson's decision not to patent the link motion, although its form was sufficiently novel. The tribunal, which denied Gray's claim, was seen by the Stephenson Company as being: 'of the utmost importance, not only to railway companies but also to manufacturers, [and] it is desirable to use every exertion to prevent a claim which I think is unjust.'[115] The link motion became the most widely used valve gear and, had it been patented, would undoubtedly have brought considerable income to Stephenson.

The extraordinary profits of the 'long-boiler' patent gave incentive to engineers, without a manufacturing affiliation, to patent their inventions, the first being Thomas Crampton.[116] A prolific and popular inventor, he took out several patents, including the first one for his 'Crampton'-type of 1842. He licensed manufacturers in Britain[117] and the United States, but the type was not seen to offer advantages over normal express types and only about twenty-five examples were built in each country. Nearly 300 were constructed by licensed manufacturers in France and Germany, however, from which Crampton received significant royalties.

The Patent Law Amendment Act, which limited each patent to a single invention rather than a cluster as hitherto, was preceded by a Parliamentary Select Committee

Fig. 4.28
Crampton-type 2 + 2-2-0 locomotive built by Tulk & Ley in 1848, *KINNAIRD*, for the Dundee, Perth & Aberdeen Junction Railway.
National Railway Museum, R.H. Bleasdale Collection.

and succeeded by an extraordinary controversy, during which a strong abolitionist movement was formed.[118] Following enactment, however, many patents were taken out for improvements to locomotive components, a small proportion being taken up, some licensed to manufacturers and earning royalties for their inventors, others prompting the establishment of new companies forming an important expansion of the component supply industry. The Act also separated British and Colonial patents, which then required separate registration.[119]

Reflecting the reduction in design and technological discretion for the manufacturers in the 1850s, there was a shift in locomotive-related patents firstly to independent engineers, motivated by potential profits, secondly to railway locomotive superintendents, following their technological development work such as for coal-burning fireboxes and compounding types, and thirdly to consulting engineers, following their work on articulated and mountain design types. The majority of patents were however, related to components, the following being a summary of the main ones.

The most important component to be introduced was the steam injector, patented throughout Europe and North America by the French engineer, Henri Giffard (1825-1882) in 1859.[120] It allowed water to be fed into the boiler at any time, a marked improvement over cross-head driven feed-pumps which required locomotives to move to pump water. It also allowed an increase in boiler size, hitherto limited by a secure water supply. In competition with Robert Stephenson & Co., Sharp Stewart & Co. secured sole rights for its British manufacture, including those for export.[121] The firm supplied thousands of injectors to British locomotive manufacturers. In 1888, when Sharp Stewart & Co. moved to Glasgow, the company sold its entire injector manufacturing division to the Patent Exhaust Steam Injector Company Limited, which later became Davies & Metcalfe Ltd.[122]

John Ramsbottom developed the water pick-up apparatus for London & North Western Railway express trains to pick up water on the move providing a significant saving in time and operating costs. Patented in 1860, it was first used only by the railway, but by the 20th century it was adopted by several other railways in Britain and a few overseas.

To overcome the route limitations of long wheel-based locomotives, lateral moving 'radial' axles were introduced in 1863 by the British engineer, William Adams (1797-1872), and about the same time, by August Wöhler (1819-1914), Locomotive Superintendent of the Prussian Niederschlesisch-Märkische Eisenbahn-Gesellschaft (Lower Silesian Railway Company). Adams' basic design was adopted extensively, but with several improvements, most particularly by Webb on the London & North Western Railway, and by a Mr. Bottomley, whose patented design of 1881 was adopted by Sharp, Stewart & Co. Ltd. and other manufacturers.[123]

The most significant of the 'proprietary' patents was a group of over twenty taken out in the United States by George Westinghouse (1846-1914) between 1869 and 1873, following his development of the continuous air brake system.[124] He experienced particular difficulty introducing the system in the United States, which required a consensus

Fig. 4.29 Locomotive fitted with patent rear radial axle-box. Sharp Stewart & Co. Ltd. 1880-built 4-4-2T (SS 2882/80) for the London, Tilbury & Southend Railway (No. 3, *TILBURY*). *Peter Wardle Collection*.

amongst railroads and industry-wide co-operation. Only through the encouragement of national standards by the Master Car-Builders' Association and a series of trials, was a standard agreed upon in the late 1880s.

Westinghouse was advised of the British requirement for automatic braking by James Dredge (1840-1906) the Editor of *Engineering*, prompting Westinghouse to develop, patent and introduce the triple-valve air brake system.[125] Reaching a consensus on a standard braking system in Britain was just as difficult as in the USA. The larger railways researched and developed a variety of alternative systems, in an endeavour both to improve upon the Westinghouse performance and to avoid being dependent upon a major component supplier. The Board of Trade directed railways to introduce automatic braking for passenger services (but not for goods services), and the two systems that were generally adopted in the 1880s and 1890s, were the Westinghouse air brake and the vacuum brake.

Many small component patents were taken out, an example being for india-rubber pads fitted into coupling rods and axle boxes, to allow freedom of movement on curved track. The patent, promoted by the Fairbairn Company, received only limited application.[126] Several components were patented in Britain by overseas suppliers, an example being the 'Jerome Patent Packing' anti-friction metallic rings that were developed in the USA and, by 1885, were licensed to five British locomotive manufacturers.[127]

The list of locomotive and component patents confirms that the conclusion of the 'long-boiler' patent in 1856 marked the end of design initiative by the independent locomotive industry. Thereafter, invention, expressed through patents, was mostly the pursuit of independent and railway-employed engineers. Of those that went on to production, the majority were manufactured under licence, whilst others gave rise to new component industries. The lack of inventive effort by the manufacturers after 1850, adds further evidence of their move away from developmental design, and their growing dependence on customer-led designs for manufacture.

The loss of design discretion had even greater impact on the industry's long-term competitiveness in the world market. The withdrawal of that discretion from the home market, in favour of railway design teams, and the overseas market, in favour of consulting engineers, left the manufacturers with little opportunity to innovate. The larger railways saw the firms as extensions of railway workshops, to which they supplied manufacturing drawings, leaving little room even for detailing work. An industry that had once benefited from exclusive manufacturing rights and licensed income from its own patents, was subordinated to a contract rôle, negotiating the best terms with which to employ the rights of other patentees.

The main-line locomotive industry was unable to persuade its railway customers, or influence their consulting engineers, to depend upon the manufacturers for technological, material and design progression. Once the railway and engineering 'empires' were developed, the manufacturers had to adopt compromise strategies which saw them play a passive role in this progression. Design work was either eliminated, where railways supplied manufacturing drawings, or was limited to arrangement and detailing work.

This lack of inventive output contrasts with the United States, France and Germany, whose locomotive industries remained prominent inventors. Patents taken out by American manufacturers provided them with important competitive advantages and were used to forestall competitors either through high royalties or refusal of manufacturing rights.[128] Baldwins alone took out fifty-three component patents between 1877 and 1900, and one major design patent for its four-cylinder 'Vauclain' compound type.

Standardisation

The progress in locomotive design in the 1830s and 1840s fulfilled the railways' requirements for greater speed, haulage and economy, but the proliferation of manufacturers resulted in incompatible components for similar locomotive types. Manufacturers were cautious to expand their production capacity too quickly due to the fluctuating demand, and railways were obliged to spread their orders across several firms to avoid lengthy delivery times. The resulting lack of component standardisation concerned the larger railways, which prompted their preparation of specification drawings. It also added emphasis to the need for standardisation of many components down to the very basics of interchangeability such as bolt threads.

The cost benefits of standardisation and component inter-changeability had been considered since Daniel Gooch's *Fire Fly* design for the GWR in 1840.[129] By the end of the 1840s, the major manufacturers were well aware of the production benefits of standardisation, and their designs generally adopted common components. As more locomotive superintendents developed specifications for their railways, however, opportunities for standardisation reduced. To discourage this trend, from 1848, E.B. Wilson & Co. offered only three standard designs, for passenger, goods and secondary duties, persuading their customers to adapt their specifications to those designs, for which there was a competitive price incentive.[130] If the customer so much as required a clack-box to be altered from the basic design, he was surcharged up to £25 for the alteration.

As railways and, later, consulting engineers developed their design capability, locomotive designs multiplied within each type, according to the perceived requirements of each line. Although each locomotive superintendent sought to standardise, his perception was limited to the internal needs of each railway, without regard to national standards. The developing intra-standardisation was thus a consequence of the growth of railway workshops, each pursuing the needs of its parent railway. This proliferation of locomotive designs from both railways and consulting engineers significantly restrained batch size and the opportunity for production economy in the second half of the century, and constrained initiatives that manufacturers may otherwise have pursued towards standardisation.

The largest number of locomotives of any one type was the DX class goods locomotive of the London and North Western Railway, of which 943 examples were made, mostly at Crewe workshops. Each locomotive was identical with the others in the class, which represented the first example of large-scale batch production.[131]

With the rise in intra-standardisation, the drift away from national standard designs became all too evident, and in 1855 Daniel K. Clark, an authority on locomotive design, wrote:[132]

> …probably five distinct classes of locomotive would afford a variety sufficiently accommodating to suit the varied traffic of railways, whereas I suppose the varieties of locomotives in actual operation in this country and elsewhere are very nearly five hundred in number. Everyone cannot be right, and

Fig. 4.30
A standard design by E.B. Wilson & Co. 0-6-0 locomotive built by Manning Wardle & Co. of Leeds in 1866 (MW 185/66), to a design principally that of E.B. Wilson & Co., originally delivered to the London, Brighton & South Coast Railway. As seen, largely unaltered, on the Manchester, Sheffield & Lincolnshire Railway (No. 132) in 1896. *Locomotives and Railways, Vol.1 (1900), p.110.*

most of them must be wrong, and it would be for the best interests of railways if the proper authorities could be unanimous in the selection of a good number of classes to uniform patterns to be adopted in future practice.

The consequences of locomotive design proliferation were small batches and high unit costs of production. It was a major factor in the economics of locomotive provision in the 19th century, and in the competitiveness of the independent industry against overseas competition later in the century. There were four contributory factors leading to design proliferation, namely:[133]

- a heavily decentralised railway system, where a minimum of workshop facilities was essential for efficient operations due to the unreliability of the first locomotives, which acted as a powerful encouragement for the proliferation of locomotive classes. This factor, however, does not take account of the fact that overseas railways were equally decentralised, each with their own workshops for the repair of similar early locomotives, but which did not venture into their own manufacture. Continental and American manufacturers moderated design proliferation by transferring innovation from large customers to smaller ones.

- the development of technology, in which the innovative process itself contributed towards product proliferation, with standardisation only becoming meaningful in the 20th century, using improved machine tools, new methods of workshop organisation and developments in metallurgy. However, thermodynamic and material innovations were important steps in locomotive evolution, which were applied to the benefit of all designs. Innovation was not, of itself, a cause of design proliferation.

- the market for railway services, in which a proliferation of passenger classes occurred to meet the changes in passenger to train weight ratios between the fastest express services and the ordinary passenger services. There was such an expansion through the introduction of fast express, fast goods, and suburban services, but it was relatively modest, and could be fulfilled by just a few standard classes. However, the development of further locomotive design types made more ambitious routes possible with steeper gradients and sharper curvature, which could be accommodated by narrow gauge or mountain type motive power.

- the independence of locomotive superintendents, who established 'private empires' in which design idiosyncrasies and proliferation of types could flourish, and who became a law unto themselves with locomotive design. This was undoubtedly true, being a consequence of the growth of railway workshop 'empires' each undertaking developmental and design work. Intra-standardisation practices gave little incentive for co-operation between railways, even though standardisation was generally encouraged through the Technical Committee of the Railway Clearing House. Exceptions were few, such as the Somerset & Dorset Joint Railway designs undertaken by the parent Midland Railway. The legal rebuff by the Locomotive Manufacturers Association, to the London and North Western Railway's manufacture of locomotives for the Lancashire & Yorkshire Railway in the 1870s, further removed incentive for design co-operation between railways.

It has been concluded that locomotive design proliferation prompted diversification in production practices amongst railway workshops and led to collusion regarding labour relations and labour market benefits.[134] It was thus partly a consequence of a perceived advantage in maintaining internal labour markets among skilled workers. The consequences of the design proliferation on the independent industry, in terms of standardised component production, batch sizes, and production practices, are considered in the next chapter.

Similar proliferation occurred with the consulting engineers' overseas designs, further diluting the manufacturers' ability to contribute to main-line locomotive standardisation, even for comparable markets, such as India. From as early as 1873, standardisation of Indian locomotive designs was considered on several occasions by the India Office, but it was not achieved until the Edwardian era, because of divided opinions between the consulting engineers in England and the locomotive superintendents in India. The consulting engineer, Sir Alexander Rendel, wrote in 1894:[135]

> It appears that the Locomotive Superintendents of India concur in one thing only, and that is, that what Consulting Engineers to the Secretary of State and to the Companies send out to them is more or less wrong.

The proliferation of British locomotive designs in the nineteenth century, and the lack of component standardisation did much to increase the unit costs of locomotive production. The continued requirement for small batches of non-standard types allowed the 'craft' manufacturers to remain in the locomotive business, where larger batch runs of standard types would have favoured the 'progressive' manufacturers. The continued

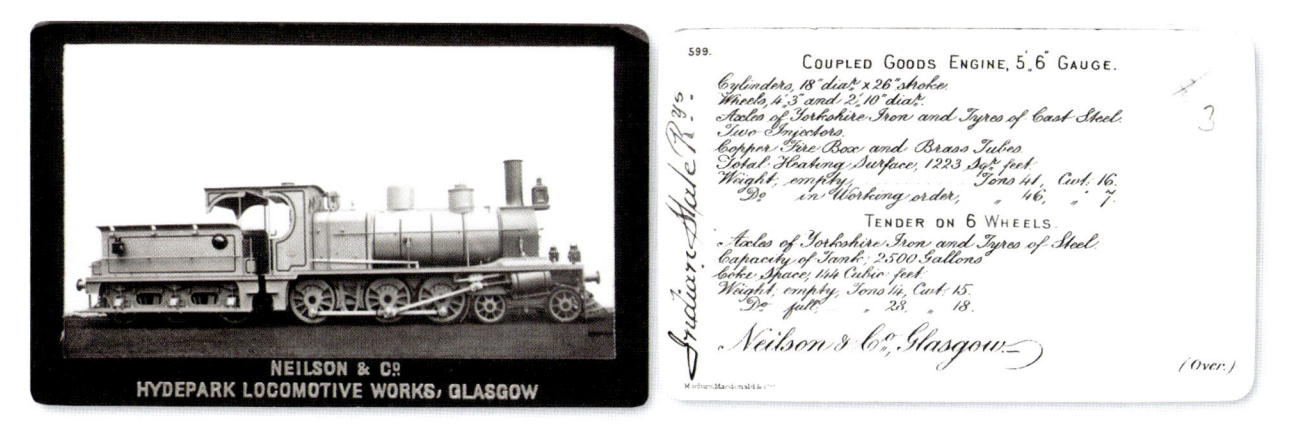

Fig. 4.31 Lack of standardisation in India in 1880. A 4-6-0 goods locomotive built by Neilson & Co. for the Indian State Railways. Data Card Obverse and Reverse views. *Author's collection.*

proliferation of independent firms was thus conditioned by the imbalance in the economics of locomotive production due to the lack of standardisation. For overseas sales, the effective cartel created by the dominant influence of the London market, largely shielded the high cost of production from comparison with the American and European industries. From the 1880s, however, the greater exposure to competition from these industries revealed both the higher costs of production and the inadequacies of design for some overseas railways.

The assessment of the 'chronic lack of standardisation' was the result of a fragmented market and an almost inevitable consequence of so many locomotive superintendents in charge of design and development teams. Had the independent industry not been forestalled in its ambitions, the standard designs of E. B. Wilson & Co., Robert Stephenson & Co., and Sharp Stewart & Co. may well have gone on to become the benchmarks for British domestic locomotive design development, emulating the German and American industries.

Figs. 4.32 Examples of non-standard 0-6-0 locomotives built for the British main-line locomotive market:
(A) Dübs & Co.-built (D 1613/82) for the Great Northern Railway (No. 737). *National Railway Museum, R.H. Bleasdale Collection*;
(B) Vulcan Foundry-built (VF 1203/87) for the Lancashire & Yorkshire Railway (No. 933). *Peter Wardle collection*; (C) Kitson & Co.-built (K 3145/89) for the Hull & Barnsley Railway (No. 56). *Kitson collection, The Stephenson Locomotive Society.*

Competing Designs in the Overseas Market

The insulation of the London locomotive market was brought sharply into focus from the 1880s as new markets opened up in countries, such as Japan, that were not dependent upon one economic influence. British locomotive designs for overseas diverged unacceptably from the best standards of American and German practice. Central to this debate, therefore, was the responsibility for designs, and the extent to which the locomotive industry was in a position to influence them, or remained as a passive sub-contractor.

Many of the overseas markets were in developing countries, often with challenging topography, and with limited funds with which to open up their respective territories. Their railways were usually laid with slow-speed track and tight horizontal radii, closer in character to American railroads than to European routes. American designs developed quite independently from European ones, the early focus for which had been the 1839-1844 attempts by the Norris company to develop European markets. Their lack of success in Britain and the early Europeanisation of the American type in Germany and Austria, were as much a lesson for the American manufacturers about the two distinct market characteristics, as for European manufacturers with export aspirations.

The introduction of railways to developing countries in the 1850s again focused attention on the divergent locomotive types in Britain and America, but it took someone of the stature of Walter Neilson (1819-1889), proprietor of Neilson & Co. of Glasgow, to understand fully the characteristics of the American locomotive. Following a visit to the USA in 1856/7, he brought back locomotive drawings showing current design practice, which he exhibited at a meeting of the Institute of Engineers in Scotland. The designs greatly impressed him, especially the 4-4-0 type which he described as:[139]

> a gay, jaunty-looking vehicle – very different from the sombre, business-like machine of the old country… The American eight-wheeled truck engine is a beautifully balanced and steady machine remarkably easy on a bad road and much safer than an English engine under similar circumstances… We may predict that the time is not far distant when we may look to our friends across the Atlantic with the expectation of learning something from them even in railway engineering.

Neilsons manufactured locomotives for the Nova Scotia Railway employing several American features.[140] For the most part, however, consulting engineers, acting for overseas railways, did not seek to learn about American practice and pursued British types with which they were familiar from their training and careers. Most particularly, by the 1870s and 1880s, many overseas locomotive designs were being precisely specified by the design teams of the large engineering consultancies such as those headed by Livesey, Fox, Rendel and Hawkshaw. In 1881, for example, Livesey's office issued a three page specification for materials alone, for two engines for the Buenos Aires Great Southern Railway. The overall specification was accompanied by at least 155 separate drawings.[141]

The continuing adoption of 'British' locomotive practices for most overseas railways became a contentious issue after 1870, following an uneasy decade of comparison with American locomotive practices in certain parts of the world, notably South America.[142] Design variations were two-fold, namely suitability for well-graded or poorly-

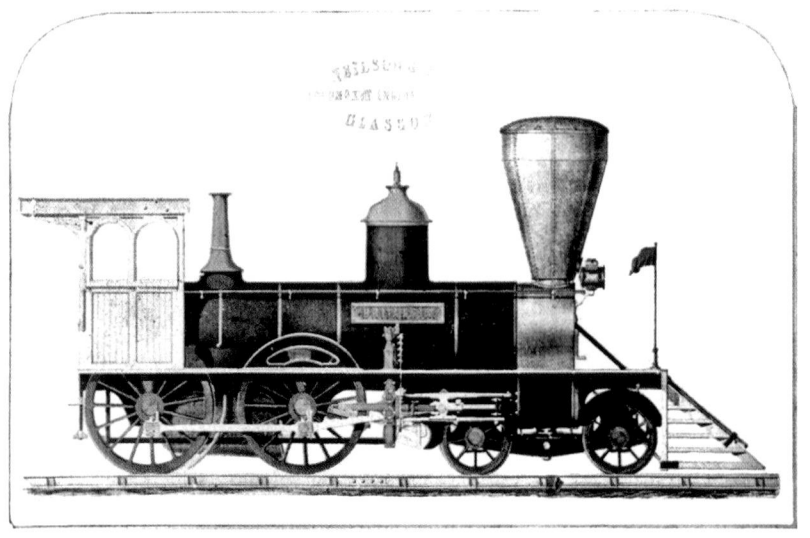

Fig. 4.33
A British-designed variant of a North American type 4-4-0 locomotive designed and built in 1857 by Neilson & Co. for the Nova Scotia Railway, BRITANNIA. *Neilson & Co. catalogue 1862 – Author's collection.*

Fig. 4.34 Designed and specified by a consulting engineer. Yorkshire Engine Co. Ltd.-built 0-6-0T (YE 382/84) for the Buenos Aires Great Southern Railway. *Author's collection.*

graded track, and the employment of short-life and long-life materials.[143] American manufacturers used bar-frame locomotives, guided by a leading bogie truck, for use on poorly-laid track with sharp vertical and horizontal curvatures, characteristic of early railways in Canada, South America, Australasia and Africa.

British designs were better suited to higher speeds on well-laid track not initially provided in the developing countries. British designers also had insufficient experience in wood-burning fireboxes, and comparative fuel consumption featured large in the debate.

The differences between American and British types were highlighted in 1878-1880, regarding the suitability of designs for railways in Victoria and New Zealand. Extensive correspondence, accompanied by several opinions, both published and private, criticising British locomotive design practice, was assembled as a seventy-nine-page booklet by an American engineer, W.W. Evans of New York.[144] The core of the booklet was a letter by Evans to the Agent-General for New Zealand in London, which preceded locomotive orders for the colony. These were largely awarded to American manufacturers, although there is no evidence of Evans being rewarded as their agent.

Evans' evidence followed the successful trial of American locomotives in New Zealand, an extract from a report by the Commissioner of Railways stating:[145]

> They [American locomotives] have now proved themselves to be both good and economical, and for attention to detail in design and general excellence in workmanship they stand out first in our catalogue of locomotives. American engines I thoroughly believe to be more suited for our lines than anything we can get built in England.

Fig. 4.35 British-built locomotive for New Zealand. 0-4-0ST locomotive built in 1873 by Neilson & Co. (N 1772/73) for the country's Government Railway. *Author's collection.*

The matter was brought to the attention of Neilson & Co. and the Vulcan Foundry Ltd. by the London agents of the New Zealand Railways, to whom Neilson & Co. responded stating:[146]

> The ordinary American type of engine, such as is in use in America, is, we have not the slightest doubt, better adapted for railways as now constructed than the engine used in this country. It is more flexible, and adapts itself better to the line than our excessively rigid engines. It has also the advantage of being less costly, though, we quite believe, equally efficient in its details, by reason of these being of simpler construction, and frequently of cheaper materials. We need not tell you that, although holding these views, it would be needless our attempting to persuade our locomotive superintendent [i.e. Consulting Engineer] to adopt even a modification of the American type, as you will be well aware of the vast amount of prejudice that would have to be overcome.

The Vulcan Foundry wrote in similar vein:[147]

> … we should like to know whether our trans-atlantic competitors built these particular engines to a specification and drawings supplied, or whether the design and carrying out of details was a matter left entirely to themselves.
>
> We suppose the latter, in which case we submit the comparison between ourselves and the American builders is most unfair.
>
> We are prepared to admit that the American type of engine… is certainly better adapted to the nature of the curves and permanent way usually prevailing in our colonies than the rigid wheel-base of our English engines; but such is the absurd conservation existing in this country that any departure from existing types would not be entertained… If English builders are compelled to adhere to a particular type and specification of an engine, they surely cannot be held responsible for its performance or failures.

Evans' booklet, published in Leeds, was clearly designed to be available to British locomotive engineers generally, and particularly those engaged with consultancies. It was also available to engineering journals, through whose columns the on-going debate was continued.[148]

The growing divergence of British and American designs was well summarised in 1893:[149]

> In the United States, long train runs, high wages, sharp competition and high rates of interest define the method of railroad operation… In other countries, generally speaking, the distances are short, competition small, wages low, and money rentals moderate… The average weight of an American goods train for level roads is not far from 1,350 tons… a foreign goods train is approximately 450 tons. So then, in the United States, heavy trains, high wages, sharp competition and high rates of interest fix the main distinctive feature of American locomotive practice, which is the greater hauling power of the locomotives. The greater power is obtained in two ways; first by using locomotives of greater weight, and second, by forcing the boilers to a degree almost unknown elsewhere…

Comparison with European locomotive designs centred on the different operating characteristics from those in Britain. Unlike the British trains of moderate length and weight, run at high speed to accommodate passenger traffic on crowded routes, continental operations were longer and heavier, and run at slow or moderate speed, often with frequent and heavy gradients.[150] British manufacturers monitored design evolution of the German, French, Austrian and Belgian industries through regular features in technical journals and the appearance of locomotives at international exhibitions. They were, however: 'filled with wonder at the multiplicity of parts and the complicated form of each piece, which seem to be features inseparable from Continental designs… which seemed to result from: the Continental system of training engineers entirely in the technical schools, where it is impossible to gain the practical workshop experience which is given by our system of apprenticeship…'[151]

From the 1850s, the eight major manufacturers in the German union[152] had maintained full control over locomotive design development in conjunction with their predominantly state-run railway customers, and in close co-operation with State and University technical departments. With minor exceptions, none of the German railways manufactured locomotives, relying entirely on the independent companies, between whom competition was intense.[153] In 1909, it was observed that: 'The State by way of assisting the manufacturers does not encourage the railways to build their own engines, but prefers to keep the manufacturers going.'[154]

The issues relating to the differing designs of locomotives in Britain, the United States and Germany were brought sharply into focus in 1902 when the three Glasgow firms, Neilson Reid & Co., Dübs & Co. and Sharp Stewart & Co. Ltd., wrote a joint letter to the *Times* newspaper. They were prompted by Lord George Hamilton, Secretary of State for India, who stated that locomotive orders for India in the three previous years had been given to American and German companies as British manufacturers were unable to provide the right engine at the right price, for which: 'they only had themselves to blame'. The manufacturers' letter responded:[155]

> With the commencement of American competition for Indian locomotives… all engines ordered for India were of the British type modified, of course, to suit local and climatic conditions, and British builders were asked to build these types only. When the American builders began to compete they were allowed to offer their own type of engine…
>
> It follows, therefore, as far as design is concerned, the Americans were allowed to supply a cheaper engine than British builders.
>
> As to materials employed in the construction, the British builders are compelled to obtain certain materials from two or three makers whose products have been found to give the most satisfactory results in working, but which are not unnaturally costly. Were the American builders in all cases restricted to the same makers?…
>
> You, Sir, treat the matter, rightly we believe, as one of national importance, and we suggest that Lord George Hamilton should send a small commission to India to inquire into the results with the American engines there…
>
> As to German competition, as the two orders recently sent to Germany are the first that have been given for locomotives it has yet to be proved that 'Germany can serve her (India) better than England in the matter of locomotives'…
>
> It would be interesting to know what makes of material are to be accepted in the case of the German engines. We can buy German tyres, axles, etc. much cheaper than we can get them in this country, but so far we have not desired and have not been invited to use these materials in the construction of engines for India…

The manufacturers' frustration, revealed in the *Times* letter, adequately sums up the effects that had been created by the consulting engineers' strong hold on design development for overseas railways. It also serves to confirm the extent to which the manufacturers had become dependent upon the specifications of their customers or their representative engineers, and to highlight the extent to which the progress of British industry had deviated from the developing practice of its competitors.

The formation of the Locomotive Manufacturers Association in 1875, should have allowed the manufacturers to speak with unanimity with the consulting engineering profession regarding overseas locomotive standardisation, particularly for market groups such as the Indian railways. There was no such dialogue,[156] however, and, in spite of its criticisms of a number of designs, the industry appears to have had no success in changing the design procedures for overseas locomotives. Although aware of the unsuitability of many British 'rigid' designs, the manufacturers were subordinated to designing locomotives according to specifications that left little room for initiative.

In spite of its dominant position in the world locomotive market at the end of the 19th century, therefore, the independent industry had been largely subordinated to the rôle of manufacturing contractors to the designs of its railway customers and their consulting engineers. Although main-line locomotive manufacturers had limited opportunity to innovate, most firms retained a design capability for other markets, such as for marine, colliery and industrial engines, and continued to offer a design

Fig. 4.36 British-built for India. A 2-6-0 built by the Vulcan Foundry Co. Ltd. in 1894 ([One of batch] VF 1387-1396/94) for the Indian Midland Railway ([One of batch] No. 156-165). *The Locomotive Magazine, Vol. 35 (1929), p.58.*

facility. Indeed, the industrial locomotive sector retained full design capability for its standard products, alongside its many other forms of capital machinery. The diminution of technological and design initiative made more difficult the development of locomotive production processes, which became the manufacturers' primary concern. The twin challenges of investment and production engineering advancement thus determined their long-term future.

Notes – Chapter 4
1. Kirby (1988), pp 24-42.
2. Fox (Ed), 1996.
3. O'Brien, Griffiths and Hunt (1996), p 158.
4. Alford (1977), p 117.
5. Picon (1996), p 49.
6. Mokyr (1996).
7. Kirby (1988), pp 287-305.
8. Bailey (Ed.) (2003), pp 167-174.
9. Bailey (1980-1981), pp 171-179.
10. Bailey (1978-1979), pp 109-138.
11. *First Report from the Select Committee on Railways* Together with the Minutes of Evidence and Appendix, Ordered by the House of Commons to be Printed 26th April 1839, Question 418, 22nd April 1839, p 23.
12. Stephenson's first patent (No 3887 of 28th February 1815) was joint with Ralph Dodds, and his second (No 4067 of 26th November 1816) was joint with William Losh.
13. Letter, Edward Pease to Thomas Richardson, Darlington, 10 M 23: 1824, Hodgkin Collection, D/HO/C 63/5.
14. Bailey, 'The Mechanical Business' in Bailey (Ed.) (2003), pp 163-210.
15. Letter, Robert Stephenson to George Stephenson, Stone Bridge, Nov.8th 1830, Robert Stephenson & Co. collection, ROB5, Folder 18.
16. Patent Specification No. 6092, Enrolled 11th July 1831.
17. Patent Specification No. 6111, Enrolled 30th August 1831.
18. Analysis of Patents in the British Library, Science Reference Library.
19. Provisional Patent No. 6372, Enrolled 26th January 1833, later a final Patent Specification No. 6484, Enrolled 3rd December 1833.
20. Alford (1977), p 117.
21. *Select Committee On Railways*, op cit (11); and *Second Report* to be Printed 9th August 1839. The third, 'Hackworth', design type had been limited to a few north-east manufacturers, and was discontinued after 1850.
22. Evidence of Hardman Earle, Director of the Liverpool & Manchester and other railways, *Select Committee On Railways*, op cit (11), Questions 5003/4, p 230.
23. Zeitlin (1997), p 244.
24. Kirby (1988), pp 287-305.
25. Channon (1981), p 59.
26. Ahrons (1927), p 94.
27. 'Specification of Locomotive Engines for the London & Birmingham Railway', July/August 1836, National Archives, Rail 384/265-269.
28. For example: 'Specification of ten 0-4-2WT locomotives for the South Eastern Railway', prepared by James Cudworth, Locomotive Superintendent, National Archives, Rail 635/225 (Locomotives built 1864 by Slaughter Grüning & Co.).
29. Darbishire (1883/4), Vol. IV, pp 217-221.
30. Usselman (1985), p 181.
31. *ibid*, pp 188-196.
32. Saul (1970), pp 146-150.
33. Warren (1923), p 410.
34. Thomas (1964), p 103.
35. Notebook prepared by 'A I', senior clerk with Neilson & Co. Author's collection.
36. Patent No. 8998, of 23rd June 1841.
37. Printed Report, 'Mr. Stephenson's Report to The Directors of the Norfolk Railway', Westminster, January 21, 1846. Retained in a scrap book thought to have belonged to George Stephenson, Institution of Civil Engineers Library, 04.
38. Warren (1923), Chapter XXVI, pp 346-357.
39. Stuart and Reed (1972).
40. Ahrons (1927), pp 70-73.
41. *ibid*, pp 60-62.
42. *ibid*, pp 83-84.
43. Rolt (1964), p 26/32.
44. Hills and Patrick (1982), p 49.
45. Ahrons (1927), pp 319/320.
46. Hills and Patrick (1982), p 74.
47. Ahrons (1927), p 153-155.
48. Patent No. 1210, Enrolled on 12th May 1864.
49. Ahrons (1927), pp 314-316.
50. The manufacturing of Fairlie's locomotives was contracted out to several firms, namely, Sharp Stewart & Co, Avonside Engine Co., Yorkshire Engine Co., Vulcan Foundry Ltd., R.&W. Hawthorn, and Neilson & Co., *Engineering*, April 9 1875, p 302.
51. Darbishire (1883/4), Vol. V, pp 178-181.
52. James Livesey's Specification Book, 1878-1882, Livesey & Henderson collection.
53. Bailey (1996/7), pp 109-138.
54. Westwood (1977), pp 132/3.

55 Robert Stephenson's *Lancashire Witch* locomotive in 1828 had been built with a primitive form of cut-off apparatus; referred to in Bailey (1978-79), p 126.
56 *Railway Magazine*, 27th November 1841.
57 Thomas (1980), p 166.
58 The locomotives were just preceded by one 'new' locomotive rebuilt from parts of older engines in the Grand Canal Street Works of the Dublin & Kingstown Railway, see Murray (1981), pp 186/7.
59 *Practical Mechanic and Engineer's Magazine*, Glasgow, 1846. Correspondence, *The Engineer*, Vol. 29, Jan-Jun 1870, pp 7-394, *passim*. *American Machinist*, February 11th 1904, p 178.
60 Letter, Robert Stephenson to Edward J. Cook (for R. Stephenson & Co.), Westminster, 31 Aug 1842, Library of the Institution of Mechanical Engineers, Crow Collection.
61 Reed (1975), pp 40-49.
62 The Liverpool & Manchester Railway Act (Geo IV, c xlix, 5 May 1826) for example directed: 'That the Furnace of every Steam Engine to be erected or built…… shall be constructed on the Principle of consuming its own smoke'. The provision was to be enforced by fines of between £5 and £20.
63 Reed (1975), pp 50-52. Also, Ahrons (1927), pp 131-136.
64 First Minute Book, Beyer Peacock & Co., p 40, 10 January 1857, Beyer Peacock & Co. collection, 0001/X. Also quoted in Hills and Patrick (1982), p 32.
65 Mr. Charles Markham of Derby, 'On The Burning of Coal Instead of Coke in Locomotive Engines', *Proceedings of the Institution of Mechanical Engineers*, 1860, p 147-176.
66 Westwood (1977), pp 107-124.
67 M. Anatole Mallet, of Paris, 'On The Compounding of Locomotive Engines', *Proceedings of the Institution of Mechanical Engineers*, 1879, pp 328-363.
68 For example, a tandem compound design patent was taken out by W.H. Nesbitt, Locomotive Superintendent of the North British Railway, No. 16,967 of 1884.
69 van Riemsdijk (1970-71), pp 1-17.
70 Ahrons (1927), pp 243-262.
71 van Riemsdijk (1994), p 9.
72 *Bulletin de la Société Industrielle de Mulhouse*, No 744, 1971, pp 79/80.
73 Ahrons (1927), pp 311/312.
74 Michael R. Bailey, 'The Oil-Burning Locomotives of the Great Eastern', *The Railway World*, Vol. 21, No. 243, August 1960, pp 238-241 & 253.
75 Westwood (1977), p 132.
76 Bailey (1984), pp 102-119.
77 Letter, James Haslam to Messrs. Jones & Potts, Vienna, 10th June 1839, Robert Stephenson & Co. collection, (ROB/5, Folder 19).
78 'Specification for 6 Wheeled Engine' (Customer not recorded), (n.d., but *c*.1847), Northumberland County Record Office, NRO 630.
79 Ahrons (1927), pp 163/4.
80 William W. Beaumont, 'The Fracture of Railway Tires', *Proceedings of the Institution of Civil Engineers*, Vol. XLVII, 1876, Paper 1453, Part I, pp 68-9.
81 Ahrons (1927), pp 163/4.
82 Larkin and Larkin (1988), pp 167/8.
83 Larkin and Larkin (1988), p 165.
84 Memorandum, Beyer Peacock & Co. to their Australian Agent, Mr. Brewster, July 10th 1884, Beyer Peacock & Co. collection, 0001/546, p 235A.
85 Ahrons (1927), pp 205-207.
86 Paper, F.W. Webb to the Institution of Naval Architects, 1886, quoted in Bowen-Cooke (1893), pp 91/2.
87 *The Engineer*, 13th August 1886.
88 Ahrons (1927), pp 206/7.
89 Darbishire (1883/4), Vol. IV, 1883, pp 245-249.
90 Ahrons (1927), p 207.
91 Brown (1995), p 40.
92 *ibid*, pp 298-301.
93 Morton (1983), pp 66/7.
94 Bourne (Ed.) (1846), p 227.
95 Vulcan Foundry Cost Book, 1870-1939, Vulcan Foundry collection, B/VF/5/6/3.
96 Sullivan (1990), pp 349-361.
97 MacLeod (1996), pp 140/1.
98 Sullivan (1990), Table 1, compiled from Woodcroft (1857).
99 Patent Specification No. 8998, Enrolled 22nd December 1841.
100 R. Stephenson & Co., Partners' Minute Book, 29th April 1841, Robert Stephenson & Co. collection, ROB/1/1.
101 Copy letter, E.F. Starbuck to Messrs. Kitson, Thompson & Hewitson, London, 17 June 1844, in letter book, Starbuck collection.
102 Copy letter, E.J. Cook (for R. Stephenson & Co.) to R.&W. Hawthorn, London, 28th May 1844, Starbuck letter Book, *ibid*.
103 Copy letter, E.F.Starbuck to E.J. Cook, London, 11th June 1844, Starbuck letter book, *ibid*.
104 Copy letter, E.F.Starbuck to E.J. Cook, 8 August 1844, Starbuck letter book, *ibid*.
105 Copy letter, E.F. Starbuck to Emile Martin, 5 Feby 1844, Starbuck letter book, *ibid*.
106 Copy letter, E.F. Starbuck, for Self & Rob't Stephenson, to Dudok Van Heel, 6th February 1844, Starbuck letter book, *ibid*.
107 Letter, J.E. Sanderson (for Robert Stephenson) to W.H. Budden (for R. Stephenson & Co.),

108 E.F. Starbuck's accounts, Starbuck collection, 131/73
109 E.F. Starbuck's accounts, Starbuck collection, 131/53.
110 Letter F.E. Sanderson (for Robert Stephenson) to E.F. Starbuck, Westminster, 13th Nov 1848, Starbuck collection, 131/81.
111 Ahrons (1927), pp 76/7.
112 Warren (1923), pp 359-370.
113 Smith (1979/80), pp 139-154.
114 Patent No. 7745, Enrolled 26th January 1839.
115 Letter, W. Weallens (for R. Stephenson & Co.) to Howe, quoted by N.P. Burgh in letter to *The Engineer*, Vol. XXIX, p 7, Jan 7 1870.
116 C. Hamilton Ellis, 'Famous Locomotive Engineers, XV, Thomas Russell Crampton', *The Locomotive*, Vol. XLVI, No.571, March 15 1940, pp 67/8.
117 Notably the London & North Western Railway, Tulk & Ley, Bury Curtis & Kennedy, Kitson Thompson & Hewitson and E.B. Wilson.
118 MacLeod (1996), pp 140-145.
119 Copy memorandum by J.W. Barlow, Ralph Peacock's notebook, Beyer Peacock collection, MS0001/70, p 18.
120 British patent No. 1665, Enrolled on 23rd July 1858.
121 Ahrons (1927), p 132, also quoting from Kneass, *Practice and Theory of the Injector*, John Wiley & Sons, New York, (n.d.). Also, John Robinson, of Manchester, 'On Giffard's Injector for Feeding Steam Boilers' *Proceedings of the Institution of Mechanical Engineers*, 1860, pp 39-51, and 'Supplementary Paper, pp 74-82.
122 Darbishire (1883/4), Vol. V (1884), pp 58-63. Also Ahrons (1927), pp 133-142.
123 Ahrons (1927), pp 160/1 & 225/6.
124 Usselman (1984), pp 30-50.
125 Article, 'The Westinghouse Brake', *The Railway Magazine*, Vol. 1, 1897, pp 362-9.
126 Patent No. 2273 of 1855, Byrom (2017), p 160.
127 Namely, Dübs & Co., Neilson & Co., Beyer Peacock & Co., Robert Stephenson & Co. and R.&W. Hawthorn Leslie & Co., see *The Railway Engineer*, Vol. VI, No.7, July 1885, pp 203/4.
128 Brown (1995), pp 60/1 & 87/8.
129 Mosse (1990-91), pp 97-112.
130 Redman (1972), p 12.
131 Ahrons (1927), p 123.
132 Clark (1855), p vii.
133 Kirby (1988), pp 287-305.
134 Drummond (1997), p 28.
135 Statement by Sir Alexander Rendel, November 1894, quoted in H.C. Hughes, 'India Office Records', *The Journal of Transport History*, Vol. VI, No.4, November 1964, p 245.

136 Rosenberg (1970), p 561.
137 Kirby (1991), p 35, citing correspondence in the Great Western Railway archives, National Archives, Rail 254/52.
138 Darbishire (1883/4), Vol. V (1884), pp 259/260.
139 Thomas (1980), p 86.
140 More detail is being learned about these locomotives from the underwater archaeological project 'Operation Iron Horse'. Artefacts now deposited at the National Railway Museum, York.
141 James Livesey Specification Book 1878-1882, Specification No 27, September 16th 1881, Livesey & Henderson collection (Locomotives built by the Yorkshire Engine Co.).
142 Summarised in a booklet, *American v English Locomotives, Correspondence, Criticism and Commentary*, No Ed., but clearly inspired by W.W. Evans, Leeds, 1880.
143 Colburn (1871).
144 *American v English Locomotives*, op cit (142).
145 Report of the Commissioner of New Zealand Railways to the New Zealand Minister for Public Works, July 24th 1878, *op cit* (142), p 23.
146 Letter, Neilson & Co., Glasgow, 27th November 1878, to Messrs. Hemans, Falkiner, and Tancred, Agents for New Zealand Railways, *op cit* (142), pp 23/24
147 Letter, Vulcan Foundry Co. Ltd. to Messrs. Hemans, Falkiner and Tancred, Lancashire, 28th November 1878, *op cit* (142), pp 24-26.
148 For example, editorial: 'Superiority of American Locomotives(?)' – *The American Engineer*, Vol. 1, No. 7, August 1880, pp 247/8.
149 David L. Barnes, 'Differences Between American and Foreign Locomotives', Paper for the International Engineering Congress of the Columbian Exposition, 1893, summarised in *The Railway Engineer,* Vol. XV, March 1894, pp 78-80.
150 Darbishire (1883/4), Vol. IV (1883), pp 217-221.
151 Darbishire (1883/4), Vol. V (1884), pp 259/260.
152 Borsig of Berlin, Maffei of Munich, Esslingen Works, Hartmann of Chemnitz, Egestorff of Hanover, Karlsruhe Works, Woehlert of Berlin and Vulcan of Stettin.
153 Karl-Ernst Maedel, *Die Deutschen Dampflokomotiven Gestern und Heute*, Berlin, 1965, *passim*.
154 *The Railway Gazette*, Vol. X, Feb.26th 1909, p 281.
155 Letter, Neilson Reid & Co., Dübs & Co., Sharp Stewart & Co. Ltd., to the Editor, *The Times*, quoted in Thomas (1964), pp 208-210.
156 Minute Book, 1875-1900, Locomotive Manufacturers Association, Railway Industry Association collection, National Railway Museum.

Chapter 5

Manufacturing

Introduction

Locomotive progress in the 19th century was as much due to developments in manufacturing and production control methods as it was to technological and design evolution. The industry was first developed by manufacturers of capital equipment, whose craft skills and production methods were applied to a wide range of industrial equipment, and for whom locomotive production was a diversification. The locomotive sector therefore not only took full advantage of these production developments, it was itself instrumental in bringing about further improvements. Investigating the motivation for improvements in this heavy manufacturing sector provides an understanding of the process of innovation and the proprietors' capital investment decisions.

The development of manufacturing during the century required major strategic decisions by the manufacturers, both in terms of accommodating the proliferating designs required by its domestic and overseas markets, and in reducing craft-dependency through the introduction of self-acting machine tools. The different perceptions of the evolution of the locomotive market by the manufacturers gave rise to the industry's growing diversity. 'Progressive' firms sought to invest in capital equipment that would decrease the cost of batch production and encourage, as far as possible, the standardisation of component design for locomotives. The 'craft' manufacturers, however, pursued a broad market base of capital goods, including locomotives, but invested much less in equipment thereby maintaining a higher dependency on craft skills.

With the major expansion in locomotive demand from the 1830s, firms quickly developed their own production processes and requirements. As designs incorporated new technologies and developed in size and power to meet the growing market requirements, the industry underwent radical and far-reaching changes. This enabled it to make components of increasing size, complexity and standardisation, to accommodate materials of increasing specification, and to reduce the unit cost of production.

Although this evolution was much constrained by the proliferation of designs from the 1850s, the manufacturers understood the scale benefits of specialisation, requiring both investment in self-acting capital equipment and the introduction of a new production culture. Specialisation in heavy manufacturing was still rare in Britain in the 1860s, but the sectors that came closest to it were locomotive, textile machinery and heavy machine tool manufacturing.[1] By the 1870s there were about ten main-line locomotive manufacturers and a similar number for industrial types, together with a number of firms who made smaller numbers of locomotives alongside other machinery for mining, shipbuilding and other heavy industries.

Economic and business historians have determined that the technological advancements of the 1830s and 1840s were followed in the 1880s and 1890s by a 'Second Industrial Revolution'.[2] Extraordinary productivity improvements took

Fig. 5.1 Neilson & Co.'s new factory site at Springburn, Glasgow, opened in 1862. *Author's collection.*

place in the second half of the 19th century, chiefly through the 'American System', in which new manufacturing and organisational processes led to mass production of finished goods with substantial unit cost reductions through economies of scale and scope. Criticism that Britain generally failed to invest in its capital-intensive industries to achieve America's high levels of efficiency, turns on the interpretation of 'mass' production, which is defined as requiring technological and organisational innovation to permit 'a small working force to produce a massive output.'[3]

By the end of the century, American firms were manufacturing large volumes of standardised equipment in contrast to the continued use of craft methods in Britain for the manufacture of customised products in small batches, which was the subject of comment on both sides of the Atlantic.[4] However, certain sectors of British industry, such as the manufacture of textile and agricultural machinery, were well used to producing very large batches of standard components on special-purpose equipment, much of it designed by the firms themselves.[5]

The key issue for specialist manufacturers, including the locomotive sector, was the organisation of their production. The Philadelphia manufacturers, including the Baldwin Locomotive Works, strove to establish 'system' in industrial practice long before its general use in management parlance.[6] The Baldwin Locomotive Works also strove to accommodate design proliferation by standardising component production, as far as possible, within the many design envelopes demanded by their customers. It has thus been noted that locomotive construction was 'systematized, but not standardized', which highlights the distinction between specialised and volume production.[7] In considering the 'systematisation' of the British locomotive industry, it is important to demonstrate that manufacturers understood production economy through specialisation and increasing batch size, and to ascertain if they sought to encourage railways to incorporate standard components in their diverse designs in the way successfully pursued by Baldwins.

Developments in the labour process were also key to production progress. There was an extraordinary advancement in 19th century manufacturing capability and production control procedures in both the heavy and light manufacturing industries.[8] Progressive development of machine tools transformed machining from a skilled activity, requiring experience and ingenuity, to an unskilled activity, allowing batching of standard components with equipment that was more robust, faster and capable of more ambitious tasks. Major advances in forging and foundry equipment extended the range of metal-forming skills and the reliability of finished components. Improving production control procedures reduced component processing time, increasing productivity and decreasing production costs. These developments have been linked with design proliferation, with particular manufacturing processes in each of the railway workshops, using different machine tools and other capital equipment, used to meet the specific requirements of each design team. This proliferation was encouraged by railway managers to deter freedom of movement by skilled and semi-skilled men, as shortages of such skills would otherwise have resulted in wage competition. It has been argued that there was collusion between the railway companies to prevent this.[9]

These findings have two implications for the independent manufacturers. First is the effect on their capital investment programmes to allow them to tender for orders from mainline railways, each of which had its own component intra-standards and machining requirements. This problem was compounded by the consulting engineers' design proliferation for overseas railway specifications. The second implication is that independent firms could have followed a similar labour strategy to that of railway workshops, which would otherwise see them entering into competition for scarce labour skills, with corresponding labour migration and increase in wage-costs.

By the commencement of the main line railway era in 1830, Britain's manufacturing industry had already divided between semi-skilled repetitive production and the craft-based production of capital goods. The concept of scale economies from volume production had been well understood and practised since the first decade of the 19th century, when Marc Brunel's sequential operation blockmaking machinery was made by Maudslay, Sons & Field.[10] The processes were well learned and adopted by the textile industry, for example, which required regular provision of interchangeable replacement components.[11]

The manufacture of capital goods, on the other hand, both custom-designed products and customised variations of standard products, had been undertaken under the 'factory' system since the 18th century. The development of their components relied on material enhancements and the progressive ingenuity of specialist machine tools and craft manufacturing disciplines. Opportunities for 'batch' production economies were limited by their small numbers and design variations. The manufacturing processes for both capital goods and standardised components were both to change substantially during the century, and their respective skills and practices followed complementary evolutionary paths.

Locomotive manufacturing was thus developed by the capital machinery industry, whose craft skills were readily adapted to the new requirements. Following the practice of that industry, each locomotive was, at first, individually made with its own uniquely-fitted components. The industry quickly learned the benefits of multiple, standardised component production but, from the 1840s, was confronted with the conflicting aspirations of volume production and design proliferation, which reduced the

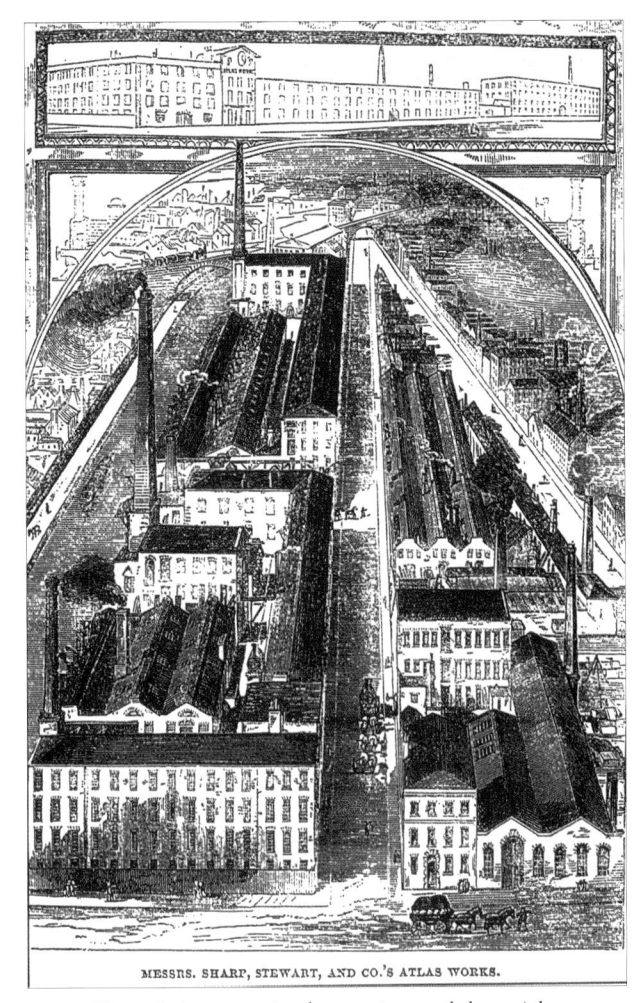

Fig. 5.2 A progressive locomotive workshop. Atlas Works, Manchester, of Sharp Stewart & Co. in c.1854. *Measom, (n.d. but c.1854), p.114.*

twenty years, however, with responsibility for design being passed to railway and consulting engineer drawing offices, the main-line locomotive sector was largely converted into a contract manufacturing business. All design work for the industrial sector, however, and detailing work for overseas main-line orders remained with the manufacturers.

The strategic decisions taken by the proprietors included consideration of what facilities to provide for in-house component production and what components to buy-in from sub-contractors. The growth of specialist sub-contracting industries, such as boiler tube manufacturers, was, itself, an important part of the development of the locomotive industry, which did much to answer the need for production economies. Although this marked the interface between the volume and capital goods industries, manufacturers moved, in part, to take over some of the volume production work, and thus diversify their activities away from being wholly capital-goods based.

The provision of specified raw materials, in sufficient quantities and at minimum unit costs, required daily negotiation and tactical decision-making by the proprietors or their managers. As the industry got under way, its material requirements created new supply industries, whilst from the 1860s, the gradual change to steel supplies required close co-operation between the industry and its suppliers. Although British railways, and consulting engineers on behalf of overseas railways, took over the responsibility for material development and specification, the manufacturers' retained the freedom to negotiate satisfactory supply arrangements, the economies of scale of their purchasing power being reflected in their tender quotations.

The development of capital equipment, particularly of machine tools, had far-reaching effects on locomotive manufacture during the century, which took it from a wholly craft-based industry to one with the potential for substantial cost benefits from standard component production. Self-acting machine tool development was motivated by a shortage of skilled craftsmen, which increasingly allowed the

opportunities for heavy and volume production practices to converge.

In spite of the promotion of standard designs by the larger manufacturers in the 1840s, the subsequent loss of design initiative and the resulting proliferation of designs, reduced batch sizes and increased production costs. The proprietors therefore had to address the strategic and tactical issues to reconcile these pressures, whilst accommodating the developing material and design technologies. The cyclical variations in the locomotive market encouraged most firms to maintain an involvement in other forms of capital equipment, in order to provide a more consistent level of work for their men and machines.

The introduction of common locomotive designs in the 1830s saw the industry pioneer the use of arrangement and manufacturing drawings, with which to ensure consistency of component manufacture. The introduction of this discipline removed an important responsibility from the craftsmen, and thus became itself, a significant part of the industry's movement away from craft-dependency. Within

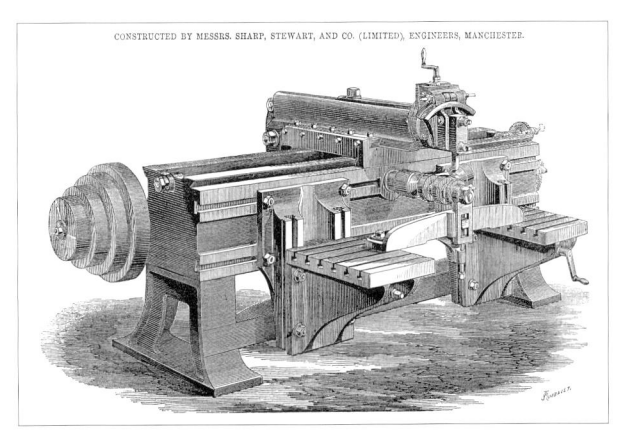

Fig. 5.3 Developments in machine tool production. Shaping machine made by Sharp Stewart & Co. in Manchester in 1873. *Engineering, Oct. 3 1873, p.268.*

employment of un-skilled men. Many of the manufacturers were themselves pioneers in the development of the self-acting tools, forging and casting equipment and throughout the 19th century several firms developed more sophisticated tools to meet their specialised requirements.

There were between 5,000 and 6,000 components in a steam locomotive,[12] with several opportunities for volume production of standard and interchangeable components. The industry well understood the economies from batch processing but, against the background of design proliferation, such opportunities depended upon railways conceding the use of standard components. The development of British railways' drawing offices, with their own intra-standardisation policies, together with the influence on overseas locomotive design exercised by consulting engineers, suppressed inter-changeability opportunities and challenged manufacturers' freedom to pursue production economies.

Production capacity was a strategic judgement for the manufacturers, for whom investment in workshops and capital equipment had to balance both a predicted loco-motive market share and a share in other product markets. The cyclical demand swings, however, with some extraordinary year on year changes (Fig.2.2), gave rise to short-term capacity shortages and surpluses. These reflected directly on manning requirements and profitability, requiring difficult tactical decisions. The options included short-term investment in, or leasing of new production facilities, sub-contracting locomotive orders and divers-ification into and out of alternative markets.

The transformation from craft to multiple production took place in different ways and at different rates, leading to a growing divergence between the manufacturers and a 'two-speed' industry by the last quarter of the century.

This divergence was a key distinction between 'progressive' manufacturers, which invested more in capital equipment to provide batch production economies, and pursued the largest orders at more competitive prices, and 'craft' manufacturers which pursued small batch orders using their craft-based skills and equipment. The divergence related as much to labour policies as to capital investment and business strategy.

In common with Britain's railways, the American locomotive industry was similarly customer-driven in terms of its design requirements but, in contrast, the Baldwin Locomotive Works succeeded in accommodating the volume production of interchangeable components with this design proliferation.[13] Comparing British and American practice therefore helps to understand the causes of the limited volume production of standard components by Britain's manufacturers.

Manufacturing Drawings

The introduction and development of drawing techniques were essential steps towards systematic design and manufacturing progress. Manufacturers embarked on this strategy both as a means to accelerate the preparation of locomotive arrangements, formerly the empirical pursuit of the millwrights, and as a means of co-ordinating and directing design strategies that had hitherto been fragmented. The evolution of manufacturing drawings, however, gave production economies through the introduction of standard components, with the potential for inter-changeability. This evolution, which took away from the millwrights the discretion for component formation, became a significant part of the manufacturers' strategy to reduce dependency on craft skills, motivated by their shortage and lengthy apprenticeship.

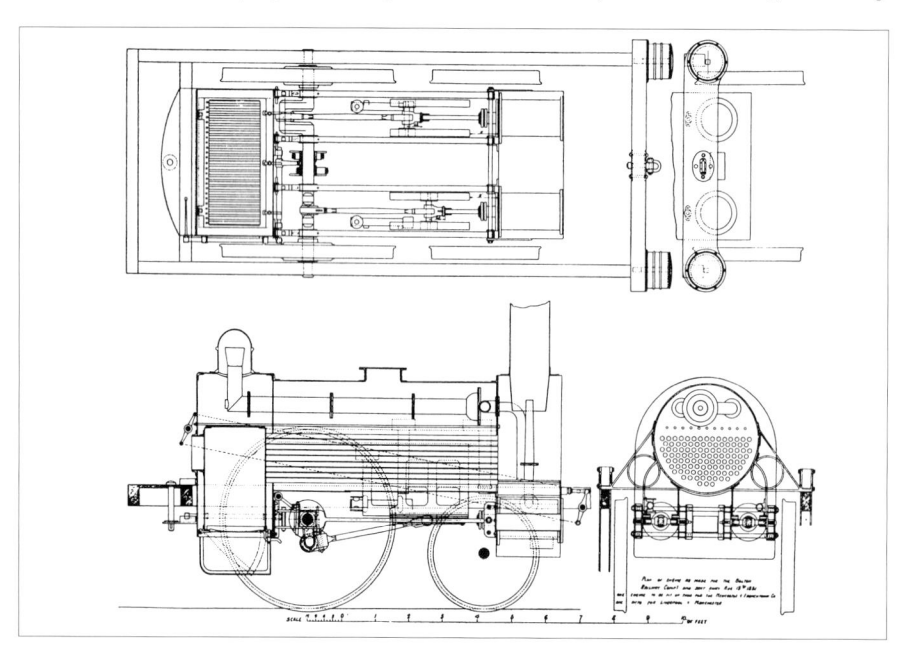

Fig. 5.4
Arrangement drawings for a *Planet*-class locomotive built by Robert Stephenson & Co. in the 1830s.
National Railway Museum Robert Stephenson & Co. collection, and reproduced in Warren (1923), p.300.

CHAPTER 5 – MANUFACTURING

Fig. 5.5 An early patent drawing. Elevation, plan and end views, with additional details, prepared by Robert Stephenson & Co., forming part of its patent 'Improvements in the arrangement and combinations of the Parts of Locomotive Engines' (Long-boiler design) dated 22nd December 1841. Patent specification No. 8998.
National Railway Museum, Robert Stephenson & Co. collection ROB/3/2.

The first requirement in the development of standard components was the progression from arrangement drawings to component manufacturing drawings. Although Robert Stephenson & Co. had introduced arrangement drawings in 1828, the millwrights, pattern-makers and fitters, who made the first locomotives, applied their craft skills without working drawings. This practice prevailed for several years with a number of early firms, and a former employee of R. & W. Hawthorn recalled the method used as late as 1845:[14]

> On ordinary foolscap sheets the work to be done was calculated and the size of boilers required, diameters of cylinders and length of stroke stated. The parts were sketched by hand alongside the calculations in sufficient detail to enable the millwright or pattern-maker to lay them down full size and to make the pattern for the founder. Copies of every part were hand sketched and afterwards the sheets were stitched together…

From the 1840s, the introduction of standard designs and components, using templates and gauges, imposed upon the craftsmen the discipline of standard parts production based on dimensioned, working drawings. The preparation of these drawings became an important part of the manufacturing process, for which teams of draughtsmen developed a new design craft. In 1844 some steam engine factories went to great lengths to protect the drawings from workshop damage:[15]

> … it is usual to have the drawings made on strong paper – not of the full size, but on a large scale – with the sizes marked on, a small portion being usually shown of the full size, when there is any thing delineated which it would be difficult to imitate by the measurements of a rule. This paper is usually fixed upon a board, and varnished, to prevent oil spots and other disfigurements, and a thin slip of wood is nailed all round the edges of the paper, both to keep the edges down and to prevent the drawing from being rubbed when the boards are piled upon one another.

A further requirement was the production of patent drawings, requiring detailed presentations of new features to be protected by the patent.

As locomotive design work for Britain's larger main-line railways was increasingly taken over by their own drawing offices, more detailed arrangement and component drawings were supplied to the manufacturers, whose draughtsmen were eventually limited to preparing tracings for workshop use.[16]

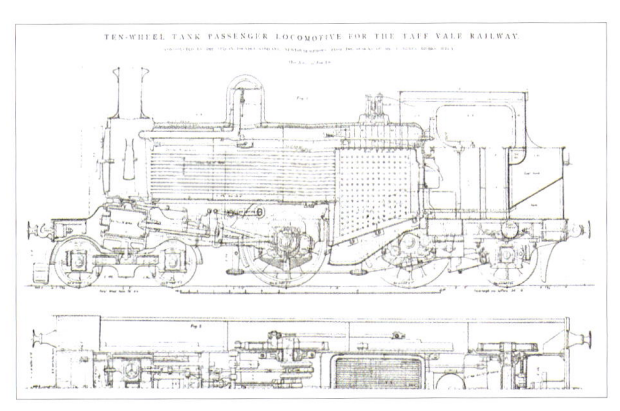

Fig. 5.6 Late 19th Century General Arrangement Drawing. Vulcan Foundry-built 4-4-2T (VF 1229-31/87) for the Taff Vale Railway (Running Nos. 170-172). *Engineering, Vol. XLVIII [Jul-Dec 1889], insert between pp.142 & 143.*

Consulting engineers representing overseas railways and, towards the end of the century, locomotive superintendents of the larger colonial railways, usually supplied only arrangement drawings to the manufacturers with their invitations to tender. Once a contract was awarded, working drawings were prepared by the manufacturers' draughtsmen, with the option of using existing templates and gauges, as far as could be accommodated by the design.

Fig. 5.7
Tracing ladies at Neilson Reid & Co.'s Hyde Park Works at the end of the 19th century.
Mitchell Library, Glasgow – NBL collection.

These drawings were submitted for approval before moving to the detailing and tracing work.[17] The requirement for traced drawings became a large and important part of the production processes for locomotive manufacture. The detailed and repetitive nature of this work was seen as lending itself to female employees and large firms such as Neilson Reid & Co. engaged a number of tracing ladies.

The evolution of British drawing practice was in marked contrast to American locomotive drawing practice, which lagged well behind. Through to the early 1850s, each of Baldwin's locomotives: 'was constructed without much reference to those which were built before or those which would come after it. Complete drawings were almost unknown.'[18]

Raw Material Supply
The locomotive industry created new demand patterns for raw materials, both in terms of innovation in the first years of production and the gradual introduction of steel later in the century. In addition, the rise and fall of the locomotive market led to fluctuating supply requirements during the century. Negotiations for the provision of adequate supplies of specified materials, in line with those fluctuations, at prices that would allow profitable fulfilment of orders, were important decisions for the manufacturers.

The employment of ferrous foundries depended on the volume of casting work to meet overall market requirements, but evidence is lacking regarding a throughput threshold above which it was profitable to establish an in-house foundry. At the start of its business in 1823, Robert Stephenson & Co.'s factory in Newcastle-Upon-Tyne was located adjacent to the Burrell foundry, of which George Stephenson was a partner, and from which it would obtain its castings.[19] As early as 1824, however, there was frustration with this arrangement and the partners minuted:[20]

> It appearing to this meeting that we labour under considerable disadvantages in not being able to found our own Cylinders & other cast metal articles, it is resolved, that an adjacent piece of ground, about 1800 square yards… be purchased… to erect a foundry upon… It is contemplated that this extension of our works, may involve a Capital equal but not exceeding the sum already invested in our Engine manufactory.

Following the foundry's opening, George Stephenson resigned from the Burrell partnership.[21]

The manufacturers that did have their own foundries bought-in supplies of several grades of pig iron for their castings. It was a rule in casting engine components, such as cylinders, 'that the greater the number of the kinds of iron entering into the composition of any casting, the denser and tougher it will be.'[22] Supplies were therefore obtained from several blast furnace companies around the country according to prevailing availability and price of the required grades. During 1832 Robert Stephenson & Co. purchased iron from twenty-one different suppliers and agencies in the West Midlands, Scotland, South Wales, Yorkshire, Tyneside, Bristol, Hull and Liverpool.[23]

However, in spite of the foundry undertaking substantial quantities of track and other general castings, its profitability was evidently marginal, and, following a drop in locomotive orders in 1832, the partners decided: 'That the Foundry shall immediately be given up, & the castings in future purchased from [various] Founders.'[24] The run-down and write-off of stock led to a loss of over £2,200, and the utensils were sold to the Burrell Foundry, from whom the Stephenson Company once again obtained most of its castings for the next thirty years.[25] Only after taking over the Burrell site in the 1860s, did the Stephenson Company again undertake its own casting.

Although most locomotive manufacturers went on to install their own foundries, and several adopted the name 'Foundry' rather than 'Works' as if to emphasise its provision, others continued through the century to buy-in their castings. The Hunslet Engine Co. of Leeds, for example, initially bought-in its castings from its adjacent competitor, Manning Wardle & Co.[26] The unusually close co-operation of these two firms probably arose from the close family ties between their respective managers, but later in the century the arrangement ended and Hunslets obtained its castings from elsewhere.

From the 1850s, as responsibility for specification and design passed increasingly to the railways, supply fragmentation reduced opportunities for predicting long-term material requirements and, correspondingly, the opportunity for negotiating the best unit prices. The problem was compounded not only by the unpredictability of the locomotive market, but also by the movement of raw material prices, reflecting the nation's wider economic health. In particular, was the manufacturers' need to balance the acquisition of materials at prices anticipated when locomotive contracts were signed, against the working capital requirements for lengthy material stock-holding. By 1880, locomotive manufacture required, typically, twelve types of materials of differing specifications, from five different suppliers.[27]

The introduction of main-line locomotives in the 1830s increased the demand for wrought iron plate for boilers, frames and tanks, the suppliers for which diversified from other markets. In addition to its forged components, Robert Stephenson & Co.'s initial requirements for wrought iron plate, bar and sections were largely met by the Bedlington Iron Company. From its origins as a nail-making centre, the ironworks had expanded in the early 19th century, under its manager, Michael Longridge, into an iron-rolling site.[28] Although there was no formal business link between the two companies, Longridge's partnership with the Stephenson Company suggests an early form of forward integration, through which he could justify his dual role to the ironworks' owners, Gordon & Biddulph of London.

Bedlington's earliest endeavours with boiler plate were not wholly successful however, as its quality and thickness often varied beyond acceptable tolerance, and from 1829 the Stephenson Company began to supplement supplies with Staffordshire and Yorkshire plate.[29] Although specific evidence is lacking, it would seem that the loss of most of the Stephenson iron business from the mid-1830s prompted Longridge to establish R. B. Longridge & Co., which began locomotive production of its own in 1837. As this company was supplied with iron from the adjacent Bedlington Ironworks, which in 1839 could roll 10,000 tons of bar and boiler-plate, it was a closer form of forward integration than the Stephenson Company arrangement had been.[30]

'Best Yorkshire Iron' was established as the best quality for boiler-plate, the first use of which was in 1830, when the Stephenson Company's Head Clerk wrote: 'I have ordered a lot of Plates from Low Moor, & have seen two gentlemen from that concern to-day, who promise to do the best they can for us –'.[31] Initially, boiler size was limited by the narrowness of the plate that could be rolled,[32] but from 1833, Low Moor rolled larger plates, allowing larger boilers to be made.[33] Plates were usually sheared to shapes and dimensions supplied by the boiler makers.[34] The tough, low sulphur Low Moor iron, 'The Best Wrought Iron in the World',[35] carried a £5 per ton premium in the middle of the 19th century.[36] Comparable iron, for high-specification components, including boilers, was produced by the adjacent Bowling Ironworks and, from the 1850s, three Leeds firms produced a near comparable quality, the Farnley Iron Co., Monk Bridge Iron Co. and Taylor Bros. & Co.

Fig. 5.8 Example of Low Moor boiler plate. Impression stamped into plate fitted to the *SAMSON* Locomotive, built in 1838 by Hackworth & Downing of Shildon. *SAMSON* is on display at the Museum of Industry in Stellarton, Nova Scotia. *Museum of Industry.*

By the 1860s, railways were specifying short-lists of boiler-plate suppliers, leaving the manufacturers to negotiate prices which would be reflected in their quotations. A Midland Railway specification for tank engines, for example, required: 'Boiler barrel, outside firebox and smoke-box plates, also rivets, stays, and hoops, to be of the best Yorkshire iron, of Low Moor, Bowling, Taylor's, Cooper's, or Farnley Iron Company's make.'[37] Other iron plate, particularly for frames and for tenders, would be specified of less expensive make, such as (Best) Staffordshire or (B Best) Glasgow make.[38]

Large quantities of wrought iron, rolled to a variety of cross-sections, were the raw material of the smiths, who forged many large and small components from wheels to handles. The iron was usually rolled from scrap iron, but the introduction, in the 1840s, of case-hardening for wearing surfaces such as tyres and slide-bars, determined the use of Low Moor or Bowling iron.[39] Wrought iron tyre bars were rolled straight, in lengths suitable for wheelwrights to form and weld into tyres for differing diameter wheels.[40]

Round iron bars for piston rods were generally converted into blistered steel from the early 1840s,[41] and cast steel motion bars and outside coupling rods were made of Bessemer steel by the 1860s.[42] The switch from wrought iron to steel was slow, determined largely by the specifications of the railways and consulting engineers, following trials of different grades. Fig. 5.9 illustrates this slow change by showing the value of materials bought-in by the Vulcan Foundry in the latter part of the century.[43] Not until 1893 was the intake of steel greater than that of wrought iron plate, and there was no perceptible change in the use of bar iron through to the end of the century.

It is thus apparent that the locomotive industry provided significant benefits to other areas of the country through the economic 'multiplier' effects of raw material supply.

New industries were established and existing industries substantially enlarged to meet the expanding requirements of both the independent and railway factories. In the absence of evidence, it is not known how accomplished the suppliers were at dealing with the fluctuating demands, or how adept the manufacturers were at negotiating prices and maintaining adequate supply arrangements. They were, however, fully aware of the prevailing prices of raw materials, and the larger manufacturers had the opportunity of their greater purchasing power to obtain better unit prices.

The introduction of steel from the 1860s required the locomotive industry to invest in new equipment, some of it developed by the manufacturers themselves, and to employ men with the requisite skills. Henry Bessemer's Sheffield works produced 'high class tool steels', which were, at first, only made available to a few Lancashire machine tool makers, but these included Beyer Peacock & Co., and Sharp Stewart & Co., said to indicate Bessemer's high opinion of them.[44]

Bought-In Components

Component provision was carefully considered by the manufacturers. Before committing major investment in specific workshops for component production, they considered the benefits of buying them, in a finished or semi-finished state, from specialist sub-contractors. Decisions to invest in foundries, forges, non-ferrous shops and boiler shops depended on their several heavy engineering markets, and not just on potential locomotive business. Although they undertook all smith, fitting and erecting work, large components could either be made in the factory or bought-in depending on the facilities available. Whilst most manufacturers had boiler shops, for example, some smaller firms found it more cost-effective to contract their boiler-making out to other firms.

In the early 1830s, Robert Stephenson & Co. bought-in boilers from the Bedlington Iron Co. to supplement those made in-house.[45] This overcame a shortage of boiler-makers but may also have been seen by the Bedlington Manager, Michael Longridge, as a form of forward integration with the Stephenson Company, of which he was also a partner. Later in the century, Andrew Barclay & Sons had bought-in all its boilers before investing in boiler-making facilities.[46]

In its earliest years of trading, Robert Stephenson & Co. bought in brass, bronze and gun-metal castings for axle and motion bearings, and boiler fittings. More than £3,000-worth of castings was acquired in the eight years to 1831, prompting the company to establish its own brass foundry the following year.[47] As locomotive production increased, the volume of brass, bronze and gun-metal castings and other non-ferrous formations that were required, persuaded most manufacturers to establish their own brass foundries.

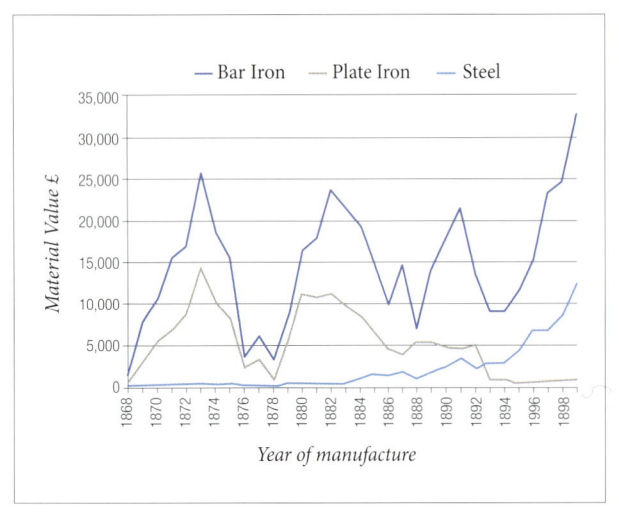

Fig. 5.9 Vulcan Foundry Co. Ltd. Value of Iron and Steel used 1868-1899. *Vulcan Foundry Co. Ltd. collection – Ledger and Journal 1864-1900.*

Heavy forgings were beyond the capability of the first manufacturers and had to be obtained from specialist ironworks. Locomotive axles, in particular, were supplied by ironworks equipped with water-driven, and later steam-driven helve-hammers, which were also used for other large marine and colliery forgings, such as ships' anchors and wagon axles. The Bedlington Iron Company was the first supplier of locomotive axles to Robert Stephenson & Co. from the 1820s, and later to R.&W. Hawthorn and Timothy Hackworth. The demanding strength and reliability requirements of main-line axles took some time to fulfil.[48] After initial problems with some of the Liverpool & Manchester Railway wagon tyres, which had been rolled and forge-welded at Bedlington, George Stephenson wrote to Michael Longridge that the tyres were:[49]

> scarcely at all welded. This says very ill for your Blacksmith's work and alarms me very much about the axletrees; For, should one of them break you are quite aware how serious an accident it would be. It is the cranked axletrees of the Engines to which I allude. The axletrees should all be numbered at your Works, and the name of the Maker inserted in a book.

The several crank-axle failures in the earliest years of main-line operation imposed upon the Bedlington Company the need to improve iron quality and forging methods. It required improvements in axle design, better forging and the use of 'best-quality' scrap iron to produce the most robust axles.[50] As the demand for locomotive axles grew in the 1830s and 1840s, ironworks in Yorkshire, Staffordshire, South Wales and elsewhere equipped themselves accordingly. The Brunswick Iron Works near Wednesbury patented and made locomotive axles by forging 'central and radial bars together, at a single heat, their hardness, tenacity, safety, and durability are incomparably greater than those produced by the old process.'[51]

Until the invention of the steam-hammer by James Nasmyth (1807-1890) in the mid-1840s, other forgings were also provided by the ironworks and forwarded to the manufacturers for machining. Recalling his time at Bedlington, Robert Rennie wrote:[52]

> I had worked at the little forge, or water wheel, as it was called, before going to the big forge. We made all the small forgings for locomotives, such as straight axles, large and small links, double eyes, motion bars, connecting rods, cross-heads and piston rods.

Although steel castings were also bought in, the hesitant introduction of cast steel generally for locomotive components gave no incentive to the manufacturers to invest in their own steel foundries. Even with the introduction of cast steel wheels after 1884, the manufacturers bought them in from specified foundries in Sheffield and elsewhere.[53] Only in 1899, did Beyer Peacock & Co. become the first independent manufacturer to establish a steel foundry, over thirty years after the LNWR's pioneering plant at Crewe Works.[54]

The introduction of forged Bessemer steel for axles and motion parts in 1866 saw the sourcing of axles return to outside suppliers, beginning with Naylor & Vickers. Cost comparison between bought-in steel axles and in-house forged wrought iron axles initially gave only limited incentive to the introduction of the former. In spite of their high first cost, some railways and consulting engineers became keen on the strength and long-life advantages of steel axles. In 1872, for example, the Midland Railway specified steel crank-axles from Taylor & Co. of Leeds for a batch of forty locomotives to be built by Dübs & Co.[55]

As specifications by railways and consulting engineers became more detailed during the century, they increasingly determined which components and raw materials were to be sub-contracted, together with their suppliers. By the 1860s, these included proprietary components, such as safety-valve spring-balances or injectors, to be bought-in from a patentee or licensee. Buying-in large quantities of components and materials required manufacturers to have a good working relationship with suppliers, as prices depended upon the wider number of orders and their consequent negotiating strength. By the 1880s, locomotives could typically require thirty-two separate orders for components and raw materials bought in from 16 different suppliers and sub-contractors.[56]

Boiler tubes were a major requirement for the locomotive industry. Initially, copper and wrought iron tubes were formed in the manufacturers' workshops using sheet materials,[57] but from 1834, sheet brass tubes were introduced, the formation and soldering of which was a repetitive, mass-production activity.[58] As boilers became larger and locomotive orders increased, the Birmingham copper and brass industry was encouraged by the manufacturers to develop its brass-rolling and tube-making skills by diversifying into locomotive tube manufacture, and reduce unit costs.[59]

Seamed and soldered tubes with flanged ends were produced from 1834/5, whilst the first seamless, brass tubes were made in Birmingham by Charles Green in 1838. To ensure accuracy of dimension for its boiler tubes, Nasmyths Gaskell & Co. sent a template to the Cheadle Copper & Brass Co., near Birmingham:[60]

> through which you will please to pass the Locomotive tubes which we lately ordered & they must be made so close to the gauge that one cannot rattle them upon it.

Fig. 5.10 Example of boiler tube from the 1850s. Recovered from a ship-wrecked site off Islay, Hebrides. Made by John Wilkes & Co. of Birmingham and fitted to Neilson & Co.-built 4-4-0 locomotive (N 386/57). *Author.*

By 1860, 4,500 tons a year of seamless tubes were being made in Birmingham for the locomotive industry.[61] One of the largest manufacturers was Thomas Bolton & Sons who made tubes in their Birmingham premises until 1858, when manufacture was transferred to Oakamoor in Staffordshire. Recent archaeology has shown that Boltons' close neighbours in Birmingham, John Wilkes & Co. supplied boiler tubes to Neilson & Co. in 1857.[62]

Early locomotive specifications left the manufacturers to negotiate for tube supplies. In 1852, a Midland Railway specification merely stated that the tubes should be: 'Brass, about 200 in number and 2 inches diameter, No 10 Wire Guage [sic] at the Fire Box end, and No 13 Wire Guage [sic] at the Smoke Box end…'[63] However, in the fast-expanding boiler-tube market, copper producers approached railway companies directly to negotiate supply arrangements. In 1860, the Midland Railway received a 'letter from the Broughton Copper Company Manchester asking to be allowed to supply Tubes for the Engines Messrs. Fairbairn & Co. are building for this Company.' The company resolved 'That this request cannot be complied with.'[64] Railways later provided a list of approved tube suppliers with whom the manufacturers were to negotiate. The South Eastern Railway, for example, stated that:[65]

> The tubes shall be of brass solid drawn, made by Allen, Everitt & Sons, [John] Wilkes [& Co.], Mapplebeck & Co., the Broughton Copper Co., or the Birmingham Battery Co …

Brass safety-valve spring-balances were adopted from the 1830s, for which the manufacturers turned to the makers of industrial spring weighing balances. George Salter & Co. of West Bromwich, manufacturers of springs and weighing machines since the 1770s, supplied the earliest spring-balances, and developed a succession of designs, including patented features, in step with boiler developments. The patent 'Salter' spring-balance was the most widely used type on British made locomotives.[66]

'Clock-face' boiler-pressure gauges were patented and introduced in 1847 by Sydney Smith of Nottingham, with the public endorsement of George Stephenson.[67] A more successful clock-face gauge was patented in 1849 by Eugène Bourdon of Paris, manufacture in Britain being licensed to John Dewrance.[68] Whilst other manufacturers also supplied to the industry, the 'Bourdon' gauge became the most widely used and was often specified by railway companies.[69]

Fig. 5.11 Early illustration of a dial pressure gauge. Footplate of an East Lancashire Railway Locomotive, built in the 1840s showing a dial pressure gauge. *American Museum of Photography.*

Other patented components specified by railway companies and engineering consultants, included injectors, introduced in 1860 and manufactured by Sharp Stewart & Co., licensee to Henri Giffard, with subsequent competition from Gresham & Craven and Davies & Metcalfe.[70] Manufacturers of proprietary continuous brake equipment were the Westinghouse Brake Company, established in 1881,[71] and Gresham & Craven, licensees of the Vacuum Brake Company's equipment.[72] Proprietary steam sanding apparatus was also fitted.

Crucible steel springs and keys, introduced during Robert Stephenson's development programme in 1828, were bought-in from specialist tradesmen. Laminated steel springs for horse-drawn carriages had developed from c.1770, and the spring-making techniques were 'coveted' by the spring-makers.[73] The Newcastle firm of French & Donnison supplied the earliest parts to the Stephenson Company, although other craftsmen later supplied boiler-tube ferrules, springs and keys.[74] Most manufacturers had insufficient volume of business to employ their own spring-makers, and specialist firms usually supplied locomotive and rolling stock manufacturers. Thomas Turton & Sons of Sheffield became a major supplier, although in 1848 it appeared to employ only few tradesmen when it apologised to Jones & Potts for a delayed delivery: 'which has arisen from the circumstance of our foreman having been the subject of a severe attack of the prevailing [un-readable] as well as one of the others of our best workmen in this department of our business.'[75]

As specifications became more detailed, railways set down particular manufacturers from whom springs should be obtained. In the 1870s, for example, Hudswell Clark & Rodgers made industrial locomotives with springs from 'the best Sheffield spring makers'.[76] In 1880, the South Eastern Railway specified springs from Charles Cammell & Co., Thomas Turton & Sons or John Spencer & Sons, which allowed manufacturers to negotiate prices, obtaining scale economies through combination of orders.[77]

The first manufacturers to buy-in Krupp's cast crucible steel tyres included Robert Stephenson & Co. which fitted them to a Great Eastern Railway locomotive exhibited at the London International Exhibition of 1862. Naylor & Vickers of Sheffield began supplying cast steel tyres in 1859, and other companies followed. From 1864, rolled Bessemer steel tyres were introduced by George & Co. of Rotherham, but with on-going reliability problems, some railways and consulting engineers preferred high-quality cast-steel tyres. The Patent Shaft and Axletree Company of Wednesbury was still supplying cast tyres at the end of the century.[78] The high cost of tyres led larger railway companies to purchase them directly and make them available to the manufacturers for fitting. The annual tyre requirements for a fleet of locomotives became so large that the railways' own purchasing power could lead to significant price reductions.[79]

The component supply industry was, therefore, an integral part of locomotive production, and an important factor in determining the costs of locomotive production. Although cost studies of component supply were under-taken from time to time, there is no evidence that they formed part of business plans considering investment in workshops and capital equipment. From the 1830s,

Fig. 5.12 Sub-contractor for wheels and axles. Marketing photograph of the Patent Shaft & Axletree Co. Ltd. of Wednesbury in 1898. *The Railway Magazine, Vol. 2 (1898), p.176.*

however, the volume supply of interchangeable, mass-produced components, such as boiler-tubes and springs, was a clear recognition of the economies of scale that could be achieved from specialist suppliers. This example of volume component production supplying to a capital equipment industry highlights the distinction between the heavy and light manufacturing sectors.

Capital Equipment

Investment in capital equipment allowed the manufacturers to expand both their volume of production and production capability. From its earliest years, the industry was faced with the twin pressures of needing increased production capacity to meet rising demand, and the reducing availability of skilled craftsmen. In its first twenty years, the industry, as part of the wider heavy manufacturing sector, was in the forefront of the design and manufacture of machine tools and other capital equipment, which made possible employment of semi-skilled or un-skilled labour.

Up to the 1820s, progress had been made by manufacturers, such as Henry Maudslay (1771-1831), Joseph Clement (1779-1844) and James Fox (1789-1859), in developing machine tools to undertake basic metal-cutting tasks.[80] In 1823, Robert Stephenson & Company's factory was equipped with machine tools and a factory-engine, made by its own millwrights to enhance their craft skills.[81] They began with simple machine tools,[82] but ingenuity was required for certain machining requirements, including crank axles.[83] Finishing work, as James Nasmyth noted, was undertaken by hand:[84]

> nearly every part of a machine had to be made and finished… by mere manual labour; that is, on the dexterity of the hand of the workman, and the correctness of his eye, had we entirely to depend for accuracy and precision in the execution of such machinery as was then required.

Fig. 5.13 Skilled machining to semi-skilled production. Drawing by James Nasmyth showing an early lathe requiring a hand-held tool, compared with a later lathe with the tool secured in a slide rest.
Nasmyth in Buchanan, 1841, p.396

Reflecting on his apprenticeship at the Stephenson factory in 1837, George Bruce recalled that:[85]

> wheels were driven on to their axles by sledge hammers… and only hand labour was available for the ordinary work of the smith's shop and boiler yard… riveting by machinery… was unknown… there were shear-legs in the yard, by which a boiler could be lifted on to a truck, and there were portable shear-legs in the shop, by skilful manipulation of which… wonders were done in the way of transmitting heavy loads from one part of the shop to another.

The rapid expansion of the locomotive market in the early-1830s posed extraordinary production problems as, although arrangement drawings were used, each locomotive was individually made through the millwrights' finishing skills using components which were not necessarily interchangeable.[86] To overcome the immediate shortage of skilled craftsmen, manufacturing was transformed by the introduction of self-acting and improved machine tools, which could be operated by un-skilled machinists.

Manchester manufacturers, notably Richard Roberts (1789-1864), Joseph Whitworth (1803-1887), William Fairbairn (1789-1874) and James Nasmyth were the primary innovators, contributing to both their versatility and standardised use.[87] This practice provided opportunity for material and layout improvement trials, and a quick response to component failure problems. 'Duplicates' of larger components, such as driving wheel sets, were made to suit each locomotive order, but otherwise components

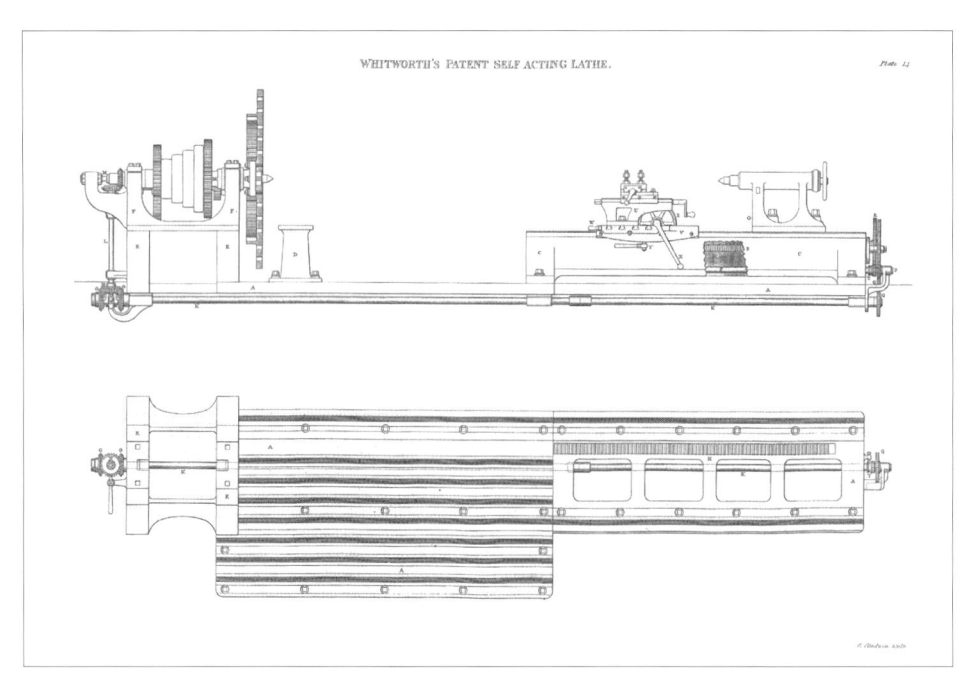

Fig. 5.14 Introduction of self-acting lathes. Joseph Whitworth's design *Buchanan, 1841, Plate 14.*

that broke in service were not easily replaced and had to be re-manufactured or replaced by railway maintenance teams.

In the early 1830s, locomotive manufacturing played as important a part in the development of machine tools, as marine and industrial engines, and textile, paper-making and mining equipment. Nasmyth later wrote that: 'Shortly after the opening of the Liverpool & Manchester Railway there was a largely increased demand for machine-making tools. The success of that line led to the construction of other lines concentrating in Manchester; and every branch of manufacture shared in the prosperity of the time.'[88]

The slotting machine, in particular, developed by Roberts through Sharp Roberts & Co., reduced considerably millwrights' time on locomotive components.[89] Whitworth noted that the planing machine reduced the labour cost of cast iron surface preparation from twelve shillings per square foot by hand, to just a penny per square foot.[90] Several manufacturers, such as Sharp Roberts & Co. and Nasmyth Gaskell & Co., were also engaged in locomotive manufacture, and were particularly innovative in machine tool development to reduce machining time for locomotive components. In 1840, Nasmyth: 'contrived several special machine tools, which assisted us most materially… to effect the prompt and perfect execution of this order [of twenty locomotives]'.[91]

Labour shortages in other crafts, notably boiler-making, and the requirement for greater precision, also led to significant developments in other forms of capital equipment. During the 1830s, boiler-plate preparation was made easier and more accurate by improvements to shearing, punching and plate-bending machines.[92] One of Sharp Roberts & Co.'s shearing machines could 'cut in two, iron plates five inches broad and five-quarters thick, as if they had been as soft as butter.'[93] Boiler plates were first hand-riveted but, in 1838, William Fairbairn patented his steam-riveting machine, allowing boiler-making to be speeded up and more accurately assembled, and with which he claimed a typical boiler barrel could be assembled and riveted in four hours, a fifth of the hand-riveting time.[94]

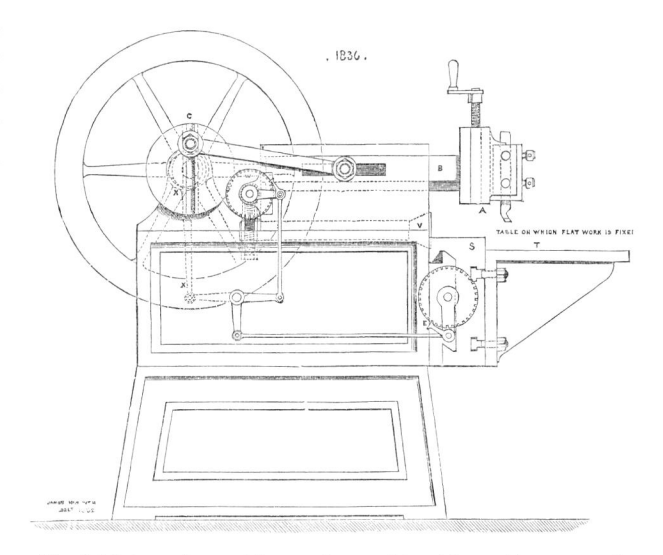

Fig. 5.16 An early precision cutting machine. Nasmyth's segmental cutting machine of 1836. *Smiles (Ed.), 1883, p.416.*

Light line-shafting, wheels and belts, powered by a central factory engine, became the usual form of machine-shop power transmission.[95] Because of the risk that engine or boiler breakdown would shut down the machine shop, Nasmyth preferred small steam engines to power individual machine tools,[96] and in 1854, Beyer Peacock & Co. installed wall engines connected directly to shafting.[97] Joseph Beattie (1808-1871), Locomotive Superintendent of the London & South Western Railway patented a wheel lathe, although Sharp Brothers developed an improved version.[98] In 1841 it was officially acknowledged that the rapid development of machine tools had: 'introduced a revolution in machinery, and tool-making has become a distinct branch of machines, and a very important trade, although twenty years ago it was scarcely known.'[99]

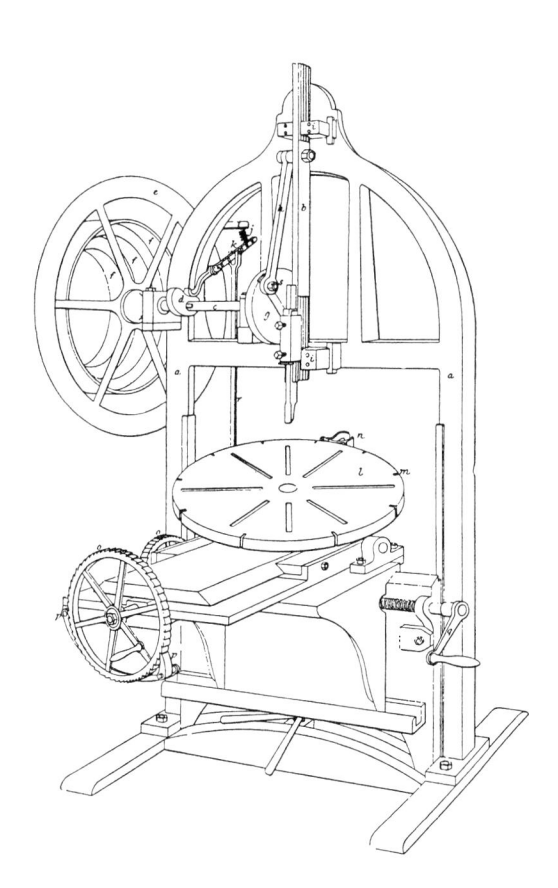

Fig. 5.15 An early form of slotting machine. Sharp, Roberts vertical version of 1835. *Buchanan, 1841, Plate XXXV.*

Fig. 5.17
Expansion of machine shops.
Example for Robert Stephenson &
Co.'s Newcastle-Upon-Tyne factory.
*Penny Magazine, 1844, p.379,
reproduced in Warren, 1923, p.90.*

In the economic boom years of the mid-1840s, an independent machine tool industry developed in major manufacturing areas, notably Lancashire and the West Riding of Yorkshire, to meet the requirements of textile, colliery, ship-building and steam engine manufacturers, as well as the locomotive industry. The strong demand for locomotives, in particular, led to: 'an increased stimulus to the demand for self-acting machine tools', which arose because the demand for skilled labour was greater than the supply and because of the men's: 'exorbitant demands… irregularity and carelessness.'[100] The sudden shortage of skilled labour led to a rise in wages, which: 'increased the demand for self-acting tools, by which the employers might increase the productiveness of their factories without having to resort to the costly and untrustworthy method of meeting the demand by increasing the number of their workmen.'[101]

From the 1850s, diverging strategies between 'progressive' and 'craft' manufacturers formed a key theme in the industry's corporate development. The former companies, such as Beyer Peacock & Co. and Sharp Stewart & Co., were sufficiently encouraged by the long-term market potential to provide improving equipment for multiple locomotive production, whilst the 'craft' manufacturers maintained a broad manufacturing base, for which the inherent skills of their workforce would continue to use more basic equipment.

Manufacturers introduced machine tools to speed up repetitive machining of standard parts, such as nut-cutting, and nut and bolt-head shaping, which became widely used in locomotive factories. These included Nasmyth's 'ambidexter' lathe, which machined two identical pieces simultaneously,[102] and the self-acting milling machine.[103] Other milling machines, including several types by Sharp Stewart & Co., were produced in the 1850s,[104] and by 1860, they were well established in locomotive factories. Their work time was much reduced from that taken by shapers and planers, which were then freed for more suitable work.[105] Wood-working machine tools were also much improved, including precision tools required for pattern-making. The use of self-acting tools was widespread by 1861, when Fairbairn reflected:[106]

> Now everything is done by machine tools, with a degree of accuracy which the unaided hand could never accomplish… For many of these improvements in 'self-acting' machine tools, the country is indebted to the genius of our townsmen, Mr. Richard Roberts and Mr. Joseph Whitworth.

Fig. 5.18 provides a good indication of the machine tools necessary for the production of locomotives and other steam engines and machinery provided by three factory inventories from about 1860.

	E.B. Wilson & Co.	R.&W. Hawthorn	Neilson & Co.
Lathes	66	}55	22
Screw-Cutting Lathes	13		2
Planing Machines	32	19	6
Drilling Machines	38	28	?
Boring Mills	5	4	?
Shaping Machines	13	5	6
Slotting Machines	20	?	5
Screwing Machines	8	?	?
Cutting Machines	7	?	?
Nut Milling Machines	None	?	1

Fig. 5.18 Examples of machine tools in use at three locomotive factories in *c.*1860 [107]

At the time of its closure in 1857, E.B.Wilson was one of the three biggest locomotive manufacturers, together with Sharp Stewart & Co. and Robert Stephenson & Co. All three, however, were much engaged with other than locomotive production, and the equipment reflects their overall manufacturing requirements.

Riveting was further speeded up in 1865, when Ralph Hart Tweddell (1843-1895), an apprentice at R. & W. Hawthorn, introduced hydraulic machines, the portable version of which, from 1871, could be moved around boiler shops.[108] The LNWR's Crewe Works were equipped with Tweddell machines in 1875 and later on the larger independent manufacturers, such as Beyer Peacock & Co., were also so equipped.[109] Some companies bought in ready-made rivets, but most rivet iron was supplied in long bars and cut to required lengths.[110]

In contrast to the extraordinary mechanisation in Britain, the American locomotive industry in the first half of the century had advanced only slowly, and was under-capitalised by comparison. In 1850 for example, Baldwin still required 'the hand skills and muscle power of its workers' aided by basic machine tools.[111] With the rapid expansion of the American machine tool industry from the 1860s, however, Baldwin were thoroughly re-equipped to overcome its shortage of labour. The accent was largely on multiplication of basic equipment rather than on innovative ways of achieving self-acting operations. The tools therefore still required 'substantial skill from their operations to achieve acceptable work'.[112]

Until the 1860s, the commonly-used 'spear-headed' carbon-steel drills wore out quickly and lost precision. In 1862, the American tool-maker, Brown & Sharpe, developed its 'Universal' milling machine for milling flutes for the production of longer-life twist drills. They were exhibited at the International Paris Exhibition in 1867 where they caused a 'sensation' and were soon adopted throughout Britain.[113]

Whilst early shear-legs lifting equipment generally gave way to pillar cranes, some manufacturers in the early 1840s, such as Maudslay, Sons & Field, developed overhead travelling cranes. When John Bodmer (1786-1864) set up his factory in Manchester in the 1840s, he installed: 'small overhead travelling cranes, fitted with pulley blocks, for the purpose of enabling the workmen more economically and conveniently to set the articles to be operated upon in the lathes, and to remove them after being finished.'[114] E.B.Wilson & Co. had five overhead cranes by 1857, possibly installed during its major expansion of 1847,[115] and Beyer Peacock & Co. had three or four by 1860.[116] Neilson & Co. appears to have been the first manufacturer to install a traverser, with which to transfer locomotives from one shop to another, when its new factory was opened in 1862.[117]

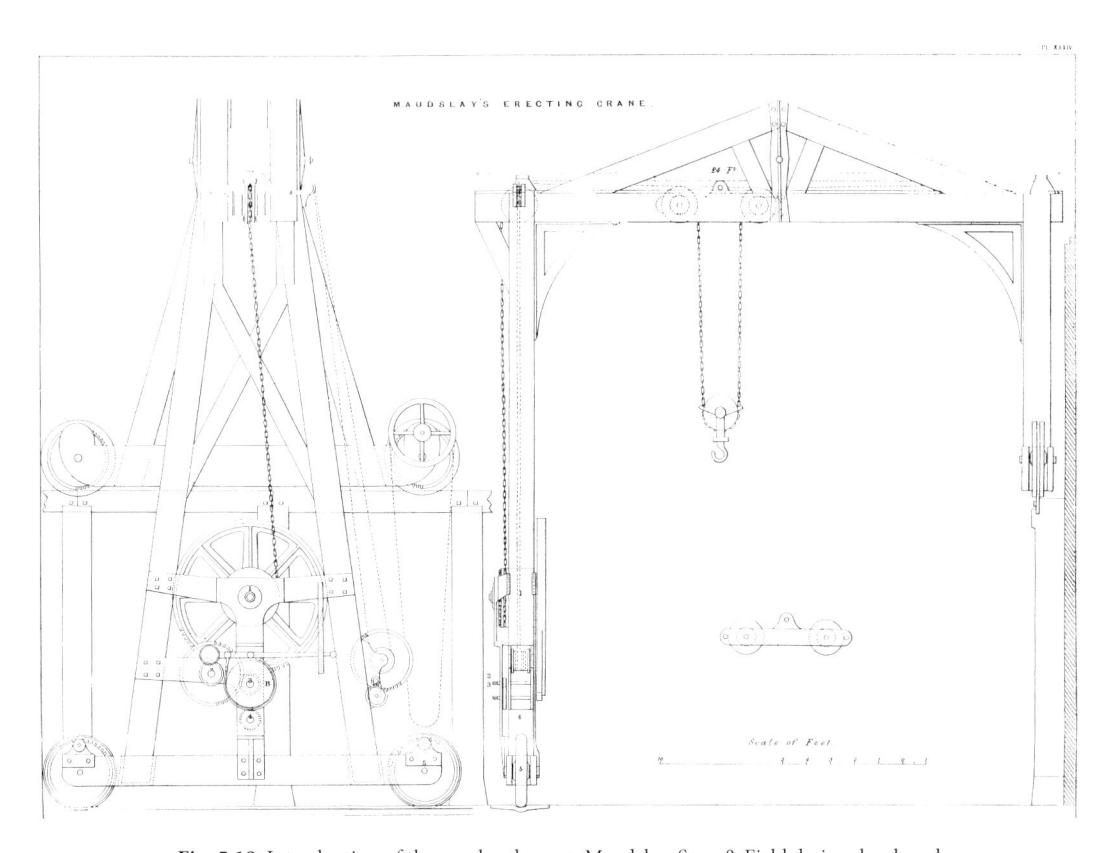

Fig. 5.19 Introduction of the overhead crane. Maudslay, Sons & Field design developed in the early 1840s. *The Artizan, Issue XI, November 30th 1843, Plate XXXIV.*

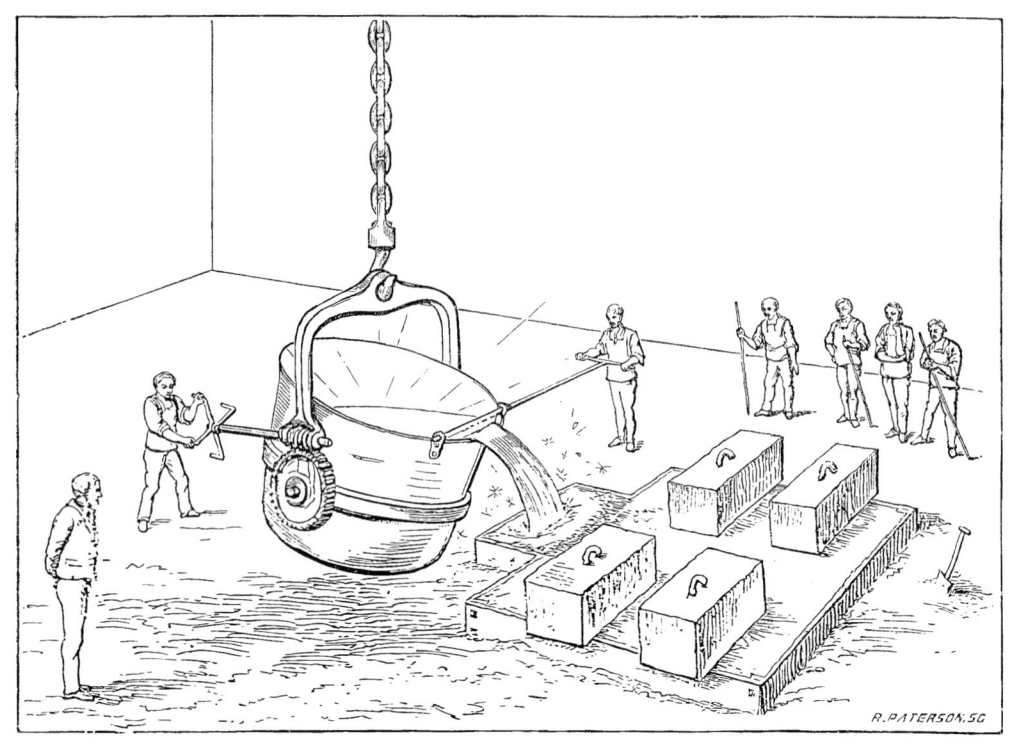

Fig. 5.20 Foundry innovation. Nasmyth's Safety Foundry Ladle adopted at his Patricroft Workshop in 1838. *Smiles, (Ed.), 1883.*

SAFETY FOUNDRY LADLE.

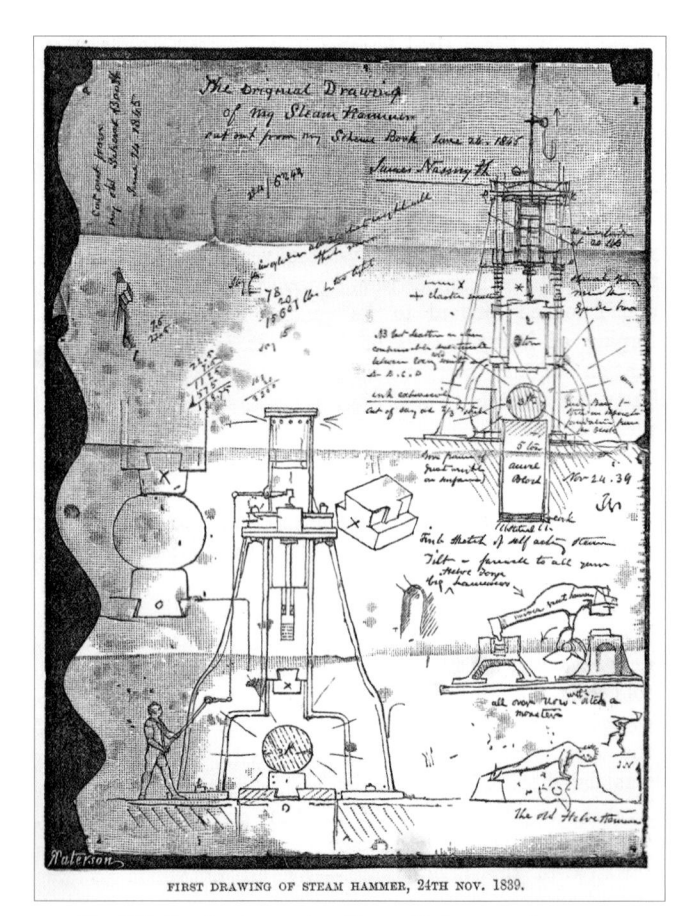

Fig. 5.21 Forging innovation. Nasmyth's first steam hammer sketch 24th November 1839. *Smiles, (Ed.), 1883, p.241.*

FIRST DRAWING OF STEAM HAMMER, 24TH NOV. 1839.

The introduction of steam-hammers, following Nasmyth's patent of 1843, was particularly important for the locomotive industry, as manufacturers were able to undertake their own axle and motion forgings, as well as work for their other markets.[119] Not only had they much work for the hammers, they also used up scrap iron recovered from other components. Several steam-hammers made by Nasmyths Gaskell & Co. were bought by locomotive factories, including Sharp Brothers and E.B. Wilson, before the patent expired in 1856.[120]

Thereafter, locomotive manufacturers, such as Kitson & Co. and Hudswell & Clarke, themselves began to compete in the steam-hammer market.[121] Beyer Peacock & Co. began making small steam-hammers for light forging work, and by 1860, the company was using three for motion forging: 'the cost was about £175 exclusive of the anvil.'[122] Locomotive crank-shafts were forged from 'faggoted' scrap iron, particularly plate shearings, which Peacock claimed was not only an economic use of scrap, but such axles would last longer than other materials.[123]

Nasmyth first manufactured and sold hydraulic presses for pressing wheels onto axles in 1839, an activity formerly undertaken by manual hammer blows.[124] Other manufacturers, including E.B. Wilson & Co. and Beyer Peacock & Co., were using them by the mid-1850s,[125] by which date hydraulic power was also used for jacks and cranes.[126] By the late 1870s, hand-flanging of firebox and boiler plates in the larger factories was replaced by hydraulic presses, using cast iron 'bending blocks',[127] a 300-ton version being used by Beyer Peacock & Co. by the end of the century.[128] Some presses were made by the manufacturers themselves, including Andrew Barclay Sons & Co. which installed one when equipping its own boiler shop at the end of the century.[129]

An indication of the capital equipment (other than machine tools) necessary for the production of locomotives and other steam engines and machinery in about 1860, is provided by the three factory inventories shown in Fig. 5.23:

Fig. 5.22 Forging progress. Nasmyth steam hammers installed at the Robert Stephenson & Co. factory in Newcastle in 1864. *The Illustrated London News, October 15th 1864, p.392.*

	E. B. Wilson & Co.	R. & W. Hawthorn	Neilson & Co.
Boiler & Tender Shops			
Punching Machines	} 6	7	} None
Shearing Machines		3	
Plate-Bending Machines	2	3	None
Rivetting Machines	2	1	None
Wheel Shop			
Hydraulic Presses	1	2	None
Forge			
Steam Hammers	5	2	3
Cranes			
Overhead Travelling	5		
Other Travelling	12	23	?
Pillar Cranes	41		

Fig. 5.23 Examples of other capital equipment in use *c*.1860.[130]

R. & W. Hawthorn was much involved with marine and stationary engine manufacture, as well as locomotives, as reflected by its boiler-making equipment. Neilson & Co. obtained its boilers from its nearby sister company in Glasgow.

In the second half of the century, improvements were made to each type of machine tool, which reduced set-up and operating time, and accommodated more demanding tasks, including steel machining. Specialised machine tools, including larger driving-wheel and crank-shaft lathes,[131] were developed by the manufacturers themselves, particularly Beyer Peacock & Co., Sharp Stewart & Co. and Neilson & Co. The introduction of rolled steel plate for locomotive frames in 1867 required a combination drilling, slotting and planing machine, the first example of which was developed by Sharp Stewart & Co., and soon adopted by other manufacturers.[132] Improvements continued and in 1872 Fairbairn, Kennedy & Naylor, machine tool makers of Leeds, developed a slotting machine 'to dispose of frames 33 feet long at one setting.'[133]

Fig. 5.24 Multi-task machine toolmaking. Sharp Stewart & Co.'s slot drilling and grooving machine displayed at the London International Exhibition of 1862. *Clark, 1862, p.140.*

Fig. 5.25
Radial drilling machine with 'universal' table designed by George Crow for Robert Stephenson & Co. and constructed by Messrs. Fairbairn, Kennedy and Naylor of Leeds.
Engineering, December 17th 1869, p.402.

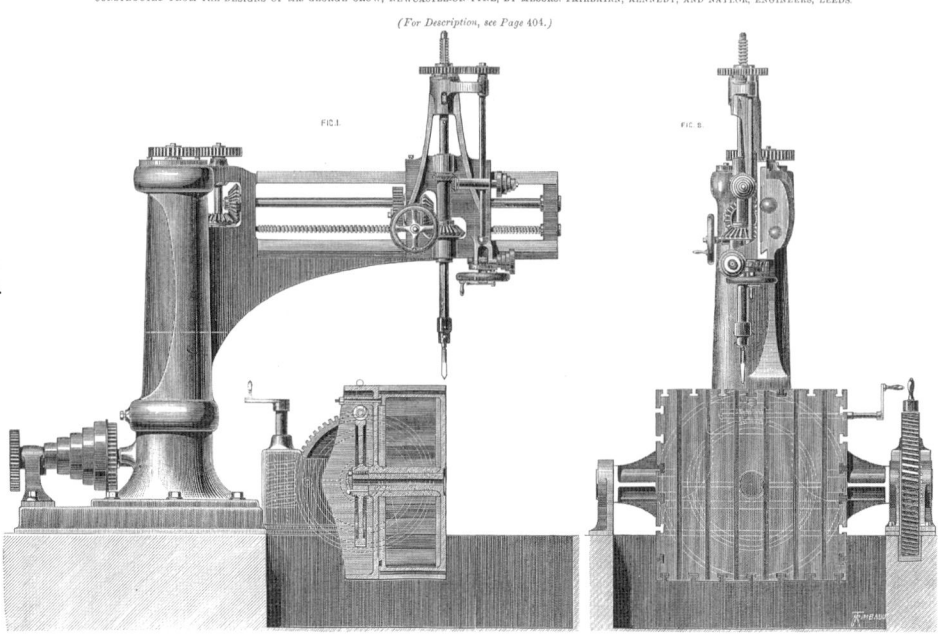

Craft manufacturers which had not proceeded with machine tool development, such as Robert Stephenson & Co., were obliged to buy-in the fewer examples of capital equipment that they did acquire. In 1869, when the company's Head Foreman, George Crow, developed and patented a radial drilling machine with a 'Universal' table, he entered into a 'sole' licensing arrangement for its manufacture and sale with Fairbairn, Kennedy & Naylor.[134]

In the last three decades of the century, the independent and railway factories developed several specialist machine tools, suitable only for locomotive purposes, and reflecting the more demanding design characteristics initiated by railway companies. Significant advancements were made at Crewe Works and in 1873 Sharp Stewart & Co. exhibited, at the Vienna International Exhibition, an example of Francis Webb's patented 'curvilinear' machine for machining the insides of driving wheel rims.[135]

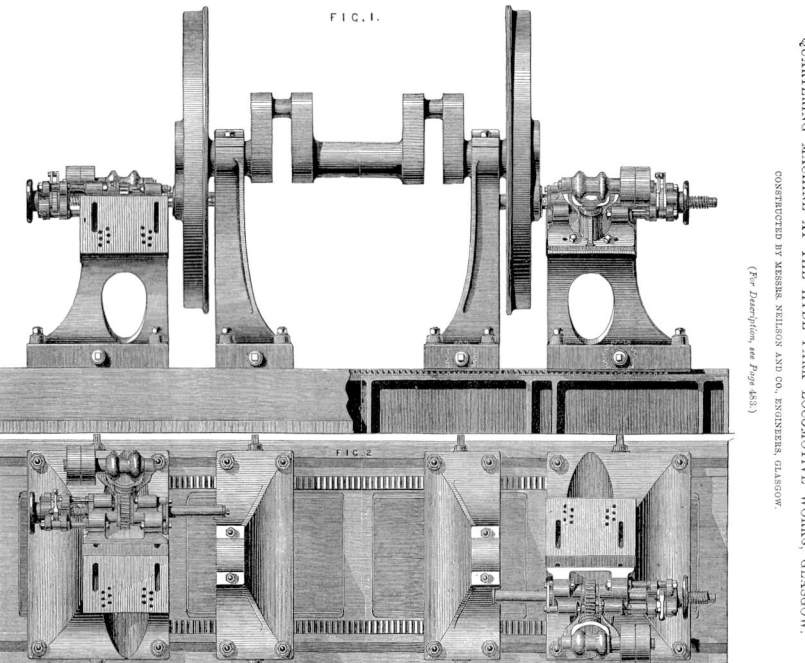

Fig. 5.26
Quartering Machine for boring crank-pin holes in driving wheel-sets made by Neilson & Co. at its Hyde Park Works, Glasgow in 1875.
Engineering, June 11th 1875, p.484.

Sharp Stewart & Co., Beyer Peacock & Co. and Neilson Reid & Co. were all prominent in the development, patenting and manufacture of machine tools for larger pieces, which accomplished more complicated machining with greater precision, and minimised manual finishing requirements. They were developed to increase productivity by reducing the need for multiple machining.[136] The Stephenson Company also developed an hydraulic locomotive weighing machine in 1886.[137]

By the end of the century, the progressive manufacturers, including Neilson Reid & Co., introduced further machine tools, such as capstan and turret lathes, which were designed to divert further work away from skilled craftsmen, and to increase the proportion of their unskilled labour. Although no details have been ascertained regarding the extent of unskilled machining, the policy did allow Neilsons much more flexibility in employment with which to cope with the fluctuating locomotive demand in the 1890s. Its unskilled labour was hired and discharged according to demand, which served to protect continuity of employment for its craftsmen. The extent of the policy is demonstrated by its workforce total, which from a low of 1,500 in 1894, had more than doubled by the end of the century when demand was high.[138]

It is thus evident that the manufacturers played a pivotal role in the development of machine tools and other capital equipment until the 1850s, and that through their further involvement with machine tool design and construction, the progressive manufacturers continued to demonstrate their ingenuity until the end of the century. The locomotive industry benefited directly from this ingenuity, which significantly reduced component production time and allowed more ambitious tasks to be undertaken by the increasing proportion of un-skilled labour, thus overcoming the shortage of craftsmen.

The continuing ingenuity with capital equipment provides evidence of the ways in which the manufacturers fulfilled the proliferation of manufacturing processes in relation to the railway workshops. There is no evidence of particular manufacturers having 'railway-specific' capital equipment or being favoured by main-line railways for having particular machine tools or skills, when contracts were awarded. The tendering system encouraged multiple tenders from ten or more firms, all of whom would have had the necessary equipment or skills, albeit with varying levels of productivity, reflected in their tender quotation.

In comparison, investment in machine tools by the American manufacturers was much more intense. By re-investing profits, the Baldwin Locomotive Works reordered its production processes with a new generation of machine tools, the value of its fixed capital increasing from $2.8 million in 1880 to $5.7 million in 1890.[139] Between 1882 and 1891, the firm took out fourteen patents related to mechanisation. Learning from the 'American System' of volume production, the radical changes prompted comment about the American locomotive industry in the mid-1880s:[140]

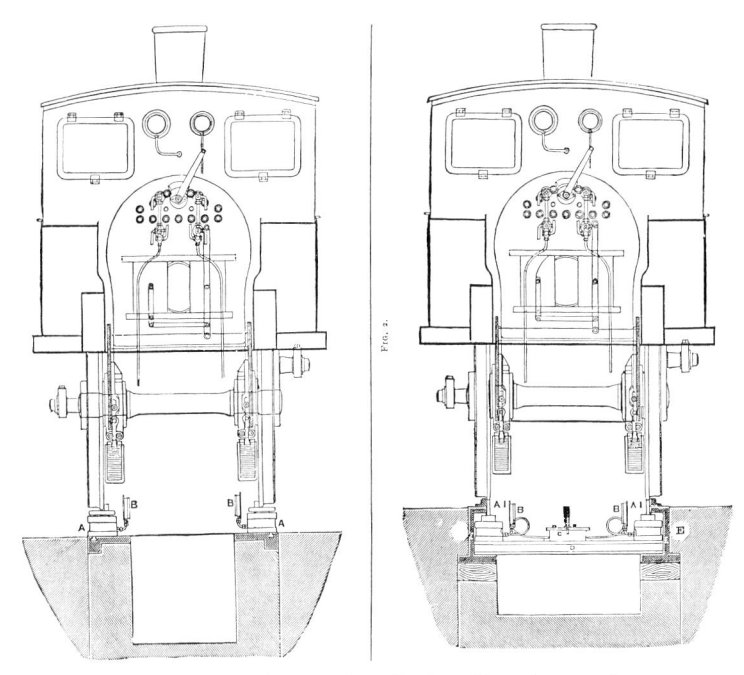

Fig. 5.27 Hydraulic weighing machine developed by Robert Stephenson & Co. of Newcastle in 1886.   *The Railway Engineer, Vol. VII, 1886, p.332.*

> As the conditions of fire-arms manufacture introduced the interchangeable system and improved machinery into a great range of small manufactures, the conditions of locomotive building are exercising a like influence in the introduction of uniform and labor-saving methods in the manufacture of marine engines and other heavy work.

As part of its mechanisation, Baldwin adopted electric drive motors from 1890, for overhead travelling cranes in its erecting shop.[141] In 1893, it applied electric power to its wheel-shop and, encouraged by 50% savings in power costs, it proceeded 'headlong' into electric drive for machine tools and cranes. Desperation drove the company into electrification, the increase in productivity allowing it to keep up with the demand in product size and quantity.[142] The larger British locomotive manufacturers also began a programme of workshop electrification at the end of the century. Beyer Peacock & Co. made an electrically driven wheel lathe in 1899 and its works were fully electrified by 1904.[143]

Comparison between the progressive manufacturers and the Baldwin Locomotive Works from the 1880s, demonstrates the extent to which economies of scale benefited the latter. It not only had a much larger domestic, and later foreign market, it also pursued a programme of component standardisation which significantly increased the opportunities for larger batch production. The proliferation of designs, and the consequential lack of standardisation denied the British industry this opportunity, and the limitations of the potential market dissuaded the manufacturers from the level of investment embarked on by the Baldwin company.

Standardisation

Standardisation of components lay at the heart of the comparison between small batch and mass production. Continuous production of standard components allowed machine tools to remain set up for common machining, increasing output by both machines and machinists, because of the time saved on re-setting for other components. There was, thus, incentive for manufacturers, which sought to specialise in locomotive production, to promote standard components and fittings as far as possible, adopting a range of templates and gauges to ensure inter-changeability.

The individuality of early locomotives gave much concern to railways, as incompatible components made maintenance difficult and expensive. Standardisation, which Richard Roberts had begun in the 1820s using his self-acting machine tools for manufacturing textile machines, was later applied to locomotive manufacture.[144] Roberts was the first to achieve, to a limited extent, standardisation of locomotive components,[145] with his system of templates and gauges, by means of which every part of an engine or tender corresponded with that of every other engine or tender of the same class.[146] James Nasmyth also introduced gauges for locomotive manufacture in 1838, as illustrated by his order for boiler tubes.

Gauges and templates helped maintain the level of precision necessary for the inter-changeability of parts between locomotives of the same class. Tool-settings and components were frequently checked against the gauges, and adjustments for wear made as necessary. As manufacturers took up their use, each made its own 'standard' and interchangeable components. With their limited manufacturing capacity, however, and the prospects of long delivery times for large orders, railways spread their contracts over several manufacturers. The separate standards of each manufacturer therefore limited component inter-changeability, and overall standardisation soon became the railways' major aim.

The first design to be accompanied by working drawings and templates, in pursuit of component inter-changeability, was prepared by the Great Western Railway for its *FIRE FLY* class in 1840. Daniel Gooch, the railway's Locomotive Superintendent recorded: 'when I had completed the drawings I had them lithographed and specifications printed, with iron templates for those parts it was essential should be interchangeable, and these were supplied to the various engine builders with whom contracts were placed.'[147]

Fig. 5.28 Initial standardisation. Great Western Railway *FIRE FLY* locomotive built by Jones & Potts of Newton-le-Willows to Daniel Gooch's design. *R.E. Bleasdale Collection, National Railway Museum.*

Particular inconvenience for all types of machinery, including locomotives, was occasioned by the incompatibility of bolt and nut sizes and their screw threads. Accurate screw-cutting and uniformity of thread had been first undertaken by Maudslay in the 1810s,[148] and Roberts, who improved precision from 1821.[149] However, new manufacturers began making screw-cutting machines with their own threads and bolt sizes, further negating component inter-changeability. In 1841, the problem was highlighted by Joseph Whitworth:[150]

> Great inconvenience is found to arise from the variety of threads adopted by different manufacturers. The general provision for repairs is rendered at once expensive and imperfect… This evil would be completely obviated by uniformity of system, the thread becoming constant for a given diameter. The same principle would supersede the costly variety of screwing apparatus required in many establishments, and remove the confusion and delay occasioned thereby. It would also prevent the waste of nuts and bolts… It does not appear that any combined effort has been hitherto made to attain this object. As yet there is no recognized standard.

Although several other Manchester manufacturers,

notably Roberts, Nasmyth and Bodmer, had advocated standard threads, it was only through Whitworth's strong advocacy, and the size and success of his own machine tool factory, that his standard 'Whitworth' threads were put into effect. The standards were soon adopted by most railways which specified for their locomotives: 'All bolts and studs to be screwed and chased to Whitworth's thread', or similar.[151]

Larger manufacturers, such as Sharp Brothers, extended their range of inter-changeable components during the 1840s. Following its major factory extension in 1847, E.B. Wilson & Co. standardised its components, and charged a premium for alterations.[152] However, the Nasmyth factory customers for large capital equipment insisted on designs tailored to their specific requirements.[153] From the 1850s, intra-standardisation of components led the larger railways to provide manufacturing drawings in order that components from their workshops and from the manufacturers were inter-changeable. Standards varied between railways, however, as locomotive superintendents pursued their individual design programmes. Although they had a frequent dialogue at meetings of the Institution of Mechanical Engineers, this largely dealt with design principles, materials, production methods and locomotive performance, whilst standardisation of components was ignored.[154]

Manufacturers pursued standard components for overseas railways as far as they could, and adopted them on designs for which they were responsible. Where locomotives were designed by consulting engineers they would adopt as many standard components as possible within the constraints of specified designs. As detailing work and production of working drawings was largely left to the manufacturers' own drawing offices, they could pursue limited standardisation, sometimes providing distinctive component characteristics.[155]

The industrial locomotive market meanwhile provided good opportunity for interchangeable components. The standard designs developed by manufacturers, such as Hudswell Clarke & Co. and Manning Wardle & Co., meant that significant economies of scale could be achieved for many components, even though customers sought a variety of arrangement variations, including track gauge. Such components as pistons and connecting rods were little changed over many years, allowing the manufacturers to produce batches and stock-pile them for subsequent use, or as part of a spares service. At times of slack demand, whole locomotives could be assembled for 'stock' for quick sale, using these standard components.

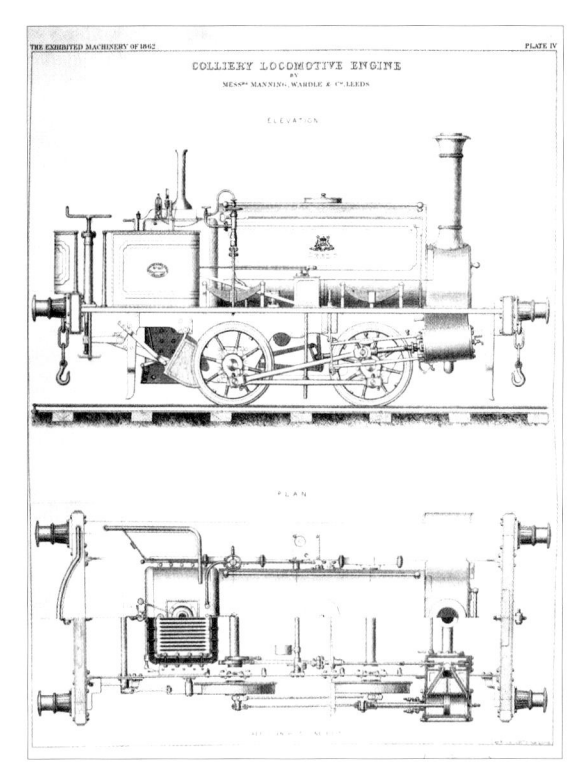

Fig. 5.30 Example of a 'standard' industrial locomotive design offered by Manning Wardle & Co. in 1862. *The Exhibited Machinery (of the London Industrial Exhibition) of 1862*, Plate IV.

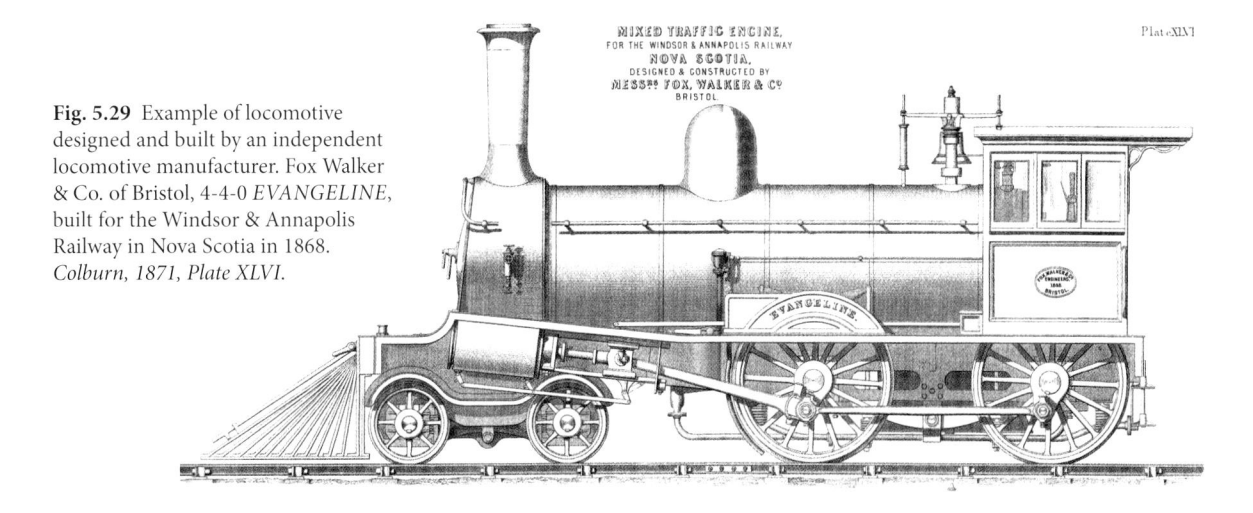

Fig. 5.29 Example of locomotive designed and built by an independent locomotive manufacturer. Fox Walker & Co. of Bristol, 4-4-0 *EVANGELINE*, built for the Windsor & Annapolis Railway in Nova Scotia in 1868. *Colburn, 1871, Plate XLVI*.

In marked contrast to these British approaches to standardisation, it is surprising that the immediate benefits of standard sizes for bolts, nuts and threads were not equally quickly taken up by American industry. The Baldwin company in America made a determined effort from 1855 to move wholly towards standard component production.[156] This was largely achieved by 1865, and by the 1870s the company had carried the principle of interchangeable components 'to great lengths'.[157] It was then able to offer main line and industrial locomotives with many standard components, whilst accommodating the customised arrangement requirements of the railroad master mechanics. Baldwin were able to achieve this because the master mechanics largely lacked the large drawing office teams of British railways and provided general rather than detailed specifications.

Baldwin thus 'leap-frogged' over the British main-line locomotive manufacturers in terms of standardisation and achieved volume production of many components emulating the 'American System' of other industries. Production costs of such common components as piston rods, cross-heads and slide-bars were substantially reduced. Inter-changeability meant that railroad maintenance requirements were also significantly aided through supply of replacement components from Baldwin, and which, in turn, became an important additional business for the manufacturer. By the 1890s, there was a large measure of standardised, inter-changeable parts, described as being among the most notable feats of 19th century American industry.[158] Although Baldwin maintained its locomotive prices at 'market' levels, its profitability from batch production of components was so good that it allowed it to maintain a high level of re-investment in capital equipment, through the remainder of the century.

Production Engineering
The reduction in locomotive manufacturing costs, and the evolution of component size and complexity during the century was achieved as much through improving production engineering procedures as by the advancements in machine tools and standardisation. The manufacturers pursued improved factory lay-outs and batch production procedures, having to judge the capacity growth they would require to make provision for expanding demand, whilst also accommodating major fluctuations in workload. The fluctuations in their capacity requirements ranged between insufficient work to keep men and machines employed, to periods of high demand when additional capacity was required to supplement the long hours being worked by an enhanced workforce, to minimise the deterioration of delivery times.

Such decisions were thus amongst the most crucial taken by the proprietors in terms of risk through over-capitalisation or under-provision of capacity. Because of the uncertainties of the locomotive market, and the different policies relating to diversification between the progressive and craft firms, these decisions were taken in relation to alternative market opportunities as well as the potential locomotive business.

The beginning of the 'factory system' in the early 19th century brought with it the recognition that matching production facilities to demand and maximising the employment of capital equipment to provide a satisfactory return, would require improved production control procedures. From 1823, Robert Stephenson & Co.'s factory undertook the manufacture of one stationary or locomotive engine every two months, before undergoing incremental increases in area and equipment to meet increased demand. Occasional downturns in demand left equipment and manpower under-used, and although capacity was doubled by 1829, the company was cautious about further investment to meet the peak locomotive demand for the Liverpool & Manchester and other railways in 1830/1.[159] It therefore concentrated on locomotive manufacture, which, even with detailed variations, benefited from increased output through batch production. Manufacturing time fell to about four months and production increased to two a month, but demand still exceeded capacity, and three locomotives were sub-contracted to Fenton, Murray & Co. of Leeds, for which arrangement drawings were made available by the Stephenson Company.[160]

From 1834/5, locomotive demand grew substantially coinciding with the rising demand for machine tools and textile machinery, and by 1840 annual output had increased by more than 500% (Fig 2.2). Manufacturers such as Sharp Roberts and Nasmyths Gaskell & Co., understood that higher output would be achieved with lower costs, if standard components were produced in batches without re-setting the tools, which Nasmyth defined as 'the production of standard interchangeable parts by means of power-driven machine tools.'[161] Nasmyth's first catalogues in 1838/9, however, contained no less than 126 entries,[162] which, at best, infers 'batch' production rather than 'mass' production.[163] The Great Western Railway's order for twenty *FIRE FLY* class locomotives from Nasmyth's in 1840, which was accompanied by drawings and templates, lent itself to batch production with which he was experienced. As Fenton, Murray & Jackson also made twenty *FIRE FLY*s, it may similarly have employed batch production techniques, perhaps learned from its long experience with textile machinery.

The practice of laying out a factory to minimise component handling and time between production processes was also developed in the 1830s. The earliest example of a planned 'work-flow' lay-out was by Nasmyth for the production of machine tools and applied to locomotive production.[164]

CHAPTER 5 – MANUFACTURING

Fig. 5.31 Cross-sectional view of factory layout drawn in December 1840 by James Nasmyth. Believed to be Nasmyths Gaskell & Co.'s erecting shop, Patricroft, Manchester. *Institution of Mechanical Engineers – NAS/4/9.*

Just prior to his factory's construction in 1836, he proposed that the buildings should be '*all in a line*… In this way we will be able to keep all in good order.'[165] On completion, the factory was described in a small booklet:[166]

> With a view to secure the greatest amount of convenience for the removal of heavy machinery from one department to another, the entire establishment has been laid out with this object in view; and in order to attain it, what may be called the straight line system has been adopted, that is, the various workshops are all in a line, and so placed, that the greater part of the work, as it passes from one end of the foundry to the other, receives in succession, each operation which ought to follow the preceding one, so that little carrying backward and forward, or lifting up and down, is required… By means of a railroad, laid through as well as all round the shops, any casting, however ponderous or massy, may be removed with the greatest care, rapidity, and security. The whole of this establishment is divided into departments, over each of which a foreman, or responsible person, is placed, whose duty is not only to see that the men under his superintendence produce good work, but also to endeavour to keep pace with the productive powers of all the other departments. The departments may thus be specified:- The drawing office, where the designs are made out; and the working drawings produced… Then come the pattern-makers… next comes the Foundry, and the iron and brass moulders; then the forgers or smiths. The chief part of the produce of the last named pass on to the turners and planers… Then comes the fitters and filers… in conjunction with this department is a class of men called erectors, that is, men who put together the framework, and the larger parts of most machines, so that the last two departments… bring together and give the last touches to the objects produced by all the others.

Although this arrangement was a far-sighted attempt to minimise handling and reduce manufacturing time, it made no provision for expansion, which would have required existing equipment to be moved with additional cost and production loss. Factory layout from the 1840s was generally undertaken on the specialist 'shop' system, which both provided for easy transfer of components between each production process and allowed for re-arrangement and enlargement with minimal disruption to overall production. Bodmer's factory for building machine tools, textile equipment and locomotives, also employed a work-flow system. It was equipped with an overhead travelling crane, and laid out for sequential machining 'according to a carefully-prepared plan'.[167]

When Beyer Peacock & Co. established its factory in 1854, it acquired land sufficient for later expansion of its manufacturing business. It was laid out both for easy movement of components between shops and for minimal disruption as the site expanded. Its single storey shops were fitted with roof lights to ensure good use of all the working areas, instead of limiting them to side windows as had been the case with earlier factories.[168] The last major new locomotive factory development in the century was the Clyde/Atlas Works in Glasgow, constructed in 1884, which was fully laid out in the shop system.

Manufacturers faced critical decisions when demand rose beyond prevailing capacity, and delivery times lengthened unacceptably. In the extraordinary mania events of 1844-47, delivery times increased to as much as three years. The then largest manufacturer, Robert Stephenson & Co., sought to accommodate the demand without risking its longer-term market standing. When it began to experience the surge of orders, it sub-contracted half of the order for 30 locomotives it had won from the Chemin de Fer Marseilles-Avignon in France to L. Benet of la Ciotat, near Marseilles, which was to supply:[169]

> on the latest and best system for which he had obtained a patent, half of the engines to be built in the works at la Ciotat under his direction and responsibility, and on condition that the engines so built in France may be in all respects equal to those which come from his works at Newcastle.

By November 1844, the Stephenson Company had decided to increase its manufacturing capacity, but, learning from experience of previous demand surges, it opted to take a seven-year lease on an additional site, half a mile from its main premises. Although this 'West Factory' was equipped with new machine tools requiring additional capital, the lease was not renewed after 1851, due to the much reduced market circumstances of that time. The Stephenson Company was very profitable in the late 1840s and seems fully to have vindicated the decision for the short-term lease. There is no evidence that other manufacturers adopted the same strategy, but E.B. Wilson & Co. interpreted the mid-1840s demand as being long-term and built its extensive new erecting shops at its Railway Foundry in Leeds. The shops were opened with much publicity in December 1847, just at the beginning of the sharp decline in orders accompanying the recession.

In spite of increasing their capacity between 1845 and 1847, the largest manufacturers, particularly Robert Stephenson & Co., were unable to meet the demand without lengthening delivery times. It therefore sub-contracted some orders by forming alliances with other manufacturers, whose reputation was acceptable to the railway customers, but whose longer-term interests would not conflict with its own. Nasmyths Gaskell & Co. made twenty-seven patent locomotives, and Jones & Potts also undertook a number of orders, for which the Stephenson Company received its due royalties.[170] Slaughter Grüning & Co. of Bristol also sought orders from the Stephenson Company, but was rejected as unsuitable.

At times of slack demand in the 1840s and 1850s, some manufacturers of main line locomotives sought continuity of work for their men and equipment by making some for 'stock', in the expectation of finding customers for them at a later date. This was commercially risky because of potential cash-flow problems and the main-line railway practice of specifying precise requirements rather than accepting stock designs.

Improved capital equipment, standardisation of components, and factory lay-out to expedite component production and assembly, were all pursued by the larger manufacturers in the 1840s and 1850s. E.B. Wilson & Co.'s 1847 plant was particularly well equipped and laid out for batch production of standard locomotives at a potential rate of six per month.[171] The new factories built by Peto Brassey & Betts and Beyer Peacock & Co., the first stages of which opened in 1853 and 1854 respectively, were similarly laid out for easy progression of components through specialist shops for their formation and machining processes, sub-assembly and final erection phases.[172]

By the 1860s therefore, there had been a divergence between 'progressive' manufacturers, whose premises were equipped for batch production and ease of work-flow, and

Fig. 5.32 A small batch machine shop. Robert Stephenson & Co.'s machine shop as depicted in 1864.
The Illustrated London News, October 15th 1864, p.393.

'craft' manufacturers which, although they had partially re-equipped with certain items of capital equipment, still relied on the inherent skills of their workforce, and had limited opportunities for production economies. The latter included the Stephenson and Hawthorn firms, which pursued small batches of marine, factory and colliery engines and other industrial equipment, whilst also pursuing locomotive orders when the market was favourable.

As the 'progressive' manufacturers, such as Neilson & Co. and Dübs & Co., expanded and acquired more specialised machine tools, forging and foundry equipment, the additional expense was further incentive to maximise productive use through the minimisation of tool and workpiece setting up and re-setting time. The organisation of the Hyde Park Works, Glasgow, of Neilson & Co. in the 1880s can be seen in the following list:

List of workshops for Neilson & Co (1881-1884)

Finishing Shop	Pattern-Makers and Joiners Shop
Machine Shop	Iron Foundry
Boiler Shop	Copper Shop
Tender Erecting Shop	Brass Foundry
Heavy Tool Shop	Paint Shop
Light Tool Shop	Millwrights
Grinding Shop	Yard
Smithy	Erecting Shop
Wheel and Frame Shop	Boiler Mounting Shop
Vertical & Horizontal Drillers	

Fig. 5.33 List of Workshops at Hyde Park Works, Glasgow, of Neilson & Co. *Data recorded in a notebook of 'A.I.', a senior clerk at Hyde Park Works. Author's collection.*

They were, however, unable to take full advantage of batch production of standard components due to the proliferation of standards of British and overseas railways. With several railways ordering small batches of locomotives, which were not always urgently required, there was little cost advantage for the well-equipped firms over the less well equipped. Full benefits of batch production were therefore restricted to the largest locomotive orders, which took advantage of the scale economies of multiple production. Fig.5.34 illustrates the growth of locomotive batch size orders in the second half of the century.

The average batch for the five largest orders each year grew from ten in 1850 to thirty-forty by the end of the century. The largest orders increased from thirty locomotives in 1850 to seventy-five by the end of the century, which allowed the larger manufacturers to reduce their unit costs. In 1896, for example, Neilson & Co. received an order from the Midland Railway for seventy-five goods locomotives and tenders at a contract price of £2,200 each. This was won against competing quotations ranging between £2,340 and £2,565.[174]

Differences in interpretation of the long-term growth of the locomotive market, and in strategic investment policies by the manufacturers were key issues which resulted in the industry's growing divergence between 'progressive' and 'craft' firms. By the end of the century this had resulted in market diversification and a wide diversity of skills and capital equipment, giving equally wide variations in manufacturing costs and productivity levels.

In the last twenty years of the 19th century, the 'craft' firms found themselves increasingly uncompetitive in main-line locomotive markets. In the continued absence

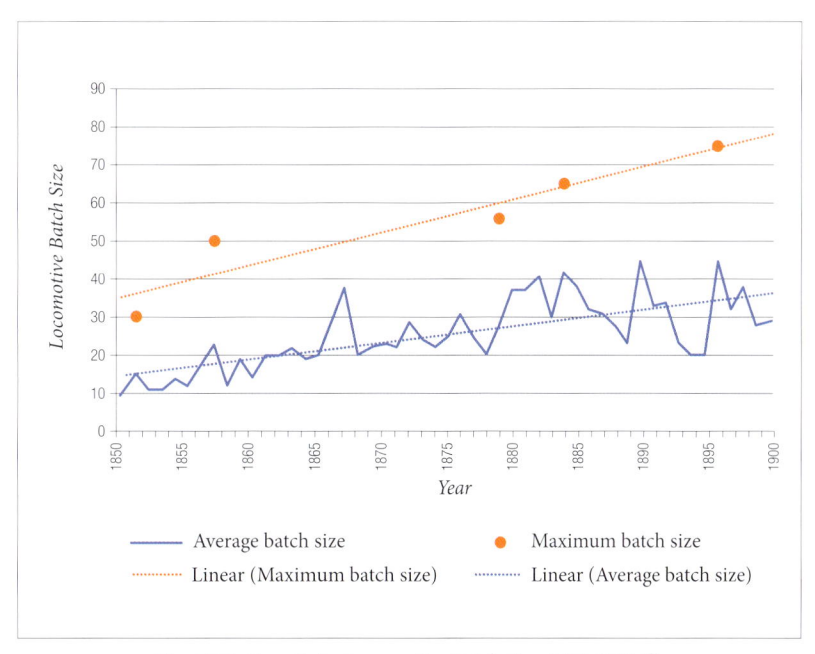

Fig. 5.34 Growth in Locomotive Batch Size 1850-1900.[173]

of major investment, they pursued small-batch manufacture of marine and industrial engines and other machinery, including replacement boilers, whose complexity benefited from their craft skills, whilst maintaining their locomotive interests through occasional small-batch orders for secondary railway and industrial customers, including re-buildings and component replacements.

In marked contrast to the 'progressive' manufacturers, the 'craft' firms undertook far less re-investment in premises or capital equipment, and their production costs were higher by comparison with their competitors. However, the 'craft' culture allowed older established firms to continue making locomotives in spite of inadequate equipment. By 1880, Robert Stephenson & Co. was obtaining orders well below its one-time capacity, but, in a buoyant market that year, it obtained an order for twenty locomotives from the Midland Railway. The Railway's inspecting engineer, Robert Weatherburn, later recorded his memories of the Stephenson site in that year:[175]

Fig. 5.35 Robert Stephenson & Co.'s Machine Shop, as seen in 1899.
Stephenson Locomotive Society, McDowell Collection/Tyne & Wear Record Office.

> ...one of the most striking personalities of the North... Geordie Crow... the works manager... was then a man over sixty years old... He was at that time, and for years, the most redoubtable mechanic in the north, and such men, once met, register an impression that never fades. Responsible, at a time of cutting prices, for the success of works more than half a century old, and destitute of modern machinery, and railway communication, and having to compete with firms equipped with the latest labour-saving tools and machinery, and so planned as to give the most rapid transit incoming and outgoing; yet with this enormous discrepancy... the firm undertook to build and deliver locomotives to the satisfaction of the Midland Railway Company. Everywhere decrepit old lathes, slotting and drilling machines that no other firm would have harboured, and few would have speculated in except for scrap; yet no finer work ever left a firm than was turned out through the care, aptitude and genius of Geordie Crow... I always felt as though I was witnessing a last desperate effort against overwhelming odds... made by one gallant man almost unaided – to regain that which should never have been lost.

In the 1880s and 1890s, the diversity between progressive and craft manufacturers was partly concealed by the cartel agreements of the ten manufacturers forming the Locomotive Manufacturers Association. These agreements, which were designed to assist the weaker members of the industry, had the effect of concealing the more costly productions of the craft manufacturers and diluting the benefits of the more productive firms.

The manufacturers' tactical response to the demand was therefore to use their equipment more intensively, made possible through major recruitment campaigns and multiple shift working. The ten members of the Locomotive Manufacturers Association increased their workforce by two-thirds from 8,250 in 1894 to 13,600 in 1900,[176] but there is no evidence of short-term leasing of additional premises to provide extra manufacturing capacity. The lengthening delivery times caused such frustration amongst some British and overseas railways that some orders were lost to American and German manufacturers.

In contrast to the railway workshops, there is no evidence of any policy by the manufacturers to make their capital equipment 'firm-specific' in order to deter free labour movement and suppress wage claims at times of skilled labour shortages. Even following the establishment of the Locomotive Manufacturers Association in the 1870s, there is no evidence of collusion with capital equipment, or in any other way to deter free labour movement. On the contrary, the discussion in the next chapter, on employment and industrial relations, indicates an extraordinary loyalty to their firms by skilled personnel.

Industrial locomotive manufacturers on the other hand, in their very different and more competitive market place, required standard component designs to remain competitive, and came closest to 'American system' manufacture. Firms, such as Hudswell Clarke & Co. and Manning Wardle & Co., had learned the benefits of volume production to produce competitively priced 'standard' locomotives even allowing for gauge and other specification changes to suit particular requirements.

Not until the late 1880s did R.&W. Hawthorn Leslie modernise its site. Its Works Manager, William Cross, reported:[177]

> The whole of the erecting shop has been remodelled, the floor lowered, and a proper floor put down, new craneways and powerful power cranes; the roof, which I found in a most dangerous condition, has been almost remade… The Wheel shop, which was formerly a collection of tumble down sheds, in various stages of dirt and decay has been entirely rebuilt. The boilerpower was formerly so bad, owing chiefly to worn-out boilers, that it was by no means uncommon for the whole place to be laid off for a day or two while they were being timbered up. One new boiler, with one man, now does easily what formerly took five boilers and four men to do with great difficulty… we now have a (boiler) yard well adapted for the class of work it is intended for, and capable of turning out a very much larger quantity of work than the older one ever could have done.

With the extraordinary expansion of the British economy in the mid-late 1890s, leading to the unprecedented increase in demand for locomotives, production doubled from a low of 500 main line and industrial locomotives in 1894 to over 1,100 in 1899 (Fig.2.2). The surge in demand was compounded by the 1897/98 seven-month national industrial dispute, which severely disrupted locomotive production (Chapter 6). Although the manufacturers were well used to demand surges, they were quite unable to accommodate the full demand, and delivery times lengthened considerably. With railway workshops at full capacity, large orders were received from the home market to add to those from overseas railways. Prices rose and, with the 'progressive' manufacturers working to capacity, orders were obtained by the 'craft' manufacturers, including Robert Stephenson & Co., which won the largest order in 1899, for forty locomotives for the Midland Railway.[178]

Fig. 5.37 Product of the late 19th century surge in orders. 4-4-0 locomotive for the Nippon Railway of Japan built by Sharp Stewart & Co. Ltd. of Glasgow (SS 4433/98). *Author collection.*

This slow improvement in productivity was in marked contrast to the American locomotive industry, which capitalised on its manufacture of standard, interchangeable components. By 1865, so soon after the Civil War, it was

Fig. 5.36 Standard design of industrial locomotive. Manning Wardle & Co. 1894-built 0-6-0T (MW 1174/1894) for Puerto de Buenos Aires (No. 3) in Argentina. *Author's collection.*

125

incorporating 'sophisticated adaptations' of New England armoury practice.[179] The drawing office developed 'shop cards' of standard components for use with gauges and templates. Master gauges, kept in a gauge shop, were used each evening to verify working gauges used in production. Each factory shop contained 'hundreds of all kinds of standard gauges and templates for boring, turning and planing'.[180] By 1881, almost all parts with machined surfaces were 'accurately fitted to gauges'.[181]

Baldwins' standardisation allowed for good batch production runs with its increasing number of semi-automatic machine tools. Further use was made of this equipment, with increased productivity, when electric lighting was introduced from about 1881, allowing double shifts and round-the-clock operations.[182] Until then, as in the British workshops, all lighting had been by gas, and productivity was poor after dark.[183] British manufacturers did not adopt electric lighting until the introduction of electric power-operated machine tools at the end of the century. Although the LNWR's Crewe workshops first used electric powered machine tools in 1890, it did not install a power-house for general electric supply, until 1903.[184] The first independent manufacturer to install a power-house to supply electric power and lighting, in about the same year, was Beyer Peacock & Co.[185]

Deliveries

Locomotives were delivered to customers using transport that was best suited for their safe delivery, incurring costs that were reflected in their quoted prices. In the earliest years all transport was by coastal shipping to the port nearest to the customer's rail-head. As rail networks expanded more use was made of rail movement for deliveries, although varying track-gauge was an important consideration in planning delivery moves.

Movements from factory gates to the shipping quays or longer distance overland movements required the provision of horse-drawn heavy lift trailers. These were firstly provided by contractors, but as production increased the factories manufactured their own 'low loaders', together with teams of horses to haul them and their attendant handlers.

Overseas customers received their newly-built locomotives by shipping movement directly from an appropriate port near to the factory where they were built, to the receiving port nearest to the customer's rail-head. Often transhipment to or from coastal vessels took place at major shipping ports, such as Liverpool, that offered long-distance vessel routes, particularly to North America. In Europe locomotives were often transhipped to smaller craft sailing up rivers and canals to reach customers' rail-heads.

For most of the century craneage at ports, or derrick cranes on many ships, were of insufficient capacity to lift complete locomotives and tenders, requiring them to be dismantled and re-assembled on arrival. Such movements required the attendance of a knowledgeable foreman from the factory to oversee the discharge at the receiving port and to supervise their re-assembly. Towards the end of the century cranage at ports and the growth in the size of ships allowed many complete locomotives to be lifted and stowed on board vessels without the need for dismantling.

Fig. 5.38 Horse-drawn heavy lift trailer conveying a new marine boiler from Robert Stephenson & Co.'s Newcastle upon Tyne factory in a newspaper montage about the company in 1881. *The Graphic, 4th June 1881.*

Figs. 5.39 and 5.40 Shipment of locomotives in 1880. Partially dismantled Vulcan Foundry-built 4-4-0s (probably VF 851-858/80), for the Bombay & Baroda Railway in India (Nos. 101-108) being loaded on to a vessel, probably in Liverpool or Birkenhead Docks.
Anon, *The Vulcan Locomotive Works 1830-1930*, Loco. Publishing Co., 1930, p.114.

Fig. 5.41 (Left) Shipment of locomotives in the late 1890s. 2-4-0 tender locomotive built by Neilson Reid & Co. Ltd for the Chemin de Fer du Nord (Northern of France Railway) (No. 690) being hoisted aboard ship in Newhaven Harbour.
The Railway Magazine, Vol.1 (1897), p.304.

Notes – Chapter 5

1. Saul, *op cit* (42), p.186.
2. Chandler (1990), p.62.
3. Chandler (1990), Part III, 'Great Britain: Personal Capitalism'; also Chandler (1977), p 241.
4. Zeitlin (1997), p 241.
5. Zeitlin (1997), p 248/9.
6. Zeitlin (1997), p 99.
7. Zeitlin (1997), p 81-107.
8. Rolt, 1965.
9. Drummond (1997), pp.32/3, Note 22.
10. Zeitlin, (1997), p 242. Also, Gilbert, (1965), and Gilbert, (1971-72), p 121.
11. Cookson, (1997), p 4.
12. Drummond, (1995), p 105, and Brown, (1995), p 173.
13. Brown, (1995), p xxviii.
14. Clarke, (nd but 1979), p 14.
15. *The Artizan*, November 1843, p.261, Article XII – 'The Management of Steam-Engine Factories'.
16. The introduction of women tracers was begun by Dübs & Co. in the 1860s. Moss and Hume, (1977), p 46.
17. Douglas Gordon, 'The Building of a Locomotive', *The Railway Magazine*, Vol IX, 1901, p 113.
18. Obituary of Charles T. Parry, *Report of the Proceedings of the Twentieth Annual Convention of the American Railway Master Mechanics Association*, Chicago, 1887, p 200, quoted in Brown, (1995), p 170.
19. Bailey (1984), p 17.
20. Minute Book of R. Stephenson & Co., 1823 - 1848, R. Stephenson & Co. collection, ROB/1/1, pp 19/20.
21. Advertisement, *Newcastle Chronicle*, 15th October 1825.
22. Bourne, (Ed.), (1846), p 227.
23. Bailey, (1984), Appendix IX.
24. Minute Book of R. Stephenson & Co. (*op cit* 20), p 35, 20 Oct 1832.
25. Foundry Account and Balance Sheet, R. Stephenson & Co., March 30th 1833, Pease-Stephenson Collection.
26. Rolt (1964), pp 32/38.
27. Analysis of Order E515, of Neilson & Co., February 1880, in Fittings and Materials Order Book, Vol.17,

28. Evans (1992), pp 178-196. Also Martin (1974).
29. Letter George Stephenson to Robert Stephenson, Liverpool, 8 January 1828, Institution of Mechanical Engineers Library.
30. Report by Thomas Davison, *op cit* (18), p 35.
31. Letter, Harris Dickinson (for R. Stephenson & Co.) to Edward Pease, Newcastle Upon Tyne, 10 Mo 21.1830, Pease-Stephenson Collection, D/PS/2/52.
32. R. Stephenson & Co. ledger 1823-1831, R. Stephenson & Co. collection ROB/4/1, *passim*.
33. Notebook by Christopher Davy, Vol.2, 1836, p 70, author's collection, recording details of the Stephenson Company's products as recorded on a visit to the Newcastle factory in 1836.
34. Gale (1966), p 97.
35. Dodsworth (1965), pp 122-164.
36. Mot, (1977-78), p 157.
37. Advertisement, 'Tank Engines for the Midland Railway', *Engineering*, Vol.VIII, Oct.8 1869, p 239.
38. For example, Contract between Messrs. Sharp Stewart & Co. and the South Eastern Railway Company for the construction of twelve goods engines, 17th May 1878, PRO, Rail 635/229.
39. Bourne (ed), (1846), p 227.
40. For example, order placed by Benjamin Hick to the Bowling Iron Works, 25th July 1836, was for '8 Engine tire bars 16.0 ft long for 5.0 wheels', Benjamin Hick Order Book, 1833-1836, Manchester Science & Industry Museum Archives.
41. Bourne (Ed.), (1846), p 227.
42. Advertisement, *op cit* (37).
43. Ledger, 1864-1889, and Journal, 1864-1900, Vulcan Foundry Co. collection, B/VF/4/1/1 & B/VF/4/2/1.
44. Lord (1945-1947), p 170.
45. Bailey (1984), pp 296/7. Also, letter, Harris Dickinson (for R. Stephenson & Co.) to Edward Pease, Newcastle upon Tyne, 10 mo.21.1830, Pease-Stephenson Collection, D/PS/2/52.
46. Moss and Hume (1977), p 77.
47. Bailey (1984), Appendix IX
48. Report by Thomas Davison, 'Description and Valuation of Bedlington Ironworks in the County of Durham and Northumberland', quoted in Martin (1974), p 35.
49. Letter George Stephenson to Michael Longridge, Liverpool, Oct.11th 1830, Phillimore Collection, Institution of Mechanical Engineers Library.
50. *Description of The Patent Locomotive Steam Engine of Messrs. Robert Stephenson and Co.*, London, 1838, p 35.
51. Article, *Railway Times*, Vol. I, August 4th 1838, p 413.
52. Letter, Robert Rennie, *Weekly Chronicle*, 24th October 1908, with description of crank-axle making in 1848.
Neilson Reid & Co. Collection, UGD10/3/1.
53. Ahrons (1927), p 286.
54. Hills and Patrick (1982), p 80. Also, Drummond (1995), p 48.
55. *Engineering*, Vol.XIII, January 26th 1872, p 63.
56. Analysis of Order E515, of Neilson & Co., February 1880, in Fittings and Materials Order Book, Vol.17, Neilson Reid & Co. collection, UGD10/3/1.
57. Bailey (1984), pp 289/290.
58. Patent Locomotive, *op cit* (50), p 10.
59. Morton (1983), p 35.
60. Copy letter, Nasmyths Gaskell & Co. to the Cheadle Copper & Brass Co., Patricroft, 13 December 1838, Nasmyth Gaskell collection, Letter Book 3, p 351.
61. Samuel Timmins (ed), *Birmingham and the Midland Hardware District*, 1866, quoted in Morton (1983), p 35.
62. Tubes from ship-wrecked Neilson-built locomotives recovered during 'Operation Iron Horse' project off the coast of Islay, Scotland in 1980s. Now retained in the National Railway Museum, York.
63. Midland Railway, Specification for Goods Engines and Tenders inserted in the Minutes of the Locomotive Committee, 3rd February 1852, PRO, Rail 491/168.
64. Midland Railway, Minutes of the Locomotive Committee, 20th March 1860, PRO Rail 491/170.
65. South Eastern Railway, Specification of a 4-Wheeled Coupled Bogie Tank Engine, Oct 18 1880, PRO, Rail 635/230.
66. Patent Specification 7724, July 9th 1838.
67. Open letter by George Stephenson, Chesterfield, Oct 15th 1847, published in 'the daily papers', quoted in Smiles (1857), pp 449-450, and subsequent editions.
68. Patent No 12,889, enrolled in UK on December 15th 1849, quoted in Colburn (1871), p 79.
69. For example, South Eastern Railway, Specification of a 4-Wheeled Coupled Bogie Tank Engine, Oct 18 1880, PRO, Rail 635/230.
70. Article, 'Messrs. Gresham and Craven's Works', *The Railway Magazine*, Vol. I (July-December 1897), pp 252-255.
71. Article, 'The Westinghouse Brake', *The Railway Magazine*, Vol. I, (July-December 1897), pp 362-369.
72. Article, 'The Vacuum Automatic Brake', *The Railway Magazine*, Vol. XII, (January-June 1903), pp 104/5.
73. Gordon S. Cantle, 'The Steel Spring Suspensions of Horse Drawn Carriages (Circa 1760 to 1900)', *Transactions of the Newcomen Society*, Vol.50 (1978-9), p 25.
74. Bailey, (1984), Appendix IX.
75. Letter, Thomas Turton & Sons, Sheffield, to Messrs. Jones & Potts, Oct. 19 1848, Jones & Potts collection, IMS 246/7.
76. Redman (1972), p 38.

77 Specification of a 4-wheel Coupled Bogie Tank Engine, South Eastern Railway, Ashford, Oct.18 1880, PRO, Rail 635/230.
78 Anon, 'A Modern Forge of Vulcan', *The Railway Magazine*, Vol. II (January-June 1898), pp 171-178.
79 For example, Specification of Ten Tank Engines for the South Eastern Railway, 9th February 1864, PRO, Rail 635/225.
80 Rolt (1965), Chapter 5, pp 92-121.
81 Bailey (1984), p 31.
82 Bailey (1978-79), pp 109-138.
83 Bailey, (1984), p.306.
84 J. Nasmyth, 'Remarks on the Introduction of the Slide Principle and Machines Employed in the Production of Machinery', in Buchanan (1841), pp 393-418.
85 (Sir) George B. Bruce, Presidential Address, *Proceedings of the Institution of Civil Engineers*, 8th November 1887.
86 Bailey (1984), Chapters 10 and 11.
87 Rolt (1965), Chapter 5, pp 92-121. Also Hamilton (1957-59), pp 184-187.
88 Smiles (Ed.) (1883), p 199.
89 Steeds (1969), p 69.
90 *Proceedings of Institution of Mechanical Engineers,* 1856, p 130.
91 Smiles (1883), p 238.
92 For example, R. Buchanan, (1845), Plates XLVIII, XLIX and L and pp 459-462.
93 Chaloner (1968-69), p 43.
94 Pole (ed) (1877), pp 163/4. Also advertisement, *Railway Times*, Vol. I, October 13th 1838, p 601.
95 Rolt (1965), p 125. Also Hamilton (1957-59), p 186.
96 Rolt, (1965) p 124.
97 Hills and Patrick (1982), p 21.
98 Steeds (1969), pp 52/3.
99 Select Committee on Exportation of Machinery, *Parliamentary Papers*, 1841, Vol. VII, Second Report, p vii.
100 Smiles (1883), pp 199/200.
101 Smiles (1883), p 307.
102 Rolt (1965), p 112.
103 Bradley (1972), pp 65/6.
104 For example, the double-spindle milling machine patented by Sharp Stewart & Co., in *Proceedings of the Institution of Mechanical Engineers*, 1856.
105 Steeds (1969), p 75.
106 William Fairbairn, Presidential Address, 1861, *Report of the British Association*, 1862, pp lxiii - lxiv.
107 Analysis of Sale Catalogue of E.B. Wilson & Co., commenced 15th day of August 1859, Hardwicks & Best, Auctioneers, Leeds. Also, for R.&W. Hawthorn, *The Artizan*, October 1863, quoted in Clarke (n.d. but 1979), p 14. Also, for Neilson & Co., see Thomas (1964), pp 88/9, inventory for the firm's move to Springburn in 1861/2.
108 Lineham (1902), pp 313-318.
109 Hills and Patrick (1982), p 110.
110 Letter J. Laird (for the Patent Rivet Co.) to Messrs. Jones & Co., Liverpool, Nov. 22 1839, Jones & Potts collection, IMS 245/4.
111 Brown (1995), pp 167/9.
112 Brown (1995), pp 167/9.
113 Steeds (1969), p 102.
114 Rolt (1965), p 126.
115 E.B. Wilson & Co. Sale Catalogue, *op cit* (107).
116 Hills and Patrick (1982), p 23.
117 Moss and Hume (1977), p 45.
118 Cantrell (1984), pp 123/4.
119 Smiles (Ed) (1883), pp 239-251. Also Cantrell (1984), Chapter VI, pp 134-180.
120 Cantrell (1984-85), pp 133/165.
121 Redman (1972), pp 24/5.
122 Richard Peacock of Manchester, 'Description of a Steam Hammer for Light Forgings', *Proceedings of the Institution of Mechanical Engineers*, 1860, pp 284-292.
123 Richard Peacock in discussion following paper by W.L.E.McLean, 'On The Forging Of Crank Shafts', *Proccedings of the Institution of Mechanical Engineers,* 1879, pp 461-483.
124 Cantrell (1984), pp 75/6.
125 Hills and Patrick (1982), p 21.
126 McNeil (1974-1976), p 153.
127 Neilson & Co., Fittings and Materials Book, Vol.17, 1880, Neilson Reid & Co. collection.
128 Hills and Patrick (1982), p 109.
129 Moss and Hume (1977), p 56.
130 Analysis of E.B. Wilson & Co. Sale Catalogue, *op cit* (107). Also, for R.&W. Hawthorn, *The Artizan*, October 1863, quoted in Clarke (n.d. but 1979), p 14. Also, for Neilson & Co., Thomas (1964), pp 88/9, inventory for the firm's move to Springburn in 1861.
131 For example Beyer Peacock & Co. installed a 7ft wheel lathe in 1854. Hills and Patrick (1982), pp 23 & 111; and a Whitworth lathe supplied to Brassey, Peto & Betts could also turn a 7ft wheel, Millar (1976), p 46.
132 *Engineer*, 9th August 1867. Also Steeds (1969), p 135.
133 *Engineering*, Vol. XIII, January 19th 1872, p 49.
134 *Engineering*, Vol. VIII, December 17th 1869, pp 402/4.
135 *Engineering,* Vol. XVII, January 9th 1874, p 30. Also, Reed (1982), pp 77-79.
136 *Engineering*, Vol. XVI, October 3rd 1873, p 268 and October 31st 1873, p 351; *Engineering*, Vol. XIX, June 11th 1875, pp 483/4; Hills and Patrick (1982), pp 55-81 *passim*. Also, Steeds (1969), pp 91, 111 &

136 155. Peacock's Patent No. 696 of 1887. *Engineer*, 21st August 1896.
137 *The Railway Engineer*, Vol. VII, 1886, p 332.
138 Reports by Neilson & Co. to the Locomotive Manufacturers Association 1890-1900, LMA Minute Book No.1, Search Engine, National Railway Museum.
139 Brown (1995), p 187.
140 Charles H. Fitch, 'Report on the Manufactures of Interchangeable Mechanism, in US Department of the Interior, Census Office', *Tenth Census of the United States*, 1880, Vol. 2, *Report on the Manufactures of the United States*, Washington DC, 1883, pp 58-59; quoted in Brown (1995), p 187.
141 Brown (1995) pp 191/2.
142 Brown (1995) p 196.
143 Hills and Patrick (1982), pp 81/105.
144 Dickinson (1945-47), p 127.
145 Rolt (1965), p 107.
146 Smiles (1863), p 271. Also *Engineer*, 13th February 1863 ascribed the use of standard gauges to Roberts as being employed for all the work on his self-acting mules and locomotives.
147 Daniel Gooch diary (1839), quoted in Mosse (1990-91), p 101.
148 Evans (1994-95), pp 157/8. Also Brooks (1992-93), pp 107/8.
149 Dickinson (1945-47), pp 125/6.
150 Whitworth (1841), pp 157-160.
151 Specification of Tank Engines for the South Eastern Railway, 9th February 1864, PRO, Rail 635/225.
152 Redman (1972), p 12.
153 Zeitlin (1997), p 243.
154 *Proceedings of the Institution of Mechanical Engineers*, passim.
155 Douglas Gordon, 'The Building of a Locomotive', *The Railway Magazine*, Vol. IX (July-December 1901), p 115.
156 Brown (1995), pp 170-183.
157 Brown (1995), pp 164 & 174-183.
158 Brown (1995), p 172.
159 Bailey (1984), pp 93-100 & 265-275.
160 Bailey (1984), p 150.
161 Musson (1982).
162 Cantrell (1984), p 67.
163 Cantrell (1984), p 76.
164 Cantrell (1984), pp 64/5.
165 Letter James Nasmyth to Holbrook Gaskell, 11th July 1836, in possession of the Gaskell family, quoted in Musson, (1957-58), p 126.
166 Booklet, *Manchester As It Is*, 1839, quoted by Musson (1957-58), p 127.
167 Rolt (1965), p 126.
168 Hills and Patrick (1982), pp 15/16.
169 Rapport à l'Assemblée Générale des Actionnaires de la Cie d'Avignon Marseilles, 29th April 1844, quoted in Warren (1923), p 96.
170 Cantrell (1984), pp 200/1.
171 Redman (1972), p 10.
172 Millar (1975), pp 44-46; and Hills and Patrick (1982), pp 15-24.
173 Analysis of locomotive orders from manufacturers records &c., (Chapter 2).
174 Midland Railway, Minutes of the Locomotive Committee, 3rd January 1896, Minute No. 4663, PRO, Rail 491/182. This quotation was, however, subject to the 'cartel' pricing arrangements of the Locomotive Manufacturers Association (Chapter 3), and probably understated the true savings from batch production.
175 Robert Weatherburn, 'Leaves from the Log of a Locomotive Engineer', No. XIX, *The Railway Magazine*, Vol.34 (1914), pp 294-300.
176 Minute Book of the Locomotive Manufacturers Association 1875-1900, Railway Industries Association collection, Search Engine, National Railway Museum.
177 Clarke (n.d. but 1979), p 55.
178 Analysis of locomotive orders from manufacturers records &c., (Chapter 2).
179 Brown (1995), p 176.
180 Article, 'The Baldwin Locomotive Works', *American Artisan*, June 5 1867, pp 482/3, quoted in Brown (1995), p 177.
181 Burnham, Parry, Williams & Co., *Illustrated Catalogue of Locomotives*, 1881, p 57, quoted in Brown (1995), p 177.
182 Brown, (1995), p 189.
183 For example Beyer Peacock & Co. had gas lighting installed in its new factory from 1854. Hills and Patrick (1982), p 24.
184 Drummond, (1997), p.52.
185 Hills and Patrick (1982), p 105.

Chapter 6
Management, Employment and Industrial Relations

Introduction

The growth of locomotive manufacturing firms during the 19th century took them from small proprietorial concerns, with less than 100 craftsmen and labourers, to large private and public enterprises with workforces up to 3,000 strong. To accommodate such expansion the manufacturers pursued new labour policies, which both developed management skills and responsibilities, and saw the transition of work skills from a fully craft-based to a partly 'factory'-based system. This was introduced against a background of improving terms of employment nationally, often arising from industrial relations disputes.

The growth of the locomotive industry through the century required an evolution in managerial responsibilities to direct employment policies and deal with industrial relations issues. The strategic decisions that were taken by the manufacturers to increase their workforces, against a background of craft shortages and sharply fluctuating demand, and the tactical decisions on employment terms, paternalism and industrial relations, were amongst the most difficult that the proprietors had to make.

To enable the firms to expand and be competitive, their evolving policies centred around both the growth of their skill-base and the employment of un-skilled labour who could undertake repetitive tasks when demand was high. They also required the flexibility to recruit and dismiss unskilled labour, according to demand, in order to preserve continuity of employment for their skilled personnel. These policies, which varied in their proportion

Fig. 6.1 Manager and Staff of Neilson & Co. *c.*1860s. *North British Locomotive Co. archive – Mitchell Library, Glasgow.*

of unskilled labour, were closely related to the progressive and craft firms' strategies on market specialisation and investment for specialised production.

The locomotive industry inherited the proprietorial management practices of the early 19th century heavy manufacturing industry, which it developed to meet its own circumstances. The vertically-integrated structure of the industry meant that it already had experience of administrative co-ordination. Each smithy, foundry, machine and erecting shop had a foreman who was responsible to a 'Head Foreman', who may or may not have been a partner of the firm. Their experience with multi-activity operations led firms to develop their administrative functions, firstly through delegated responsibilities from the partners to experienced workshop foremen and head clerks and, subsequently, through employment of specialist managers.

As management and manufacturing technology evolved, so did the division of labour, into 'time-served' skilled craftsmen, semi-skilled artisans, premium and craft apprentices and un-skilled 'labourers'. The changing technologies had both cause and effect on the composition of the workforce and employment policies. The all-embracing skills of the millwrights, foundry-men and boiler-makers were replaced by the hierarchical 'factory' system, which passed their former decision-making discretion to the proprietors, head clerks and head foremen who determined technological progress, designs, material development and manufacturing programmes.

The management and motivation of a mixed labour force, against the background of increasing 'de-skilling' of repetitive production, called for evolving employment policies to administer, encourage and discipline the changing workforce. For major firms in the capital goods sector, such as locomotive manufacturers, these policies included further cost-reducing measures, notably productivity-related incentive piece-work and the imposition of overtime when the locomotive market was strong.[1] However, the extent to which differential wage rates and other employment incentives were necessary between manufacturers, to recognise the relative attraction of their different sites and the need to develop paternalism to encourage loyalty and longevity of service, were important issues faced by the industry.

At the commencement of the railway era, the employment of millwrights and other skilled craftsmen brought an inherent discipline to locomotive manufacturing, born from respect through the apprenticeship system. The commitment of young men to be trained as junior engineers by their experienced elders, or to learn a 'craft' from experienced 'journeymen', and to 'bind' themselves to be trained for a period of between four and seven years, had been established in the 18th century. In the 19th century craft training remained the central route to skilled employment and manufacturers maintained an ongoing commitment to apprentice training.[2] Although the manufacturers sought to employ unskilled men for routine machining tasks, their requirement for the craft skills remained high, and the numbers of skilled 'journeymen' exceeded half the country's manufacturing workforce by the end of the century.[3] This draws into question whether firms generally regarded apprenticeship training as an important part of their long-term labour policies, or whether they were seen as a form of inexpensive 'bound' labour.

As experience in locomotive construction developed, the on-going shortages of skilled labour brought the potential for competing claims for manpower between the locomotive firms, and to wage escalation. Paternalism in railway workshops, such as Crewe, was seen as important to prevent workers migrating to other railway towns.[4] Paternalism and patronage not only created the basis for managerial strategy, but it also formed the basis of its system of industrial relations. The employment policies of the independent manufacturers therefore determined whether they were similarly motivated towards paternalism and the extent to which it was felt to be of benefit in avoiding migration.

The improving employment terms mostly arose from industrial disputes, which were as much to do with the retention of craft skills and the preservation of 'closed shops' to minimise the erosion of bargaining power, as they were to do with the basic hours of work or rates of pay. The evolution of these issues from local to national disputes brought about an increased representation by the manufacturers on regional and national employer federations, which does not indicate any particular accommodation with the skilled workers.[5] In testing to see the extent of any accommodation, therefore, the part played by the manufacturers, in dealing with both their fellow employers and the trades unions, provides an indication of how influential their role was in the country's overall changing employment scene.

Whilst delegation of proprietorial responsibilities to managers was generally adopted throughout the industry, the evolution of employment policies most distinguished the growing diversity between the progressive and the craft enterprises. The shortage of skilled labour throughout the century, with attendant higher wage expectations, was a major pre-occupation for the manufacturers. They resolved this shortage with varying emphasis on the development and acquisition of self-acting machine tools and other capital equipment, resulting in a divergence in the numbers of 'time-served' craftsmen employed by progressive and craft firms.

The millwright, the smith and the foundryman had already developed skills to manufacture many forms of heavy machinery by the commencement of the railway era. The breadth of their practical and intellectual skills was essential for the development of early locomotive technology, but the extraordinary rate of the market's expansion soon led to a shortage. The industry sought to overcome this by productivity improvements through radical changes in machine tool technology and manufacturing processes. Basic component preparation converted to 'factory'-based employment of un-skilled machinists.[6] However, the locomotive industry remained, on the whole, more dependent on skilled labour than repetitive manufacturing industries, with boiler-makers, foundrymen and fitter-erectors adapting their skills to improving techniques and equipment.

The nature of skill and its application to the heavy manufacturing sector, had to accommodate the on-going requirement for several craft skills whilst reducing the costs of routine machining tasks.[7] The first issue for the manufacturers was to identify those repetitive activities which could be undertaken with new equipment by labour recruited for the purpose, and for whom training could be completed in just a few days. Much of their activities, however, would remain 'skilled', because no machine could be devised to carry them out.

The second form of de-skilling, however, was motivated by different considerations on the manufacturers' part, namely the effective administration of design and technological developments, co-ordination of activities, planning of work programmes, and bulk ordering of raw materials and components – all work formerly undertaken by 'time-served journeymen'. Skill was more than just experience of metal-working and fitting, it included discretion and freedom to make decisions on the selection of materials and how the work was carried out.[8] The move towards a 'factory' system in a competitive environment meant that these responsibilities and functions would pass to the proprietors or their managers. The adaptation of craftsmen's skills, and the introduction and development of the factory system of production were therefore important considerations in the evolution of the locomotive industry.

A further important issue was the retention of staff. The move into 'factory' work required a new culture among workers, which required them to adjust to the regularity and discipline of factory work.[9] Factory workers were said to have a 'restless and migratory spirit' and a stable, rather than a better labourer was usually worth more to a manufacturer. Firms in the heavy manufacturing sector therefore sought ways to develop employment policies for un-skilled staff. However, the shortage of skilled craftsmen made it important for the locomotive firms to maintain loyalty, discipline and longevity of service through attractive employment terms. Hours of work, wage rates and productivity pay incentives in manufacturing establishments were all better than the provisions of the 1867 Factory Acts Extension Act, which was the first legislation to affect the manufacturers.[10] The requirement for employment incentives, including the provision of housing, extra-mural and other community benefits, and whether employees enjoyed a higher level of paternalism than in light industries, are therefore important indicators of the manufacturers' response to the shortage.

Paternalism was an important issue with the locomotive companies. 19th century employers in certain factory towns secured a position of both ideological and cultural authority over their workforce.[11] It is likely, however, that this applied more to volume industries, employing largely un-skilled labour, in a relatively free market. By contrast, railway companies shared many of the workplace strategies of industrial paternalism. Their workshop towns, such as Crewe, benefited from the provision of welfare, recreational and learning opportunities, and facilities which were partly the means by which workers created structures of self-help.[12] Unlike several of the railway workshops, however, the independent locomotive workshops were usually in existing urban areas, allowing most employees the opportunity of alternative employment. These manufacturers were thus required to treat their workforce with equanimity, with the necessity for, and consequences of, company paternalism.

In general, these matters were not, of course, wholly within the employers' control. 19th century industrial relations saw organisational growth, by both trades unions and employers' federations, as pressures grew to preserve craftsmen 'closed shops' and introduce a shorter basic working week and higher wage rates. 'Friendly' societies amalgamated into regional and national trades unions, allowing pooled resources in support of local actions, presenting the manufacturers with industrial relations issues quite unlike the local ones to which they were accustomed. Urban manufacturing craftsmen enjoyed real bargaining power in the mid-century, other groups being either too geographically diffuse or too mobile to be effective.[13] The unions also provided unemployment benefit, sick pay and superannuation, which further reduced dependence on employers and increased union control.

Labour relations were a severe test of managerial expertise. Between 1850 and 1890 there was an implicit accommodation between engineering employers and their skilled workers, all parties benefiting from British economic supremacy and relatively stable craft-labour employment.[14] It is however important to consider the role of the locomotive manufacturers in national employment disputes. A comparison with the railway-owned workshops is also appropriate as these workshops, whose workforce was closest to the independent manufacturers in skill and experience, had relatively trouble-free industrial relations, there being only three or four sectional strikes between 1838 and 1914.[15]

Management and Supervision

Strategic decisions relating to employment largely rested with managing partners and, from the 1860s, with General Managers and Managing Directors. Whilst the proprietors of the smaller firms supervised the manufacturing and administrative functions directly, the larger firms delegated the tactical decision-making responsibilities to head foremen and head clerks respectively. At the start of locomotive manufacturing, the vertical integration of the heavy manufacturing industry had already instigated the supervision of employees by shop foremen, and as the firms grew in size and in number of employees, their responsibilities for recruitment, discipline and work programmes grew accordingly. Much reliance was placed by the manufacturers upon policies which sought both to preserve employment for the skilled craftsmen and pursue a growing proportion of un-skilled, but increasingly experienced machinists and other repetitive workers. The head foremen and their respective workshop foremen therefore played key roles in carrying out those recruitment policies in accordance with the prevailing work programmes.

The salaried responsibilities of managing partners were usually divided between manufacturing and administration. The partners of the Jones & Potts firm, John Jones and Arthur Potts, so divided their activities. Potts spent much time marketing, selling and chasing payments, whilst Jones supervised the manufacturing work at the Viaduct Foundry in Newton-le-Willows.[16] The largest manufacturer in the early years, Robert Stephenson & Co., was exceptional, as the consulting activities of both George and Robert Stephenson allowed them only occasional appearances at their Newcastle factory. From the 1820s, therefore, although Robert Stephenson, as Managing Partner, made policy decisions through frequent correspondence, the factory was administered by a 'Head Clerk' (Office Manager) and a 'Head Foreman' (Works Manager).

After Stephenson's death in 1859, his cousin, George Robert Stephenson (1819-1904), became the new Managing Partner and directed the firm's policies from his Westminster consulting office until the end of the century. Although the firm did not obtain limited company status until 1886, it was the first manufacturer to delegate managerial responsibilities to a salaried General Manager, George K. Douglas, in 1862.[17] No evidence has been found to confirm the motivation for the appointment, but it would seem that, unlike his cousin, George Robert Stephenson felt unable to devote sufficient attention to the management of a firm of 1,200 men at such a distance.

Some firms converted to limited companies following the 1862 Companies Act and passed the responsibility for strategic planning to boards of directors attended by managing directors, whilst others remained as partnerships. The conversion of Charles Tayleur & Co. into the Vulcan Foundry Co. Ltd. in 1864, for example, saw the appointment of William Frederick Gooch (1825-1915), as Managing Director reporting to the Company's Chairman, Edward Tayleur.[18] Gooch had been the GWR's Swindon Works Manager for seven years before his appointment, and his experience of employment and organisational procedures, as well as production processes brought about the modernisation of the Vulcan Foundry.

In contrast, the conversion of Slaughter Grüning & Co. into the Avonside Engine Co. Ltd. in 1864 saw its Managing Partner, Edward Slaughter (1814-1891), appointed as Managing Director.[19] Although this provided continuity of management, it lacked the injection of new direction into the 28 year-old establishment. New direction was belatedly introduced in 1871, when Slaughter became non-executive Chairman of the company, and engaged Alfred Sacré (1841-1897) as Managing Director.

Financial incentives were occasionally negotiated to retain the services of competent managers. When Heinrich (Henry) Dübs (1816-1876) was appointed Assistant Manager of Beyer Peacock & Co. in 1857, with the expectation of being appointed Manager of its proposed factory in Vienna, he was paid a salary of £500, the same as Beyer and Peacock themselves, and an 'additional sum of one half per cent upon the amount of work turned over on trade account or upon a minimum of £100,000 per annum.'[20] Although appointed for an initial two years, Dübs was given notice after just six months, the reasons for which were not recorded. He was subsequently taken on by Neilson & Co. and, became a partner, but he also left that company after disagreement in 1864, to found his own company in Glasgow.

The stimulus to the recruitment of specialist managers from outside the locomotive firms was the Companies Act of 1862. These managers brought with them particular knowledge of capital equipment, production processes, organisational and employment procedures, as well as the administrative skills of cost and financial accounting. By the late 1860s, the number of salaried managers (or executive directors) with the largest firms had increased to, typically, six men. The number could be larger, depending upon the breadth of the firm's activities. In 1869, for example, Robert Stephenson & Co. had nine salaried managers, for its workforce of 1,200 men.[21]

By the last decade of the century, the breadth of responsibilities varied widely, between full control by executive partners, such as Neilson Reid & Co. and Dübs & Co., executive directorships of private limited companies, such as Beyer Peacock & Co. Ltd., and executive directorships of public limited companies, such as R.&W. Hawthorn Leslie & Co. Ltd. which became a public company in 1886. It had a Works Managing Director, William Cross, to represent its engine works, whilst employing a General

Fig. 6.2 Product of partnership. Slaughter Grüning & Co. 1859-built 4-4-0 for the Ferrocarril de Tarragona-Barcelona-Francia (Tarragona Barcelona and France Railway) (SG 369/59).   *The Locomotive Magazine, Vol. 33 [Feb. 1927], p.261.*

Manager for its shipyard.[22] The Vulcan Foundry had five salaried directors by 1891, reporting to William Gooch.[23] There were also General Managers of limited companies, such as Robert Stephenson & Co. Although some limited companies were amongst the leading firms in the country in the development of the new administrative order, the Neilson and Dübs partnerships were the most 'progressive' in terms of manufacturing and employment strategy.

The long tradition of vertical integration in the manufacturing industry brought significant workshop management experience to locomotive manufacture. This experience was developed from the 1820s into responsible workshop management by Robert Stephenson & Co., thus allowing the Stephensons to pursue their consulting careers. The employment of a salaried 'head foreman' and 'head clerk' was soon emulated by other firms, placing the locomotive industry in the forefront of the emerging multi-activity managerial practice. The firms' organisational 'pyramids' were thus low and unlike the multi-level hierarchy of main-line railways although they were similar to their workshop organisations.[24]

The employment of head clerks by Robert Stephenson & Co. from 1823 represented an early example of sectional management through delegated responsibility. The practice followed the example of the firm's partner, Michael Longridge who was, himself, employed as manager of the Bedlington Iron Works. One of the first appointees, Harris Dickinson, was 'a very pushing young man',[25] whose actions sometimes caused embarrassment. Robert Stephenson was unhappy that Dickinson had such a prominent position, and disliked the idea of a salaried head clerk, preferring instead the traditional link between risk and reward. When consideration was being given to appointing a new head clerk in 1836, he wrote:[26]

> …neither do I believe that the management will be much impressed by employing a manager of the description named in your letter – Dickinson was precisely the kind of man you allude to – he was active, intelligent and what is usually termed a man of business – but the Est. would have been ruined by this time had that kind of management not been entirely altered … if any manager is brought to Forth Street, he ought to have a share – and ought to confine his attention to the financial department, as any interference with the mechanical will I fear throw all wrong –

In 1836, Stephenson recruited Edward Cook as Head Clerk from his father-in-law's establishment in London, where: 'His occupation hitherto has always been confined I believe to accounts'[27] By 1845, Cook's successor, W. H. Budden, reporting to Stephenson in London, was responsible for external affairs (marketing, sales and purchasing) and the accounting and time offices, and supervised the work of the estimating and purchasing staff, a total of about forty personnel. In 1849, following four very profitable years for the Stephenson Company, he received a large bonus of £500 as reward for services, dedication and success in achieving profits.[28]

Other manufacturers similarly employed head clerks, and, by the end of the century, it had become normal practice to employ these departmental managers reporting to general managers, managing directors or managing proprietors. Their responsibilities included supervising the preparation of cost and management accounts. Andrew Barclay Sons & Co., which was several times in financial difficulty through poor management practices, was obliged by its creditors to recruit a Financial Manager to place it on a sounder financial footing, after the firm's second sequestration in 1882.[29]

From the opening of their factory in 1823, the Stephensons appointed William Hutchinson as their 'Head Foreman', there being a close harmony between them which allowed him full executive status over manufacturing matters. When, in 1834, Edward Pease proposed that his son, Joseph, might have some executive responsibility, Stephenson wrote with some concern:[30]

> I learnt from Hutchinson that you had spoken to him in reference to the occasional superintendence of your son – To me, he has expressed an apprehension that he might not be allowed in my absence to follow up such arrangements within the walls of the manufactory as he now has the power of doing – This apprehension is quite natural, and I embrace this opportunity of mentioning it, in order that I may express to you, my strongest conviction that Hutchinson is trustworthy – talented & assiduous with the success of the concern at heart – The energies of any man in such a situation as Hutchinson's are I believe maturely influenced by the degree of independence both in thought and action, which he is permitted to experience … I feel from a long and thorough acquaintance with him, that the strong interest which he now takes in the economical arrangement in the working department would be lessened by any limitation of his powers –

In 1839, Hutchinson received a £150 bonus from the partners for his services to the firm,[31] and, in 1845, Robert Stephenson recommended to his fellow partners that he should be made a partner. Stephenson's London assistant, J. E. Sanderson wrote:[32]

> Mr. Robert Stephenson has talked the matter over with his father and they think that from Mr. Hutchinson's talent & assiduity he is entitled to have a permanent interest in a concern that he has so ably supported for so many years.

Senior managers were occasionally promoted to become partners after several years with a firm, their equity being built up annually out of profits. By that time, Hutchinson managed 850 employees, and was responsible for all production matters, including design, capital equipment, recruitment and industrial relations.[33] Stephenson's faith in Hutchinson, whom he referred to as the 'oracle',[34] was rewarded by a well-managed factory until his death in 1853. He was succeeded by the equally competent George Crow, who supervised the factory until his death in 1887, and who was, in turn, replaced by his son, W. H. Crow.

Fig. 6.3
Product of a well-managed factory. Robert Stephenson & Co. 1853-built 0-6-0 (RS 918/1853) for the Newcastle & Carlisle Railway No. 37 BLENKINSOP, as running un-named on the North Eastern Railway as No. 485.
R.H. Bleasdale Collection, National Railway Museum.

CHAPTER 6 – MANAGEMENT, EMPLOYMENT AND INDUSTRIAL RELATIONS

Fig. 6.4
Nasmyths Gaskell under good management. The 1849-built 0-6-0 (NG 86/1849) for the York, Newcastle & Berwick Railway, No. 193, as modified to an 0-6-0WT by the North Eastern Railway. *R.H. Bleasdale collection, National Railway Museum.*

Works Managers supervised several workshop foremen. In 1839, Nasmyth's works was:[35]

> divided into departments, over each of which a foreman, or a responsible person, is placed, whose duty is not only to see that the men under his superintendence produce good work, but also to endeavour to keep pace with the productive powers of all other departments.

Giving his opinion on obtaining a suitable candidate, Nasmyth again enforced the contemporary view of the importance of linking risk and reward:[36]

> If a first [rate] man, a fair salary say £150 a year & a per centage on the work turned out is the only way to secure him for a permanency and if he has a little money of his own get him to put it in the business.

When Nasmyth engaged Robert Willis as his Works Manager in 1852, he paid him a salary of £200 and provided a house, rent free. The following year Willis was also provided with an annual bonus incentive:[37]

> For every £1,000 of work ordered … you shall receive a bonus of £1 per thousand of value up to £50,000 and for every £1,000 worth of work in excess … you shall receive a double rate of Bonus namely £2 for every £1,000 output.

By 1856, Willis' basic annual salary had been raised to £300, in addition to the bonus incentive.

Other proprietors were equally aware of the importance of recruiting good works managers as a means of importing both managerial skills and technical capability. Works Managers with good reputations were occasionally 'head-hunted' by rival firms. Henry Dübs, who had shown considerable talent in the drawing office of Sharp Roberts & Co., was recruited as Works Manager for Charles Tayleur & Co. in 1842, where he spent fifteen years before, in turn, being recruited by Beyer Peacock & Co.[38] Beyer Peacock's Works Manager from 1860, Francis Holt (1825-1893), was recruited by R. & W. Hawthorn in 1871, and his successor, Robert Burnett (1838-1916), left in 1877 on ill health grounds, but later became Chief Mechanical Engineer of the New South Wales Government Railways.[39] Family connections were occasionally the means to obtain senior positions. William W. Clayton (1848-1901), for example, who was appointed Works Manager of Hudswell, Clark & Rodgers in 1876, was the son of one of the firm's non-executive partners. He was, himself, made a partner on William Hudswell's death in 1882, and became Chairman when the firm gained private limited company status in 1899.[40]

Fig. 6.5 Progress through family connections. 0-6-0ST *TATHAM*, built by Hudswell Clarke & Rodgers in 1879 (HC 214/1879) for the Leeds & Yorkshire Co-operative Coal Mining Co. Ltd. (Lofthouse Colliery). *Industrial Locomotive Society, negative 30758.*

The workshop foremen, reporting to the proprietors or 'head' foremen, had the authority of 'lower' or even 'middle' management and were not, as became the practice in the 20th century, supervisory grade personnel reporting to junior management. They were, in essence, 'Assistant Works Managers' and thus took over the decision-making discretion of the former millwrights, foundry-men and boiler-makers. Their departmental responsibilities included hiring and firing of personnel, and responsibility for the output and quality control of each shop, including the motivation and productivity of the work-force.

Foremen were best at determining applicants' qualifications, and re-hiring after lay-offs was speeded up by their knowledge of the abilities of many of the applicants, a point emphasised in relation to railway-owned workshops, such as the LNWR's Crewe Works.[41] In America, similarly, Baldwins vested the responsibility of recruitment with its foremen.[42]

Reliance upon foremen for efficient operation of their workshops was total. Identifying them as his 'vice-regents of practical management', Nasmyth:[43]

> always took care to make my foremen comfortable, and consequently loyal. A great part of a man's success in business consists in his knowledge of character. It is not so much what he himself does, as what he knows his heads of departments can do. He must know them intimately, take cognisance of the leading points of their character, pick and choose from them, and set them to work which they can most satisfactorily superintend … I always endeavoured to make my men and foremen as satisfied as possible with their work, as well as with their remuneration.

By 1869, Robert Stephenson & Co. had twenty foremen reporting to the Head Foreman, George Crow.[44]

Several foremen went out with the first locomotives to be used in a new country or railway, to erect them and train the railway personnel. They were sometimes induced, by the offer of high wages and senior positions, to stay as locomotive superintendents of the new railways or to set up engineering workshops. Being the industry leader in the 1830s, Robert Stephenson & Co. lost several experienced foremen. In 1839, for example, Joseph Hall, who accompanied the first locomotives for the München-Augsburg Eisenbahn-Gesellschaft (Munich-Augsburg Railway), was induced to remain in Munich by Joseph von Maffei to set up his locomotive factory that became one of the largest in Germany.[45] John Haswell accompanied the first locomotives built by William Fairbairn & Sons for the Wien-Raaber Bahn (Vienna-Raab Railway) in 1840. He was also given the responsibility of setting up the first locomotive works in Vienna, where he subsequently remained, and which became one of the largest in Austria.[46]

There is no evidence that the manufacturers introduced 'piece-mastering'. This was the practice of engaging sub-contracted workshop personnel under the management of a foremen or 'piece-master', who was wholly responsible for the profitable execution of work in each workshop. It was adopted by some railway-owned workshops in the 1840s, notably 'railway towns', such as Swindon, Wolverton and Derby, although not in Crewe which maintained a policy of direct labour employment.[47]

Although there is no conclusive evidence to explain fully the cost benefits of the piece-mastering system, it was a means of overcoming the shortage of skilled men at relatively isolated locations.[48] The piece-masters used social, religious and family associations in their recruitment drives, which were seen to provide family loyalties and a well-managed workforce, although there could often be a dislike of the piece-master. As workforces became more stable, piece-mastering in railway workshops was replaced by directly employed, piecework labour. Swindon works, for example, switched to direct labour in 1865, although Derby and Wolverton workshops maintained the piece-mastering practice through to the 1890s.

It is possible that piece-mastering was introduced by the larger independent firms, such as Neilson & Co. or Dübs & Co., or by sites such as the Vulcan Foundry, whose remoteness led to recruitment difficulties. The lack of evidence, however, regarding either perceived advantages or any discontent, suggests that the system was not practised. It is likely that, with the firms mainly based in Manchester, Leeds, Newcastle, Glasgow and other major manufacturing centres, workmen were easier to recruit than for the railway workshops, and there was no need to sub-contract the work. The manufacturers' workforce was generally more secure in the knowledge that their skills could be switched to other products, such as marine and stationary engines, at times of low demand for locomotives. *In extremis*, they also knew that, if they had to be laid off, or if their employer failed completely, they had the opportunity of recruitment by other firms in their town.

The lack of evidence regarding the adoption of 'piece-mastering' as a means of maintaining a satisfactory level of employment and productivity in the independent locomotive industry, is in marked contrast to the Baldwin Company in the USA. The company introduced an 'inside-contracting' system in 1872, to counter a strike threat, but which then saw the practice as an important means of raising productivity in the last three decades of the century.[49] It was seen to offer the advantages of group piecework, for which the contractor would bid by the piece or job, whilst assisting supervision in a workforce of over 2,500 men. An American railroad executive noted in 1903 that 'The contractor is a piece-worker on a larger scale. As he is paid by the job, he has the incentive to turn out his work as quickly as possible and to get as much as possible out of the men under him.'[50]

The different approach on such an important issue is most likely to have been the diverging production systems. The comprehensive adoption of interchangeable components and their volume production by Baldwins required a 'mass-production' employment policy, with a greater emphasis on routine production by an un-skilled workforce. The much smaller batch production opportunities of the British industry, and their retention of a higher proportion of craftsmen called for a continuation of direct employment.

Employment

Skilled labour was scarce before the locomotive industry began. Millwrights in the main textile manufacturing areas were in short supply in the 1820s,[51] and in 1824, William Fairbairn found some difficulty in employing a sufficient number of millwrights at his Manchester works because such 'hands are very scarce at present.'[52] Millwrights, whose skills and experience were wide-ranging and who were generally independent craftsmen, limited their own numbers in order to safeguard their negotiating strength in terms of wage levels and hours of work. Their numbers were restricted through the apprenticeship system partly because the millwrights with a 'rude independence… would repudiate the idea of working… with another unless he was born and bred a millwright.'[53]

There was no alternative to employing millwrights for locomotive manufacture, as they had the necessary skills on which the emerging industry depended. Fairbairn, himself a former millwright, later wrote that:[54]

> The millwright… was to a great extent the sole representative of mechanical art, and was looked upon as the authority on all the applications of wind and water… as a motive power for the purposes of manufacture. He was the engineer of the district in which he lived, a kind of Jack-of-all-trades, who could with equal facility work at the lathe, the anvil, or the carpenters bench… Generally, he was a fair arithmetician, knew something of geometry, levelling, and mensuration, and in some cases possessed a very competent knowledge of practical mathematics. He could calculate the velocities, strength and power of machines; could draw in plan and section and could construct buildings, conduits, or watercourses, in all the forms and under all the conditions required in his professional practice.

The shortage of millwrights threatened to curtail locomotive development, particularly as demand grew rapidly in 1835/6. Manufacturers were concerned that the shortage would lead to a general rise in wages which would make them less competitive, whilst not allowing for any significant reduction in manufacturing time. The shortage thus prompted the introduction of self-acting machine tools, which not only reduced machining time substantially, but were operated by un-skilled machinists, trained relatively quickly to operate them. The several new types of artisans thus caused the millwrights' activities to be subsumed into the residual work of fitters and erectors, where their skill and experience of tolerances and fitting remained essential. Fairbairn later described this evolution into the 'factory system' of manufacturing:[55]

> In these manufactories the designing and direction of the work passed away from the hands of the workman into those of the master and his office assistants. This led also to a division of labour; men of general knowledge were only exceptionally required as foremen or outdoor superintendents: and the artificers became, in process of time, little more than attendants on the machines.

By the time Nasmyth set up his Bridgewater Foundry in 1836, he recruited unskilled men to train as machinists, and took on men who lived near the Patricroft site:[56]

> It was for the most part the most steady, respectable, and well-conducted classes of mechanics who sought my employment… In the course of a few years the locality became a thriving colony of skilled mechanics… The village of Worsley… supplied us with a valuable set of workmen. They were, in the first place, labourers: but, like all Lancashire men, they were naturally possessed of a quick aptitude for mechanical occupations connected with machinery…

Boiler-making was another skilled craft in short supply as the early locomotive market expanded, and the manufacturers had to look further afield to recruit sufficient men. In 1830, the Stephenson Company's Head Clerk, reported that: 'The Liverpool men have arrived and got to work, and seem tolerably contented; but we have not had sufficient experience to form as yet much opinion of their abilities.'[57] The boiler-makers 'brethren' or unions maintained a tight 'closed shop' that, again, kept recruitment low and wages high. Regarding Robert Stephenson & Co., Edward Pease wrote in 1835 that: '… we sent here and there, & gave great wages for the most experienced boiler makers…'[58] Boiler-making remained a skilled craft throughout the century, although routine tasks were speeded up by new equipment, especially riveting machines and hydraulic presses.

RULES AND REGULATIONS

TO BE OBSERVED BY

THE WORKMEN

IN THE EMPLOYMENT OF

ROBERT & WILLIAM HAWTHORN.

OCTOBER, 1838.

WORKING HOURS.

1.—On and after the 8th FEBRUARY, to the 4th NOVEMBER, the Bell will be rung at Six o'Clock in the Morning, and at Six o'Clock in the Evening, for a Day's Work.

On and after the 4th NOVEMBER to the 11th NOVEMBER, at Half-past Six o'Clock in the Morning, and at Six o'Clock in the Evening.

On and after the 11th NOVEMBER to the 2nd FEBRUARY, at Seven o'Clock in the Morning, and at Six o'Clock in the Evening.

On and after the 2nd FEBRUARY to the 8th FEBRUARY inclusive, at Half-past Six o'Clock in the Morning, and at Six o'Clock in the Evening, for a Day's Work.

On SATURDAYS, and on one Working Day before Christmas Day, and the same before New Year's Day, the Day's Work will end at Four o'Clock in the Afternoon, except when Working out more than Two Miles from the Works, the Saturday's Work will end at Twelve o'Clock.

2.—MEAL TIMES, are from Eight o'Clock to Half-past Eight in the Morning, for Breakfast, except from the 11th NOVEMBER to the 2nd FEBRUARY (the Winter Quarter,) when the Breakfast Time will be from Half-past Eight to Nine o'Clock, and from Twelve to One o'Clock for Dinner at all times of the Year.

3.—OVERTIME to be reckoned at the rate of Eight Hours for a Day, both in and out of the Works.

4.—Every Workman to Write on his Time Board with his Time, the Name of the Article or Articles he has been Working at during the Day; and what Engine, or other Machinery they are for.

5.—Every Workman to be provided with a Drawer for his Tools, with a Lock and Key. The Drawer and Key to be Numbered, and all his Tools to be marked with the same Number, and the Letters R. & W. H. The Key to be left with the Foreman, on going from home for more than a Week; to be accountable for his Tools when leaving the Employ; and in case of Loss the amount to be deducted from his Wages.

                                                              s. d.

6.—Any Workman who is longer in taking his Time-Board out of the Office than Five Minutes after the first Bell rung in the Morning, to be fined (if thrice in one pay) 0 3

To be allowed to commence Work at any Time before Eight o'Clock in the Morning, having the time he is behind the ringing of the first Bell deducted off his Wages, when it shall have amounted to a Quarter-Day.

The second Quarter-Day to commence at Half-past Eight o'Clock in the Morning, except during the Winter Quarter, when the second Quarter-Day will commence at Nine o'Clock.

7.—Any Workman neglecting to return his Time-Board into the Office, when done Work, to be fined 0 3

8.—For taking out, or putting into the Office any other Time-Board than his own, to be fined for each Board 0 3

9.—For leaving his Gas Cock open, or Candle burning 0 6

10.—For enlarging or in any way altering or damaging his Gas burner... 1 0

11.—For leaving his Work without giving Notice to the Foreman, or Time Keeper 0 6

If absent at any Time during Work Hours, the Time to be kept off.

12.—Any Workman opening the Drawer of another, or taking the Use of his Tools without Leave, to be fined 0 6

13.—Any Workman taking Tools from a Lathe, or other Piece of Machinery, to be fined... 0 6

14.—Any Workman not returning Taps, or Dies, or any general Tool, to the Person who has charge of them, as soon as he is done with them, to be fined.......... .... 0 6

15.—Any Workman coming to, or returning from Work, who comes in, or goes out, at any other Door, than that adjoining the Office, or Store House in which his Time-Board is kept, to be fined 0 6

16.—Any Workman taking Strangers into the Manufactory without Leave, to be fined 1 0

17.—Any Workman interfering with, deranging, or injuring any Machinery, or Tool, to pay the cost of repairing the damage, and to be fined 1 0

18.—Any Workman putting on his Coat, or leaving his Work before the Bell is rung, to be fined 0 4

19.—For shouting, or creating any other noise, or tumult, during the Meal, or any other Hours, more than necessary for the peaceful carrying on of the Manufactory, to be fined 0 4

20.—For being intoxicated in the Works, to be fined 1 0

It is requested that no Spirits or fermented Liquors may be brought into the Manufactory.

21.—For Smoking during Work Hours 0 6

22.—For bringing a Dog into the Works 0 6

23.—It is requested that should any Workman see another breaking any of the above Rules, he will give information of the same.

24.—A Week's Notice will be given, and required, previous to leaving the Employ.

25.—Every Workman to provide himself with One Pair of Compasses, One Pair of Callipers, One Two Foot Rule, and One Plumb and Line, and any Workman not providing himself with these Tools, will have them supplied to him by his Employers, and the value deducted from his Wages.

26.—Boys or Apprentices to be fined only half the above.

N.B. All Fines to go to a Fund to be called the *Accident and Sick Fund*, (the same to have no connexion with the Sick Fund at present established in the Manufactory). Any Workman happening an Accident, or being Sick, to be allowed out of the Fund 3s. per Week, or such other Sum or Sums, while he is not able to Work, as the Stewards may determine, with the consent of the general Body, from time to time, whilst there is Money in the Fund.

The Fund to be equally divided amongst the Workmen, the Boy's receiving half that of the Men, at the end of each Year.

Newcastle-upon-Tyne: Printed by John Hernaman, Journal Office, 19, Grey-Street.

THIS DOCUMENT WAS FOUND DURING THE DEMOLITION OF ROBERT STEPHENSON & Cos OFFICES FORMERLY COCKCROW HALL, HEBBURN. 1941.

Fig. 6.6 Rules and Regulations to be Observed by the Workmen in the Employment of Robert and William Hawthorn, October 1838. *Clarke (nd but 1979), p.5.*

The introduction of the steam-hammer from the mid-1840s saw the manufacturers broaden their skill-base by re-training some of their smiths and recruit experienced forge-men from iron-works. Forging, particularly by wheel-smiths, remained a skilled craft throughout the century, as was the work of the pattern-makers and foundry-men, whose skills were enhanced as more complex casting was pursued.

The extraordinary demands for locomotives during the mid-1840s, led to a severe shortage of machine-shop craftsmen, prompting further advances to capital equipment in an endeavour to reduce manning requirements. As volume production of certain components became feasible with new milling and other improved machine tools, the repetitive nature of some of this turners' work was also passed to un-skilled machinists. Their output was passed to the turners, fitters and erectors who continued to exercise skill and judgement in the finishing work. As improved machine tools were introduced during the remainder of the century, yet further skills were removed from the turners' craft and placed with un-skilled machinists.

The shortage of craftsmen was sometimes compounded by the absence of suitable housing. Whilst most locomotive factories were located in urban areas and recruited their workforce locally, some manufacturers were obliged to provide housing and other amenities. So poor was the available housing in Patricroft, that Nasmyth 'Expended several of pounds in erecting cottages for our own workmen, because those in the district were so defective.'[59] By 1850, he had built eighty-nine dwellings, for each of which the men paid a 3/- weekly rent.[60]

The two factories established at Newton-le-Willows were too far from both Liverpool and Manchester to recruit their workforce. Charles Tayleur & Co. built the 'Vulcan Village' on the south side of the Vulcan Foundry site to accommodate many of its employees.[61] The village had its own school, inn and general stores, but was small compared to railway workshop towns, such as Swindon and Crewe. Jones & Potts' Viaduct Foundry at Newton had similarly been provided with thirty-three workers' cottages by its closure in 1851.[62] Beyer Peacock & Co. provided some housing in order to recruit a sufficient workforce to its relatively isolated Gorton Foundry, east of Manchester. Charles Beyer also paid for the building of a church and a day school close to the factory, and the re-building of the parish church.[63]

Although Neilson & Co.'s growth was constrained by lack of skilled men, it is likely that this was due to the somewhat extreme Scottish patriotism of its proprietor, Walter Neilson, reflected in his avowed intention not 'to employ a single Englishman', when he started up in the 1840s.[64] He experienced a shortage of craftsmen, however, when he embarked on locomotive work in Glasgow, and it therefore became necessary for him to recruit some workmen from further afield.[65]

Manufacturers developed employment strategies which recognised the importance of the skilled crafts, whilst avoiding dependence upon them. Their strategic decisions on levels of employment for each of the several craft skills that they employed were, as far as possible, related to a continuing level of work, to maintain their availability and loyalty in a competitive labour market. The shortage resulted from the lengthy apprenticeship system and the limitations placed upon recruitment of apprentices by the 'journeymen' craftsmen, which served to secure their employment at potentially higher wage levels.

It was, however, the very dependence upon the craftsmen, in the 1830s and 1840s, which threatened to restrict the growth of the industry and which prompted the manufacturers to pursue, vigorously, the development of capital equipment to allow routine tasks to be performed by men who were not 'time-served' journeymen. The employment of this unskilled labour for much of the repetitive machining, which was contentious throughout the century, not only overcame the shortage of skilled craftsmen when locomotive demand was high, it also reduced the wage bills.

Employment policies were, however, constantly being conditioned by the uncertainties and fluctuations of the locomotive market. As they evolved after the 1850s, therefore, these policies contributed to the divergence of the industry, between the progressive locomotive manufacturers and the craft manufacturers, with their

Fig. 6.7 Vulcan Village in Newton-le-Willows, Lancashire, built for the employees of Charles Tayleur & Co. in 1833. *Author.*

broader market base, including small locomotive batches. The larger investment in capital equipment, particularly machine tools, by the larger companies increased their proportion of unskilled labour by comparison with the craft manufacturers.

Fig. 6.8 Product of craft manufacture. Rothwell & Co. of Bolton 1854-built 4-2-4WT for the Bristol & Exeter Railway (No.46).
R.H. Bleasdale Collection – National Railway Museum.

The extraordinary fluctuations in the locomotive market led to tactical decisions by the manufacturers, which largely saw continued employment for skilled men, whilst unskilled labour was laid off and recruited according to demand. When demand was slack, manufacturers would endeavour to retain their craftsmen in order to be well-placed when orders built up again.[66] When labouring staff were laid off, craftsmen diversified, where possible, into alternative product manufacture, or were put on short-time working. James Nasmyth thought it unwise 'to let your men scatter all over the country' during depressed periods, but rather 'to wait for the good times to come.'[67] In this way, there could be said to have been an accommodation between the employers and their skilled labour, but this mutually beneficial arrangement was limited to a ready supply of skills and continuity of employment, which engendered a tradition of staying with one firm, rather than any accommodation on the terms of employment.

In planning the locations of their factories, the manufacturers had to anticipate the limitations and availability of labour in their respective areas, which resulted in significant variations in labour rates and other terms of employment. Not only had wage levels to be higher in more remote areas, but housing and other benefits had to be provided. The larger manufacturers were anxious to improve levels of productivity following their increased levels of investment in capital equipment in the latter part of the century. Several firms introduced incentive piecework payments, but success usually depended upon the system and its often contentious implementation.

The apprenticeship procedures lay at the heart of the skilled labour market and were vigorously protected by the craftsmen who saw the system as the means to preserve their job security and terms of employment. The manufacturers were equivocal in their approach to the training, particularly premium apprenticeships, which were not always beneficial to their long-term labour requirements but represented a compromise in the confrontations over un-skilled labour.

Wages rose in the early 1850s because the demand for labour exceeded the expansion of the labour force.[68] It is more likely, however, even with the greater use of self-acting tools that the shortage of craftsmen was due to the effects of the national 'lock-out' in 1852.[69] Nasmyth wrote at this time:[70]

> It will be well mean time that you be looking out for some first rate handy fellows who you know to be such and who will knock out the work in first rate style and with all speed… I allude more particularly to [locomotive] Erectors among whom should be one or two men suitable to go out with the Engines when they are ready and see them delivered in a most satisfactory manner to the company… Mean time if you happen to know of a few first rate turners you will oblige us by sending them this way as we are more in want of hands in that department than any other.

The shortage of craftsmen served to maintain favourable terms of employment, which were always superior to volume industries. Wage scales and hours of work were in advance of the minimum provisions of the 1867 Factory Acts Extension Act.

Fig. 6.9 A craft manufacturer. The Holmes Engine & Railway Works, Rotherham, of Dodds & Son. *The New Illustrated Directory; Men and Things of Modern England, 1858.*

In the peak of the 1868-1873 cycle there was a rapid rise in wages, making it the first boom actually to be constrained by labour shortage. It is, again, equally likely that the effects of the 1871 'nine-hours' labour dispute caused the increase in wage costs. The disputes and the resulting wage increases are evidence of the lack of accommodation between the employers and their skilled labour over terms of employment.

When the locomotive market was depressed, manufacturers laid off un-skilled labour. In 1842, R.B. Longridge & Co. was obliged to release its men, the local press reporting: 'It is no uncommon circumstance to see groups of men, sitting about unfrequented parts of the town [Bedlington], playing cards.'[71]

Fig. 6.10 provides an example of the effect on the Vulcan Foundry Ltd., of the close relationship between the number of locomotives manufactured and the annual wages bill. Although a direct comparison is not possible because of the several other engines and machinery also manufactured by the firm, it is clear that employment closely followed the prevailing locomotive market.

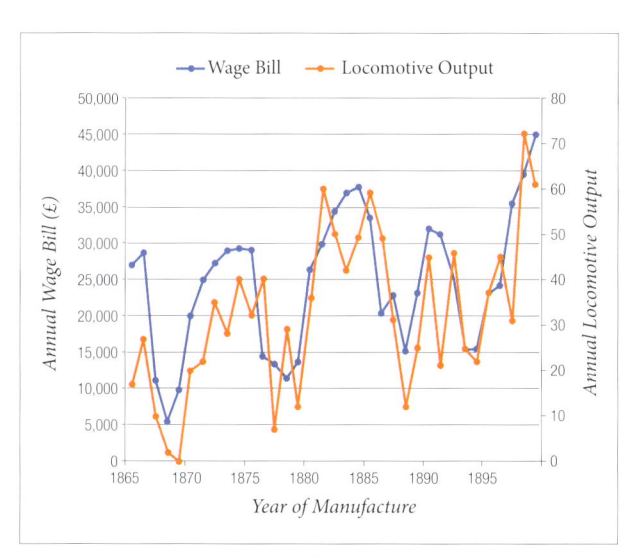

Fig. 6.10 Vulcan Foundry Co. Ltd 1865-1900. Comparison between annual wage bill and locomotives manufactured. *Data obtained from Vulcan Foundry Ltd. Ledger 1864-1889, and Journal 1864-1900, Vulcan Foundry collection, B/VF/4/1/1 and B/VF/4/2/1.*

In a similar way, Fig. 6.11 compares the total number of employees of the Hyde Park Works, Glasgow, of Neilson & Co., and Neilson Reid & Co., with its locomotive output in the same years. The numbers of labourers again rose and fell in accordance with the prevailing locomotive market. The rise and fall of the wages bill reflects the recruitment and lay-offs of labourers (and craftsmen in severe recessions such as 1868) and by overtime payments or short-time working. The proportion of labourers to skilled personnel was between 15 and 25 per cent and their dismissal was the easiest way of reducing the workforce and the wage bill, even though craftsmen then had to undertake some of the residual repetitive tasks.[72] Commitments to apprentices meant retaining them until the conclusion of their 'time'.

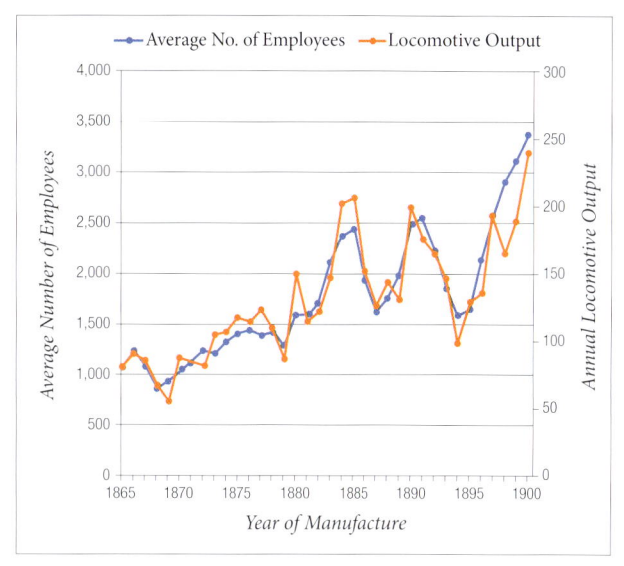

Fig. 6.11 Neilson & Co. & Neilson Reid & Co. comparison between number of employees and locomotive output 1865-1900 (*data recorded in a notebook of 'A.I.' – a senior clerk at the Hyde Park Works*). *Author's collection.*

Craftsmen usually supplied their own 'tools of the trade'. In 1846, when R.&W. Hawthorn's Forth Banks Works burnt down, there was 'a considerable loss…[including] the tools of about 200 workmen.'[73] The Stephenson Company's draughtsmen were each required to 'provide himself with all necessary Drawing Instruments.'[74] R.B. Longridge & Co., however, provided tools for its employees, each man being responsible for the tools supplied, the value of any losses being deducted from his wages.[75]

In the early years of the century, the millwrights' 'societies' had laid down a 6am to 6pm workday (daylight hours in winter), a net 10½ hour day after meal breaks.[76] R.B. Longridge & Co. adopted this working day although, following the 1836 dispute, this was reduced to ten hours, with higher rate overtime.[77] By the mid-1840s, whilst most manufacturers were working 60 hour weeks, some voluntarily reduced this figure accepting that too long hours could actually reduce productivity. In 1846, Robert Stephenson & Co.'s draughtsmen worked from 7am to 6pm, (4pm on Saturdays), a net 55 hours.[78] Some Lancashire manufacturers introduced the 'English Week' (early finish on Saturdays), resulting in 57½ or 58½ hour weeks, whilst the normal Saturday finish in the north-east became 4pm.[79] By 1861, the average working week remained at 58.8, although the normal working week in the north-east was 60-61 hours.[80] In spite of the 1852 dispute largely against the expectation of systematic overtime, manufacturers required regular overtime to be worked at times of high demand.

1858

RULES & REGULATIONS

TO BE OBSERVED BY THE WORKMEN,

IN THE EMPLOY OF

FOSSICK & HACKWORTH.

1.—The Bell will be rung at Six o'Clock in the Morning, and at Six o'Clock in the Evening, throughout the year, for a day's work; on Saturdays, the day's work will end at Four o'Clock.

2.—The Meal-times allowed are, from Half-past Eight to Nine o'Clock for Breakfast; from One to Two for Dinner; during which period: the first quarter of a day will end at Half-past Eight o'Clock; the second, at Twelve o'Clock; the third, at Half-past Three; and the last at Six o'Clock.

3.—Overtime to be reckoned at the rate of Eight Hours for a day; but no Overtime to be entered, until a whole day of regular time has been worked.

4.—Every Workman to be provided with a Chest or Drawer for his Tools, with Lock and Key.

5.—Any Workman neglecting to write on his Time-board, with his time, the name of the Article or Articles he has been working at during the day, and what Engine or other Machinery they are for, to be fined 1s.

6.—Any Workman who does not return his Time-board to the office, when done work, and on the Wednesday Evening previous to the pay not later than Half-past Ten o'Clock, should he work to that time, with his full time written thereon, to be fined 1s.

7.—Any Workman either putting into the office, or taking out of the Store-house, any other board than his own, to be fined, for each board, 1s.

8.—Any Workman leaving his job, without screwing down his gas, or on quitting work, leaving his candle burning, or neglecting to shut his gas-cock, to be fined 1s.

9.—Any Workman enlarging, or in any way altering or damaging his gas burner, to be fined 1s.

10.—Any Workman leaving his work, without giving notice to his Foreman, to be fined 1s.

11.—Any Workman opening the drawer of another, or taking his Tools without leave, to be fined 1s.

12.—Any Workman taking tools from a Lathe or other piece of Machinery, to be fined 1s.

13.—Any Workman not returning taps or dies, or any general Tool, to the person who has the charge of them, as soon as he has done with them, to be fined 1s.

14.—Any Workman taking Strangers into the Manufactory, without leave, or talking to such as may go in, to be fined 1s.

15.—Any Workman coming to, or returning from work, who comes in or goes out at any door, other than that adjoining the office, to be fined 1s.

16.—Any Workman interfering with, deranging or injuring, any Machinery or Tools, to pay the cost of repairing the damage.

17.—Any Workman washing himself, putting on his coat, or making any other preparations for leaving work, before the bell rings, to be fined 1s.

18.—Any Workman taking any kind of Wood, Chips, or Shavings, out of the Works, to be fined 1s.

19.—Any Workman creating tumults or noise in the Manufactory, at any time, to be fined 1s.

20.—Any Workman Smoking, during working hours, to be fined 1s.

21.—Any Workman using Oil, to clean his hands, or for any other improper purpose, to be fined 1s.

22.—Any Workman giving in more time than he has worked, to be fined 2s. 6d.

23.—Every Workman using the following Tools, either to provide himself with them, or to have them supplied to him by his Employers, and the cost deducted from his wages, viz.:—one pair of Compasses, one pair of Callipers, one two-foot Rule, one Plumb and Line, and one Square.

24.—No Workman to leave the employ, without giving two weeks' notice; and the same to be given by *Fossick and Hackworth*, to any Workman whose services they may cease to require, except in case of misdemeanour.

25.—Any Workman not giving proper notice to the Night Watchman, in respect to starting and leaving off Work, is subject to be Discharged, without the usual notice.

Fig. 6.12 60-hour week employment terms of Fossick & Hackworth in 1858.
Durham County Record Office CP/Shl 16/6.

The passing of the 1867 Factory Acts Extension Act brought the manufacturing industry into the same legislative framework as textile and other factories, which had been progressively subject to employment restrictions under the Factory Acts of 1831-1864. Although it was the first legislation to affect the locomotive manufacturers, it had little direct effect as their hours were generally below those required by the Act.[81]

At the beginning of the railway era, wages in the main manufacturing centres averaged about 20/- per week. Foreman millwrights could earn up to 42/- per week, including allowances for installing engines on customers' premises. R.&W. Hawthorn employed smiths and fitters for about 18/-, an experienced machinist received 22/-, whilst long-serving men and foremen could earn 25/- to 28/- weekly. Labourers were paid, typically, 12/- per week.[82] By the early 1870s, the company's wage scales had increased generally to an average of 28/- weekly.[83]

Higher wage rates were necessary in relatively remote areas to attract sufficient labour. In 1855, Beyer Peacock & Co. paid higher wages to attract a workforce from other manufacturers to its factory in the Gorton area of Manchester. Its average weekly wage was about 34/-, with all craftsmen receiving more than 30/-, and foremen receiving £5 a week.[84] Beyer Peacock's wage rates remained higher than their competitors in the main industrial centres, such as Leeds. In 1866, the Hunslet Engine Co. paid an average wage of about 30/-, with fitter/erectors being paid 26/- to 32/-, boiler makers up to 34/-, machinists 16/- to 26/-, and labourers 16/-. Overtime figures showed that wages could be increased for working up to a 72 hour week.[85]

At the conclusion of the 'lock-out' dispute of 1898, minimum weekly wage rates had been agreed with the ASE Union for skilled men working a 53 hour standard week. The wage-scales ranged between foundry-men and pattern-makers at 36/-, to machinists at 26/- per week.[86]

Incentive piecework payments were introduced by several locomotive manufacturers in the latter part of the century as they pursued larger batch production. Piecework was largely confined to volume industries where there was incentive and opportunity for productivity improvements. Surprisingly, the earliest reference to piecework in a locomotive factory was in the 'Rules and Regulations' of R.B. Longridge & Co. on its start-up in 1837, which required 'Every workman on piecework' to give an 'accurate account of time at the office when leaving work'.[87]

With greater batch production opportunities, piecework was introduced into railway-owned locomotive factories early on. By 1861, half of Swindon's union-based workforce was so employed, with smaller proportions at Brighton, Crewe, Wolverton, Doncaster and Darlington.[88] Baldwins in America had fully implemented piecework payments by 1857, as it pursued interchangeable component batch production, and demonstrated that increased productivity not only lowered unit production costs, but also unit overhead charges.[89] During boom markets piece rates were left unchanged, the workforce took home more pay by increasing output and enhanced the company profits. This gave the workforce a larger than average portion of the margin between production costs and selling prices.

By the 1880s/1890s, several independent manufacturers had implemented piecework schemes, as they sought to reduce costs in the face of increased competition. Piecework offered opportunity for higher wages through faster production, particularly for un-skilled machinists who operated several machines simultaneously, with lower supervision. Kitsons' Works Manager, E. K. Clark, introduced a form of 'Payment by Results' in the 1880s, but serious anomalies arose in the distribution of awards by the 'autocratic and ingenious Leading Hand', which led to several disputes.[90] As estimating experience was gained by the 'Price Settler', piece-times were reduced and stifled incentive. In replacing this system, Clark himself allocated a price to each 'piece' and machine process, by his own 'observation, and estimate, by trial and error'. Employees were made individually accountable for their time, a move which still met with dissatisfaction, and which Clark later determined was a failure.

In England it was more difficult to reduce piece rates, justified by increases in labour productivity, due to the introduction of labour-saving machines, which made them less attractive to manufacturers.[91] Unlike the Baldwin practice, as piecework experience was gained, manufacturers reduced standard production times, making them more difficult to improve upon. Piece-work rates, introduced by Hudswell Clarke & Co. by the end of the century, were described as 'low', the bonus for early completion being usually time and one eighth, but long hours of overtime were worked, sometimes through the night.[92] If piece times were regularly bettered, the price was re-assessed for future work, with a consequent reduction in earnings. This practice by several manufacturers led to growing discontent, building to outright opposition by the 1898 'lock-out'.

Not all locomotive works successfully introduced a piece-work system. At R.&W. Hawthorn Leslie's factory in 1889, William Cross complained of the lack of accurate cost records:[93]

> … piecework is simply guess-work on the part of the foreman aided by such imperfect costs as he has been enabled to obtain for himself. At present and for some time past, we have been working no piecework in the machine and erecting shops, owing to want of sufficient data to settle prices…

Peckett & Sons went as far as to state that: 'No piece work is allowed, as the proprietors are determined to avoid scamping, and to ensure first-class workmanship...'[94] Manning Wardle & Co. also continued to pay its workforce an hourly wage rather than by piece work.[95]

Apprenticeships

Some manufacturers engaged educated young men from families of 'reasonable means', and gave them five to seven year apprenticeships, for which their fathers paid a premium to the firm's partner as well as meeting their sons' living expenses. Such men, some related to the firm's proprietors, could expect to become engineers in due course, and their apprenticeships involved several aspects of 'civil engineering' beyond workshop craft skills. Thomas Gooch was apprenticed to George Stephenson for six years in 1823, the first two of which were 'in the occupation or business of practical engineer and in making building and fitting up Steam engines and the various branches of Machinery.'[96] In 1827, Stephenson wrote to another parent, William Salvin:[97]

> I will engage to take your son to give him instruction in my profession, for the term of five, six or seven years – My fee is 200 guineas at the signing of the indentures: his board and lodging to be paid by his friends, until I can receive such a sum for his labor, as will pay it.

The Stephensons attracted several applications from families in the 1830s, which prompted them to increase the premium, up to £300 and even to £500. Robert Stephenson wrote:[98]

> We have at present as many, indeed more, young men than we can sufficiently employ. If we increase the number (which we have frequent opportunities of doing) we should only be doing the young men injustice, because they would not have proper and sufficient experience to learn the profession... Taking young men, although it may be a profitable part of our business is one that incurs great responsibility...

In 1839, William Lawford was apprenticed to Stephenson for five years for a premium of £525, about eighteen months of which was spent at the Stephenson factory.[99] Other engineers also received many applications for premium apprenticeships. James Nasmyth agreed to take George Grundy's son in 1838 for £300, even though he could not be accepted for three or four months.[100] At that time, Nasmyth set down the distinction between premium and craft apprenticeships:[101]

> It is only our premium apprentices whom we engage to instruct 'in every branch' of the art or business of machine makers & engineers – the others are continued at one employment such as fitting or turning or removed from it to another occupation to suit our convenience.

The jump in demand for locomotives in the mid-1830s and the consequential increase in the Stephenson Company's workforce was accompanied by an increase in the number of apprentices. This was not welcomed by Robert Stephenson, as reflected in a letter to Joseph Pease:[102]

> If it be the particular wish of your father to place the youth (mentioned in your letter) at Forth Street, I shall of course not object, although I had given instructions that no more apprentices should be taken – they are an everlasting source of mischief – we were always in hot water with them, and I regret having made any arrangements for allowing some of them to come into the office, to become acquainted in every detail with our plans &c – They have no sooner done so, than they leave and carry away what has cost us a great deal of money and more thought –

Nasmyth went further in his dislike of the apprenticeship system, which he considered to be 'the fag end of the feudal system' and advocated its abolition 'in every branch of business'.[103] He also felt 'they caused a great deal of annoyance and disturbance. They were irregular in their attendance, consequently they could not be depended upon for the regular operations of the [Bridgewater] foundry. They were careless in their work and set a bad example to the unbound.'[104]

The premium apprentice system had the benefit of revealing promising talent at an early stage and allowing that talent to be developed accordingly. Edward Snowball was apprenticed to Stephenson in 1846 at the age of sixteen, and such was his intuitive design talent that, before he had completed his five-year term, Stephenson installed him as his Chief Draughtsman.[105]

Although there were limited openings for premium apprenticeships in the second half of the century, the demand largely favoured railway-owned workshops,[106] although some firms, such as R.&W. Hawthorn, did engage 'operative and scientific' apprentices, for which parents paid a premium of £400.[107] Proprietors' sons often undertook apprenticeships in their fathers' firms. In 1863, for example, J. Hawthorn Kitson started his apprenticeship at the Kitsons' Airedale Foundry.[108] Later in the century, Edwin Kitson Clark was apprenticed there, after leaving

CHAPTER 6 – MANAGEMENT, EMPLOYMENT AND INDUSTRIAL RELATIONS

Cambridge University with a first class classics tripos in 1887. He later became a foreman, Works Manager, Partner, Director and finally Chairman of Kitsons.[109]

Employment policies between the British and American industries similarly diverged over the question of apprenticeship training. Baldwins' cessation of its apprenticeship schemes in favour of machine-related training, was again in marked contrast to the perpetuation of apprenticeships by the British manufacturers. Premium apprenticeships were peculiar to Britain and reflected the practical characteristics of British engineering training, in contrast to the theoretical technical school courses of their European counterparts. The maintenance of a high proportion of craft jobs in the industry justified the continuation of craft apprenticeships, but the evidence strongly suggests that the extent of their engagement was to provide inexpensive 'bound' labour, particularly during times of dispute.

Each manufacturer took on craft apprentices in accordance with its anticipated long-term requirements. Young men were given five to seven-year apprenticeships in a specific craft, as set down on their 'indentures', which was signed by a partner or director, the apprentice's father and the youth himself. The standard and legally-binding indentures set out the nature and period of the training, together with the annual wage rates, which depended upon the nature of the training and a proportion of the craftsmen wage rates. The apprentice was 'bound' to the principal who was, in turn, obliged to see the youth through to the completion of his 'time'.

The crafts were jealously guarded within families, it being the normal practice for the eldest son of a 'journeyman' to be offered a preferential apprenticeship, which was a self-perpetuating permanent upper class within the working classes.[110] The practice of sons following in their fathers' crafts was widespread and served to reinforce the claim for 'closed shop' status adopted throughout the century, and to maintain which there were several disputes, particularly towards the end of the century.[111] With the expansion of the industry in the 1830s, the supply of eldest sons was insufficient, and second and third sons, and sons of fathers out of the trade, were encouraged to take up an apprenticeship.[112]

Apprenticeship periods and wages varied according to the craft. Three apprentices engaged by Robert Stephenson & Co. at the beginning of the railway era were, respectively, taken on for seven years as an 'Iron Founder' (wages 4/- rising to 12/-), a five year term as an 'Engine Builder' (wages 6/- to 14/-), and a four year term as a 'Boiler Builder' (8/- to 12/-).[113] The apprentices were reliant at first on families or friends for accommodation and subsistence. Robert Millward, who was 'bound' to Nasmyths Gaskell & Co., was reliant upon his father to 'provide for the said Robert Millward meat drink washing and lodging and wearing apparel and necessary Instruments and Tools of all kinds…'[114]

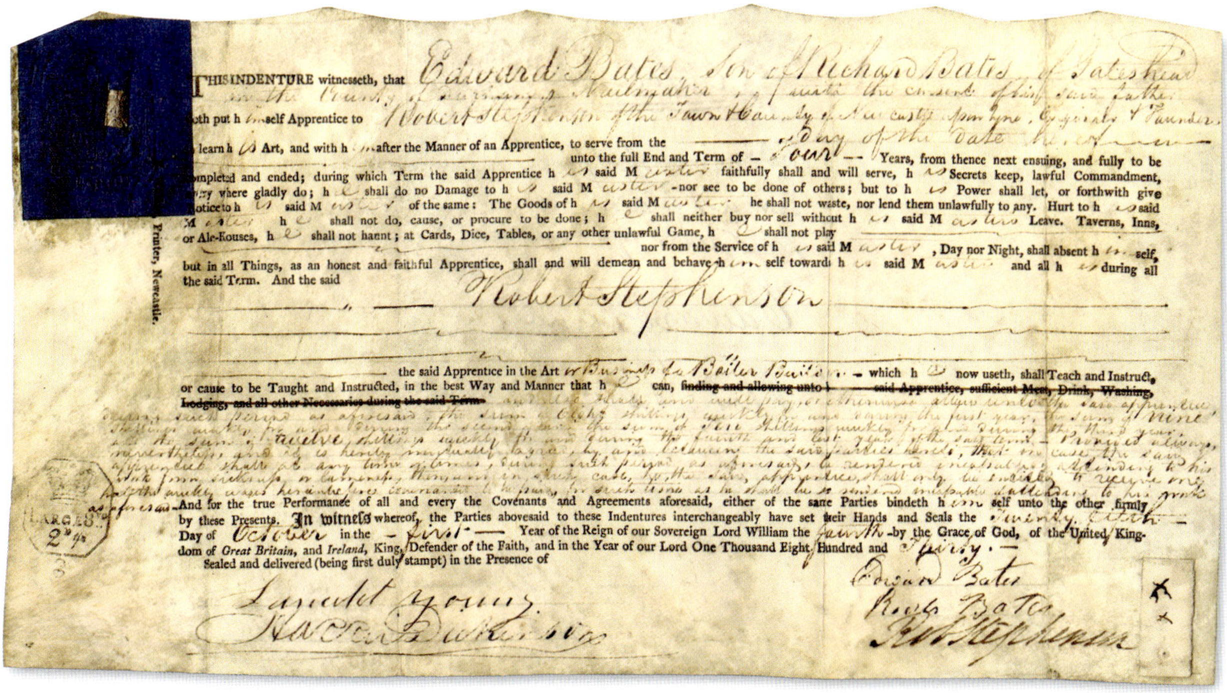

Fig. 6.13 Indentures for Edward Bates to Robert Stephenson 'in the Art or Business of a Boiler Builder', 25th October 1830. *Institution of Mechanical Engineers, Crow Collection, IMS/167/2.*

Nasmyth was much against the apprenticeship system and soon abandoned it in favour of employing unbound 'intelligent well-conducted young lads' who were given responsibility according to aptitude:[115]

> They took charge of the smaller machine tools, by which the minor details of the machines in progress were brought into exact form without having recourse to the untrustworthy and costly process of chipping and filing… We were always most prompt to recognise their skill in a substantial manner. There was the most perfect freedom between employer and employed. Every one of these lads was at liberty to leave at the end of each day's work. This arrangement acted as an ever-present check upon master and apprentice. The only bond of union between us was mutual interest.

Apprentices would occasionally be used as less expensive 'labourers', particularly when major disputes were looming. In advance of the nine-hours dispute of 1871, for example, R.&W. Hawthorn recruited 101 apprentices and, by August 1872, just over 20% of its 1,000 employees were apprentices.[116]

Although long apprenticeships were seen both as a way of developing future skills and as a justifiable form of 'bound' servitude to provide a ready means of inexpensive labour, they continued to limit the availability of skilled labour during the century. The pressure from the journeymen, in their protracted dispute with the employers over the use of non-apprenticed labour, meant that a high proportion of labour was being apprenticed unnecessarily. By the end of the century, the training requirements of manufacturing industry generally were considered to be backward by comparison with normal American practice. This backwardness was stressed in a paper in 1902:[117]

> It is wasteful and foolish that a man must serve from four to five years before he can become a proficient turner, when we know that in a few weeks a laborer can learn to operate two grinding machines and produce cylindrical surfaces that it is impossible for the most skillful turner to duplicate, either in regard to quality or to cheapness. This is not saying that it is unnecessary to train men to become good turners; it is, however, unnecessary to train them to perform many operations which are superseded by new methods of working…

and concluded that training should be confined to 'men who are necessary for supervising and keeping plant in a high state of efficiency.'

The British practice was pursued by the locomotive industry through to the twentieth century, in contrast to the Baldwin Locomotive Works in America which discontinued craft apprenticeships after 1868. As the firm moved so comprehensively towards component standardisation and batch production, the necessity for lengthy training became irrelevant.[118]

The manufacturers' employment policies were thus dominated by the relationship between skilled and unskilled labour. The craft culture which had been so dominant at the beginning of the railway era continued to dictate employment decisions in the locomotive factories in spite of the manufacturers' moves to decrease dependence on craft skills. The motivation for employing un-skilled labour, made possible through the introduction of self-acting and automatic machine tools, was three-fold. Not only were the costs of repetitive production reduced, and the shortage of skilled labour minimised, but un-skilled men provided an employment cushion, which allowed their level of employment to rise and fall in accordance with the fluctuating market. This general policy resolved one of the manufacturers' major tactical problems, whilst allowing them to preserve their requisite levels of craft skills through favourable employment terms and paternalistic incentives.

Boys aged between thirteen and eighteen were taken on by several factories. In Newcastle in the 1840s, both Robert Stephenson & Co. and R. & W. Hawthorn employed between sixty and seventy lads, about 15% of their respective workforces.[119] In addition, the Stephenson Company also employed three boys under the age of thirteen, although this practice was not taken up by the Hawthorn Company.

Paternalism

With the continuing shortage of skilled craftsmen during the century, paternalistic as well remuneration incentives were important aspects of the manufacturers' employment policies. This was as much to stimulate loyalty and continuity of service in manufacturing areas where the men had alternative employment opportunities, as it was to attract men to the more remote locations.

Paternalism was implemented from the start-up of the railway workshops in the 1840s,[120] but its origins lay much earlier, with some manufacturing proprietors, such as Michael Longridge.[121] From 1810, the men of the remote Bedlington Iron Works could send their children to 'schools', the boys receiving instruction 'in the usual branches of education', whilst the girls were 'instructed in needle-work &c'. The workmen were provided with a library and regular newspapers, and with 'surgical advice and medicine', towards which married men paid 2d a week, and single men 1d per week. A benefit society was formed into which members paid 3d per week and to which was added the value of any fines imposed on the workmen.

Any contributor unable to work through sickness received 8/- weekly from the Society, which by 1836, had 80 members. A savings bank was also set up, into which over £1,000 had been deposited by 1836. Michael Longridge was thus pioneering in his introduction of these welfare and other community benefits, the motivation for which was the welfare of his staff, almost certainly prompted by the remoteness of Bedlington works.

It is not recorded if Longridge introduced similar benefits at the Stephenson factory during his tenure as managing partner, but they were introduced for the R.B. Longridge & Co. men from 1837.[122] It was a condition of service that 'every person who enters service shall pay to the medical fund.' A pension scheme was also introduced, its terms being:

> Any agent, clerk or man in the works service, shall after ten years (not including term of apprenticeship) becoming incapable of work through old age or accident in the company's service or receiving any disease not occasioned by intemperance, receive a pension of £5 per annum for life, or after twenty years, £10 per annum, and after one year, £1 for each year.

In 1839, a Bedlington news sheet wrote: 'Few establishments can boast a better regulated set of workmen. No means have been left unemployed to make them comfortable.'[123] The provision for the well-being of the Bedlington workmen was, however, two or three generations ahead of such provision generally in the country. Not until 1880 did the Employers' Liability Act require employers to set up insurance schemes for their employees,[124] whilst only in 1897 did the Workmen's Compensation Act introduce the responsibility on employers to provide for the welfare of their employees.[125] One of this Act's provisions allowed claims for compensation to be made against employers for injuries due to an 'accident arising out of and in the cause of his employment.'

Welfare and educational opportunities were provided by some other employers. In 1853, for example, when the Canada Works at Birkenhead was opened by Peto, Brassey & Betts, a canteen was provided, whilst the three proprietors donated £75 for the construction and filling of a 600-volume library. A concert and dance in Birkenhead was put on for five hundred of the men and their wives, which raised a further £12 for the library, which was stocked with copies of local and national newspapers. The men paid a penny a week towards the reading room which was run by the workmen themselves.[126]

Prior to the confrontational industrial relations of 1871 and later years, looser forms of paternalism, such as works dinners and other collective activities, were usually perceived to engender loyalty and longevity of service.

On New Year's Eve, 1839, Robert Stephenson provided a dinner in the Newcastle factory, which was well received by the workmen.[127] A more ambitious dinner was provided by Stephenson in 1845 for his 850 workmen, who processed from the factory to the centre of Newcastle to celebrate the passing of the Newcastle & Berwick Railway Act, for which he was Chief Engineer. They then 'were provided with a Dinner at the expense of their employers, and others interested in Railways.' This was seen to be 'proof of the kindly feeling that exists between the employers and their workmen.'[128]

At the beginning of each year, from 1856, Beyer Peacock & Co. provided a 'Programme of Amusements' for its workforce 'to celebrate the [anniversary of the] opening of Gorton Foundry.'[129] The venue was Manchester's Belle Vue Gardens and a procession by the workmen was followed by a programme of sporting events, and concluded at five o'clock when the men were 'allowed to introduce Ladies for Dancing.'

In c.1864, five years after becoming Managing Partner of the Stephenson Company, George Robert Stephenson instituted an annual New Year's Eve dinner for the c.1,250 men. Over twenty taverns in Newcastle were needed to accommodate everyone.[130] The reports of the dinners reflected the strong 'collectivism' by each craft group, who dined separately.[131] They were addressed by their foremen, who were the 'Chairmen' for the evening, and whose speeches reflected a good relationship between management and workforce. It is likely however, that the dinners were discontinued after 1870 as the workforce became embroiled in the nine-hour dispute and went short of money through their contributions to the movement's strike fund.

Paternalism was rewarded by loyalty and longevity of service. An 1888 article about Robert Stephenson & Co., noted that two employees had continuous service since the late 1830s, one of whom was still employed at the age of seventy-two.[132] Neilson & Co. had one employee who began his apprenticeship in 1859 and retired in 1901,[133] but the longest service seems to have been that of Beyer Peacock's first employee, Thomas Molyneaux, who was recruited from Sharp Stewart & Co., where he had been engaged since 1831, and who retired in 1903 after forty-nine years service (and a total working life of seventy-two years).[134] Such longevity has similarly been noted in relation to railway-owned workshops.[135]

Loyalty and longevity of service for senior personnel could be rewarded by *ex-gratia* payments. In 1860, George Crow, the Stephenson Company's Head Foreman, received from George Robert Stephenson, £100 'as a token in expression of this feeling [the deep interest you have always taken in performing the duties of your department] on my part and of my sincere regard for you.'[136] In 1856, three senior employees of the Stephenson Company, George

Crow, L. Kirkup and Edward Snowball, were considered to be 'deserving of distinction' by the International Jury for the 1855 Paris Universal Exhibition.[137]

Loyalty to employers and to each other was a notable characteristic of paternalistic employment. 'The workmen themselves would occasionally contribute to persons or groups in distress. 'In 1836, for example, when a John Dixon and family lost their Newcastle home by fire, the workmen from several factories donated sums of money into a relief fund for him.'[138] 'The death of a proprietor or senior manager saw collective mourning by the workmen. The funeral of William Hutchinson, then a Stephenson Company partner, in 1853, saw a procession of over 800 workmen, 'the whole proceedings were conducted in a solemn and creditable manner.'[139] On the death of Robert Stephenson six years later, 'a great number of gentlemen from Newcastle attended the funeral [in London] – the North Eastern Railway Co. gave them return tickets.' In Newcastle itself, the memorial service in St. Nicholas Church was attended by 1,000 workmen 'from the different factories.'[140]

Paternalism was thus recognised by the manufacturing proprietors as an important policy with which to maintain their requisite levels of craftsmen in a competitive labour market. The loyalty and longevity of service demonstrably confirmed the benefits of this policy. After 1871, however, reports about paternalistic endeavours are noticeably absent. The inference is that the 'nine hour' dispute of that year marked the beginning of an era of confrontational relationships in which paternalism could no longer play such a significant part in the manufacturers' employment policies.

Industrial Relations

'The Employers explain too little; the Employees exclaim too much'[141] This exhortation by Edwin Kitson Clark summed up the industrial disputes which, in separate eras through the 19th century, severely disrupted the heavy manufacturing industry, including the production of locomotives. The main failure that may be ascribed to the manufacturing industry generally, and the locomotive firms in particular, was their inability to find an accommodation with the trades unions over the employment of un-skilled men. The issue was a running sore throughout the century, re-surfacing with each claim for improved hours and wages. The issue was the driving force behind the expansion of the 'Friendly Societies' into national trades unions, which brought about, in response, the development of the Engineering Employers' Federation. In spite of the loyalty generated by generally good terms of employment and paternalistic benefits, the confrontations were largely driven by the perceived dilution of the standing of skilled craftsmen.

The proprietors and their senior managers were central to these disputes which were as much concerned with the displacement of skilled by un-skilled men, as they were on wage and reduced hour claims. As the trades unions became more unified in their preparations for the disputes, the manufacturers' independent tactics employed in the early disputes were replaced by joint responses by employers' federations. There was, however, a lack of cohesion between them, which prolonged some disputes and failed to resolve the central issue of skilled employment by the end of the century.

The foundations of collective representation were laid at the beginning of the locomotive industry. In 1825, a 'Millwrights Benevolent Benefit Society… For the Relief Of Each Other When In Distress And For Other Good Purposes…' was formed in Newcastle.[142] Its founding committee, including at least one employee of Robert Stephenson & Co., restricted its membership such 'that no one shall be admitted a member who is not working as a journeyman Millwright or Engine Builder.' There was an 'admission-money of ten shillings and sixpence' and a monthly shilling subscription, which allowed the Society to build up sick benefit funds.

This and other societies, such as the Friendly Society of Mechanics formed in Manchester in 1826, were used to establish 'closed shops' by the millwrights. William Fairbairn complained that the millwrights had a 'rude independence… [and] would repudiate the idea of working… with another unless he was born and bred a millwright.'[143] The resulting shortage of millwrights was already evident in 1825 when the Stephenson Company was manufacturing its first locomotives. Michael Longridge wrote:[144]

> As the Darlington Rail Way Engines must be finished in three months we have no choice at present but to comply with the demands of the Men – which however will be attended with bad consequences. There have been two Meetings of the Master Millwrights – & there is to be another on Monday – but it is of no use whatever.

There were several industrial disputes in 1836, as the new trades unions sought to take advantage of the healthy economic climate to increase wages and reduce working hours.[145] Employers saw the co-ordinated action of the unions as a particular threat to be denounced and turned some of the disputes into issues of union membership and closed shop practices. Boiler-makers' 'Friendly Societies', including the Manchester-based Society of Friendly Boiler Makers, effected a closed shop. Nine Liverpool firms, including the locomotive manufacturers, Edward Bury & Co. and George Forrester & Co., jointly sought to get around this by laying off boiler-makers who had struck for higher wages, and refused to re-instate them unless they 'quit the 'union''. Several thousand men were laid off during the dispute, which the managements sought to end by recruiting men 'who will pledge themselves to remain unconnected with the present or any future boiler-makers club' from other parts of the country.[146] The strike succeeded in winning a 10-hour day for the employees of several Lancashire firms.

A strike at James Nasmyth's new Bridgewater Foundry in 1836 had a profound effect on his attitude to industrial relations. The issue was the first confrontation over the skills issue, with the Engineer Mechanics Trade Union seeking employment only for 'time-served' union members.[147] Pickets prevented most willing employees from getting to work during the three months strike, and to avoid a humiliating defeat Nasmyth travelled to his native Scotland to recruit '64 first-rate men, who had been wheelwrights and carpenters, smiths and stonemasons.' The strike effectively collapsed on their arrival, and, when regular working proceeded, the majority of Nasmyth's men were Scottish.[148] Following the conflict, he would neither tolerate trade unionism nor employ only time-served journeymen.

Against this background of closed-shop dispute and resulting shortage of craftsmen, Nasmyth, Fairbairn, Roberts and others introduced new equipment and processes that could be taken up by un-skilled workmen. With each successive demand-peak for locomotives and other capital goods, particularly in the mid-1840s, the consequential requirement for routine machining was met by further machine tool innovation. Nasmyth, who was not alone in refusing to employ trade union members, arranged a separate employment contract with each employee, determining that relations between employer and workman should be based on 'the principles of free trade, without the intervention of third parties.'[149]

Demonstrations in favour of political aims were not unusual in the 1830s, and Longridge's reputation of being a good but firm manager was tested in 1839 when the 'Chartist' movement turned on the Bedlington Iron Works and the adjacent locomotive Works of R.B. Longridge & Co. Over several days, there was a 'riotous assembly' which some of the Bedlington men were intimidated to join, about which Longridge wrote:

> I felt utterly ashamed that any of my men should thus far sully the fair name of the Bedlington Iron Company's workmen by mixing with such a rabble.[150]

Longridge's long oration to the mob, which he printed and circulated, demonstrated his firm opposition to Chartism and likened its advocates as being novices trying to repair an engine: 'We might possibly have skill enough to take the engine to pieces, but I am sure that we could never put it together again…' The cessation of the conflict was a vindication of Longridge's tough stand, which earned the respect of the workmen: '… I am very popular with all the Workmen.'[151]

Fig. 6.15 Site of Chartist dispute. R.B. Longridge & Co factory and Bedlington Iron Works near Morpeth in *c*.1840. *Uncredited view in on-line article by Alan Fryer, Bedlington Ironworks, 5th February 2018, part of Blythtown.net.community view, accessed June 2021.*

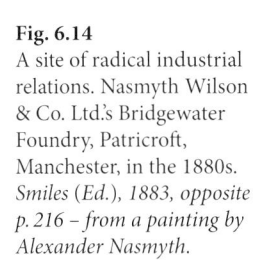

Fig. 6.14 A site of radical industrial relations. Nasmyth Wilson & Co. Ltd.'s Bridgewater Foundry, Patricroft, Manchester, in the 1880s. *Smiles (Ed.), 1883, opposite p. 216 – from a painting by Alexander Nasmyth.*

The 'Plug Riot' strike in the summer of 1842, the first general strike in Britain, spread out from the north west of England.[152] Although the country's recession had led to wage reductions, the multi-origin discontent gave fuel to the embers of the Chartist and anti-Corn Law movements. The strike and attendant rioting affected most industries, including the north-west locomotive manufacturers. A deputation of Sharp Roberts & Co.'s men organised the Manchester trades regarding Corn Law repeal, and one of their smiths, described as a 'ringleader' and an 'Owenite Socialist', was arrested during the disturbances.[153] There is no surviving record of Sharp Roberts' response, or that of any other locomotive firms, but the strike collapsed through lack of internal cohesion and the determined actions of the Magistrates.[154]

The 1842/43 recession reduced the shortage of craftsmen, but, with the rapid growth in the locomotive market from 1844 the shortage returned. The manufacturers employed further unskilled personnel who could be quickly trained to operate the new self-acting machine tools. There was strong resistance from the trades unions, such as the Journeymen Steam Engine and Machine Makers' Society (JSEM) and the Friendly Society of Ironfounders. The employment of 'adult' apprentices and 'illegal' men by Jones & Potts in 1846 led to a direct conflict with the JSEM, the resulting four-month strike at their Viaduct Foundry becoming a test case on the closed-shop issue.[155] Several strikers were arrested, including the General Secretary of the Union, which undertook nation-wide fund-raising towards the men's defence. After trial and conviction of the strikers at the Liverpool Assizes in 1847, the union raised further funds for their appeal which was successful in reversing their conviction.

The economic upturn from 1851 coincided with a re-organised and enlarged union organisation in the manufacturing sector, the JSEM having become, through amalgamation, the Amalgamated Society of Engineers (ASE). So soon after the euphoria of the Great Exhibition, the ASE sent demands to employers on three main employment issues, namely the employment of un-skilled men, the removal of piecework and automatic overtime working. Thirty-four Manchester employers, including Sharp Brothers, Fairbairn & Sons and James Nasmyth & Co., formed a 'Central Association', and invited major London employers to join their resistance.[156] They advised the ASE that its demands were 'totally inconsistent with the rights of employers of labour' and that if a strike occurred at any of the thirty-four firms they 'have unanimously determined to close our establishments.'[157] The employers duly closed their works and 'locked out' 3,500 union members, 1,500 other artisans and 10,000 labourers in Lancashire and London.

Sharp Brothers laid off over 600 men, and Fairbairns 2,500 men in their Manchester and London factories. Thomas Fairbairn took an active part in trying to resolve the dispute and wrote letters on the subject to the *Times*.[158] Nasmyth 'locked out' 300 union members from the Bridgewater Foundry, but employed all his other men, and used the opportunity to increase his quota of 'unbound' apprentices.[159] He later wrote:[160]

> I placed myself in an almost impregnable position, and showed that I could conduct my business with full activity and increasing prosperity, and at the same time maintain good feeling between employed and employer.

The lock-out lasted for nearly three months, by which time the ASE funds were nearly exhausted, and its survival threatened. Although the ending of the dispute heralded a long period of industrial peace, the ASE had learned lessons, and resolved that any future dispute would be regional rather than national, to make better use of its financial resources.[161]

Fig. 6.16
Locomotive built during the 1851 dispute. James Nasmyth & Co.-built 0-6-0 (N 99/51) for the York, Newcastle & Berwick Railway (No. 206), shown latterly as North Eastern Railway No. 206.
R.H. Bleasdale Collection – National Railway Museum.

CHAPTER 6 – MANAGEMENT, EMPLOYMENT AND INDUSTRIAL RELATIONS

Labour relations were generally harmonious for the twenty years between 1851 and 1871, in which year the issue of the basic working week erupted. The economic recovery in 1870 was the opportunity for the men of the Tyneside shipyards and factories, including the Stephenson and Hawthorn sites, to seek weekly instead of fortnightly pay, calculated to be worth 2/- a week from the resulting credit saving.[162] The request was promptly accepted by most employers, with the exception of R. & W. Hawthorn, whose 1,200 men walked out. It was the first general stoppage at the site, and the Hawthorn management hurriedly implemented weekly payment.[163]

Even after the provisions of the 1867 Factories Act Extension Act, the 1871 'Nine Hours Movement' on Tyneside pursued a shorter working week, without any political issues or fratricidal conflicts. Less than a quarter of the skilled engineering workers were members of trades unions in the build up to the dispute.[164] Although the leaders of the 'Nine Hours League' held trade union office, their influence and authority derived primarily from individual qualities of leadership, rather than from trade union organisation.

The nine hour demand was first rejected by Wearside manufacturers, prompting a strike which sent Tyneside employers into immediate conference to agree a response when the demands would also be made on them. Sir William Armstrong, whose works were by far the largest, and who was emphatically opposed to a reduction in hours,

Fig. 6.17 The commencement of the Nine Hours Movement on Tyneside involving the men of R. & W. Hawthorn and Black Hawthorn & Co., together with moves by the employees of Robert Stephenson & Co.
Newcastle Guardian and Tyne Chronicle, May 27th 1871 (digitally enhanced).

took charge of the employers' strategy, and dominated its proceedings. Armstrong needed the support of other large firms, particularly R. & W. Hawthorn and Robert Stephenson & Co., both with c.1,200 men, (and Black Hawthorn & Co. with 500 men), whose conduct was central to the dispute. Black Hawthorn was a new partnership whose inexperienced partners, faced with the conflicting demands of the nine-hour claim and the strength of Armstrong's advocacy to reject, felt obliged to prove themselves as strong proprietors and joined Armstrong's intransigent stand.

Fig. 6.18 Locomotive built during the 1871 'nine-hours' dispute. Black Hawthorn & Co.-built 0-6-0 ST (BH 130/71) for the Llynvi & Ogmore Railway. *The Locomotive Magazine, Vol.33 (1927), p.157.*

The same dilemma was faced by George Robert Stephenson, regarding Robert Stephenson & Co., who was resident in London and dependent upon his General Manager, George Douglas, for the detail and mood of the dispute. Stephenson was a sagacious and generous employer, who was not prepared to throw away years of equitable industrial relationships to suit Armstrong's unequivocal approach. He wrote a long letter to his men from London declining their request for a nine hour day, agreeing to discuss the issue without coercion, and stating explicitly that he would not join the other Tyneside employers. The men remained at work throughout the dispute but contributed substantial sums of money to the strike fund, which placated any antipathy there may otherwise have been.[165]

The crippling strike lasted from May to October 1871, when the nine hours day was conceded by the manufacturers to commence from the beginning of 1872. Stephenson also promptly conceded the nine-hour day but, characteristically sending a message of independence to the other employers, brought forward its implementation to November. There was a rapid implementation of the reduced hours in factories around the country. In December 1871, for example, when the men of the Avonside Engine Co. requested the nine-hour day, the Chairman, Edward Slaughter, replied that: 'the hour was given by the Directors ungrudgingly, and it was resolved upon by the Board even before the men applied for it.'[166]

The events of 1871 had demonstrated the successful outcome of a regional dispute, and, with further industrial disputes breaking out, the manufacturers' representatives formed the Iron Trades Employers' Association (ITEA). The Glasgow and north-east employers declined to join the latter group admitting they 'were still disunited as a district'.[167] Further disputes flared up on Tyneside in 1874/5 as a representative committee of the workforce, requested all 'The Employers of Engineering Labour in the Newcastle District' to provide a general increase in wage rate. After negotiations beset with communication difficulties, the employers, including Robert Stephenson & Co., agreed to two phased 5% increases, to be followed by a third 5% 'later'. 'Later' was open to interpretation, as Armstrongs and Stephensons first paid it, but then withdrew in the face of the recession later in the year, and Hawthorns withdrew from it completely.[168]

In devoting part of his Presidential Address to the Institution of Civil Engineers in 1876, George Robert Stephenson stressed the importance of employment terms being commensurate with business opportunities, and accepted that employers were far from always being correct.[169] In an obvious reference to Armstrong's stand in the 1871 dispute he stated: 'It cannot be denied that 'temper' has been a very potent element in causing and prolonging some of the larger strikes with which the country has been afflicted.' He appealed for understanding and reconciliation and observed that:

> The men themselves have the cure in their own hands, and it can only be effected by their strong determination to stamp out the cause of so much unhappiness… I do not speak bitterly, but in sorrowful earnest, and I know that there are a large number of good men to whom these remarks will not apply. Let me say, therefore, that it behoves us, who have had superior opportunities, better culture, more immediate means of surrounding ourselves with refining influences, to prove that we are, at least, learning to overcome prejudices, and patiently to take every reasonable method of proving to the men whom we have to employ that we entertain no animosity against them.

Industrial disputes became more widespread in the late 1870s and 1880s, however, as the workmen pursued further reductions in hours and wage increases. The survival of the Avonside Engine Co. was threatened by a bitter dispute over the directors' imposition of a 12½% wage reduction from 1878.[170]

The hours worked by the workforce of Neilson & Co. in Glasgow reduced from fifty-seven hours per week, before the nine-hour dispute, to fifty-four hours per week from January 1872 and a further reduction to fifty-one hours per week a year later. The men's call for three week's paid holiday on top of this reduction was resisted however, and an agreement was reached to return to fifty-four hours per week when the extended holiday arrangements were conceded from January 1879.[171]

Other locomotive manufacturers enjoyed better industrial relations, not least because of loyalty after many years service. E. K. Clark, later wrote that 'Among the [Kitsons'] men as a whole there was a continuous rumble of complaint as to the rates of wages, but partly because of the permanence of employment in the past, and confidence in the future, there was little active discontent.'[172] There was a similar feeling prevalent in Crewe and Derby railway workshops in the 1880s.[173]

There were several strikes in the early 1890s, for a shorter working week and higher wages, and, in 1896, the ITEA, which helped employers financially, supplying strike-breakers and black-listing strikers, was enlarged into the national Engineering Employers' Federation (EEF).[174] The central constitutive element in the formation of the EEF was the diversity and deep ambiguity which underlay the employers' insistence on managerial prerogative.[175] Although the employers, including the locomotive manufacturers, were responding to cost pressures by further capital investment and employment

Fig. 6.19 Built at a time of good industrial relations. Kitson & Co. 1886-built 0-4-2 (K 2965/86) for the Madras Railway (No. 14). *Kitson collection, The Stephenson Locomotive Society.*

of un-skilled labour, their sectoral diversity, overlaid by regional specialisation was to produce internal dissension in the confrontations of 1897/8.

Although the employment of non-union men led to the first round of inconclusive negotiations between the ASE and the EEF, the spark which led to the 1897 dispute was a further call for an eight-hour day, about which the employers had been in two minds, prompted by the recession of 1893 and the consequential availability of labour. Francis Marshall, of R.&W. Hawthorn Leslie & Co. Ltd., had suggested in that year that 'all should lead to an immediate reduction and I would support an 8-hour day – omitting breakfast time – this should be done now…' There was no response from his Chairman, Benjamin Browne, however, who was a leading member of the EEF, which had no such policy, and he probably suppressed the idea.[176]

In July 1897, before a selective strike could take hold, the EEF 'locked-out' 25,000 workers from 250 firms. By mid-September, with the core of the dispute reverting to working practices, the lock-out affected 34,000 workers, largely in the north of England, and including several locomotive manufacturers, which had far-reaching effects on the industry. In Glasgow, Sharp Stewart & Co. locked-out its men, although Neilson Reid & Co. avoided action, Hugh Reid stating there was no point in joining the EEF 'to fight for a freedom which we already possess.'[177] There was a shut-down by Fowlers in Leeds,[178] whilst John Kitson, Kitsons' Chairman, took an active part in the dispute, as a member of the EEF's Emergency Committee.[179]

Sir Benjamin Browne, R.&W. Hawthorn Leslie's Chairman, played a central role, and in October 1897, wrote to a fellow Hawthorn-Leslie Director:[180]

I *must* ask the Board to acknowledge that all Union or strike questions be left to me. I am practically the Head of the Peace Party on the Executive of the Engineering Employers and my position is a *very* delicate one…

Browne helped bring about the end to the dispute, after seven months, in January 1898. The EEF held fast to the employers' right to introduce into all workshops any condition which had previously been accepted by the union somewhere in the country, and the eight hour day was rejected.[181] The ASE's right to collective bargaining over wages was re-affirmed, in return for an acknowledgement of the employers' right to hire non-union personnel, demand up to forty hours overtime per month, employ as many apprentices as they chose, place any suitable worker on any machine at a mutually agreed

Fig. 6.20 Locomotive built during the 1897 lock-out. Kitson & Co. Ltd. built 4-4-0T (K 3727/97) for the Trinidad Government Railways (No. 13). *The Locomotive Magazine, Vol. 41 (1935), p.363.*

rate, and institute piecework systems at prices agreed with individual workers.[182] One of the 'essential agreements' was the provision for all piecework payments to be paid directly to the employees, and not determined and distributed by leading hands.[183] The lock-out was a turning point in industrial relations, following which unions gave less emphasis to craft exclusivity and more to political action.[184]

The cost of the dispute had been dear, having occurred during a buoyant economy, and the effects were serious for the locomotive manufacturers. Edwin Kitson Clark, for example, later wrote, with some understatement, that: 'in the guerrilla skirmishes before 1896, in the direct campaign that followed, the firm and personnel of Kitson & Co. have played a part sometimes wisely, sometimes unwisely…'[185] The backlog of locomotive orders was extreme and the industry as a whole was forced to quote the longest delivery times to their customers since the mid-1840s.

∼

The proprietors and their senior managers were equivocal in their handling of industrial disputes. Nasmyth's strong stance on freedom to employ un-skilled labour was too early for other proprietors, who may have felt vulnerable to the loss of scarce skilled men and preferred a more measured transfer of repetitive tasks to un-skilled personnel. The equivocation was due both to concern over losing craftsmen and to the personal characteristics of the proprietors themselves. This ranged between the confrontational approach of James Nasmyth and John Jones, and the more conciliatory approach of those other proprietors, such as Michael Longridge and the Stephensons, whose firm but fair employment maintained the men's respect and loyalty.

The effective accommodation between the manufacturers and their workforce in the twenty years before 1871 left them unprepared to face the concerted approach of the trades unions. The new generation of proprietors and senior managers retained divergent views on the major issues, especially the employment of un-skilled men. The regional federations, and the later national representative bodies, which were formed as defensive responses to the disputes, were dominated by strong personalities and were not wholly representative of employers' views. This weakened their resolve to deal conclusively with the issue of un-skilled employment, which was thus carried forward into the 20th century.

The crippling effects of the 1897 dispute revealed divisions of opinion within the Employers' Federation, representatives of the locomotive firms being directly involved in the dispute's conduct and resolution. The issue of the erosion of craft skills remained unresolved after the dispute, and in spite of their overwhelming victory the employers' internal divisions failed to make significant progress in reshaping work organisation. They carried into the next century their continued dependence on skilled labour leaving them vulnerable to future craft disputes.[186]

The divergent employment policies thus highlighted the individualism of the manufacturing proprietors. Their strategic and tactical decisions on employment represented differing views on the levels of un-skilled employment and the rate of their introduction, employment terms, paternalism and the maintenance of industrial harmony. Employers such as Michael Longridge and the Stephensons had shown extraordinary vision in attracting core workforces that remained loyal even in adversity. In contrast, Nasmyth's confrontational approach, for all his pioneering advancements in machine tool technology, had resulted in continuing problems that eventually led to his premature retirement.

There was no distinction in industrial relations between the 'progressive' firms with a high proportion of un-skilled labour, and the 'craft' firms more dependent upon their skilled workforce. Amongst the most harmonious firms were Neilson & Co. and Robert Stephenson & Co., respectively the largest and most successful firm, and the least successful which failed at the end of the century. Employment decisions, and acceptance of those decisions, were a combination of business acumen and respect.

Notes – Chapter 6

1. Zeitlin (1997), p 242.
2. Zeitlin (1997), p 247.
3. Jefferys (1945), p 207.
4. Drummond (1995), p 63.
5. McKinlay and Zeitlin (1989), pp 33-47.
6. For example, discussed by Chandler (1977), pp 269-272; Musson (1975), pp 109-149; Saul (1968), pp 186-237; and Drummond (1995), pp 40-132.
7. Braverman (1974); Friedman (1977); More (1980); Gordon, Edwards and Reich (1982); and Burawoy (1985).
8. Drummond (1995), pp 92/3.
9. Pollard (1963-64), pp 254-271.
10. Rowe (1971), p 52.
11. Joyce (1980), p 92.
12. Revill (1998). Also Drummond (1995), pp 186-208.
13. Southall (1996), p 61.
14. McKinlay and Zeitlin (1989), p 34.
15. Drummond (1989), p 8.
16. Letters, Arthur Potts to John Jones, Jones & Potts collection, IMS 248-252.
17. Warren (1923), p 416.
18. Anon, *The Vulcan Locomotive Works 1830-1930*, Locomotive Publishing Co., London, p 8.
19. Davis (1979), pp 32/3.
20. Hills and Patrick (1982), p 17.
21. Articles, *Newcastle Chronicle*, 1st January 1869 and 1st January 1870.
22. Clarke (n.d. but 1979), p 56.
23. Journal 1864-1900, Vulcan Foundry Collection, B/VF/4/2/1.
24. For example the LNWR, Drummond (1995), pp 58-65.
25. Letter, Joseph Locke to Robert Stephenson, Liverpool, Feb:25:1827, Library of the Institution of Mechanical Engineers, Crow Collection.
26. Letter, Robert Stephenson to Joseph Pease, Newcastle Upon Tyne, 12 April 1836, Pease-Stephenson Collection, D/PS/2/54.
27. Letter, Robert Stephenson to Edward Pease, London, 27 Oct 1836, Pease-Stephenson Collection, D/PS/4/5.
28. Memorandum, 16 May 1849, in Minute Book, Robert Stephenson & Co. collection.
29. Moss and Hume (1977), p 73.
30. Letter, Robert Stephenson to Edward Pease, Bedlington Iron Works, 25 Dec 1834, Pease-Stephenson Collection, D/PS/4/4.
31. Minute, 26 Feby 1839, in Minute Book, Robert Stephenson & Co. collection.
32. Letter, J. E. Sanderson to T. Richardson, Westminster, 12 May 1845, Pease-Stephenson Collection, D/PS/2/66.
33. Article, *Newcastle Chronicle*, September 6th 1845.
34. Jeaffreson (1864), Vol. I, p 13.
35. Booklet, *Manchester As It Is*, 1839, quoted by Musson (1957-58), p 126.
36. Letter J. Nasmyth [for Nasmyths Gaskell & Co.] to J. Dundas, Patricroft, 23 April 1839, Nasmyth Gaskell Collection, Letter Book 4, p 375.
37. Letter, J. Nasmyth to R. Willis, 1 June 1853, Bodleian Library, Oxford, quoted in Cantrell (1984), p 237.
38. Hills and Patrick (1982), p 17.
39. Hills and Patrick (1982), p 56.
40. Redman (1972), pp 43 & 50.
41. Drummond (1995), p 62.
42. Brown (1995), p 135.
43. Smiles (Ed) (1883), pp 311-312.
44. Articles, *Newcastle Chronicle*, *op cit* (21).
45. Article, *The Engineer*, Vol. 84, 1897, pp 31/2.
46. Byrom (2017), p 111.
47. Drummond (1995), p 69.
48. Drummond (1995), pp 72/3.
49. Brown (1995), pp 153-157.
50. John W. Converse, 'Some Factors of the Labor System and Management at the Baldwin Locomotive Works', *Annals of the American Academy of Political and Social Science*, Issue 21, Jan. 1903, p 6, quoted in Brown (1995), p 156.
51. Jenkins & Ponting (1975), pp 109/110.
52. *Fifth Report from Select Committee Respecting Artisans* &c., Parliamentary Papers, 1824, Vol. 5, pp 566-569.
53. *ibid*.
54. Pole (Ed) (1877), pp 26/7.
55. Pole (Ed) (1877), p 47.
56. Smiles (Ed.) (1883), pp 216/7.
57. Letter, Harris Dickinson to Edward Pease, Newcastle Upon Tyne, 10 Mo.21.1830, Pease-Stephenson collection, D/PS/2/52.
58. Letter, Edward Pease to Thomas Richardson, Darlington, 7 Mo 4 1835, Hodgkin Collection, Durham County Record Office, Darlington Public Library, D/HO/C/63/18.
59. Board of Health Report, Barton-upon-Irwell, 1852, p 22, quoted in Cantrell (1984), p 230.
60. Nasmyths Gaskell & Co. Day Book, 8th February 1850, quoted in Cantrell (1984), p 230.
61. *The Vulcan Locomotive Works*, *op cit* (18), p 14.
62. Reed (1982), p 26.
63. Hills and Patrick (1982), p 54.
64. Vamplew (1972), No 3, p 328.
65. Mark O'Neill, 'Walter Montgomerie Neilson and the Origins of Locomotive Building in Glasgow', in Nicolson and O'Neill (Eds) (1987), p 5.
66. Rowe (1971), p 79.
67. *Royal Commission on Trades Unions, Tenth Report,*

67 Parliamentary Papers, 1868, Vol. 39, Q.19,155, quoted in Cantrell (1984), p 242.
68 Habakkuk (1962), p 195.
69 Byrom (2017), pp 164-69.
70 Letter J. Nasmyth to R. Willis, 29th April 1852, Bodleian Library, Oxford, quoted in Cantrell (1984), p 236.
71 Martin (1974), pp 16/7.
72 The proportion of labourers to skilled artisans was 20-25% at Crewe Works. Drummond (1995), p 49.
73 Article, *Newcastle Chronicle*, 3rd Feb 1846, quoted in Clarke (n.d. but 1979), pp 9/10.
74 'Regulations for the Drawing Office of R. Stephenson & Co.', quoted in Warren (1923), p 100.
75 Martin (1974), p 45.
76 Rowe (1971), p 41.
77 Martin (1974), p 45.
78 Warren (1923), p 100.
79 Rowe (1971), p 45.
80 Jeffrys (1947), p 32.
81 Rowe (1971), p 52.
82 Clarke (n.d. but 1979), p 4.
83 Clarke (n.d. but 1979), p 29.
84 Hills and Patrick (1982), pp 19/20/285.
85 Rolt (1964), p 34.
86 Lane (1980), p 166.
87 Martin (1974), Appendix, p 45.
88 Drummond (1995), p 70.
89 Brown (1995), pp 149/150.
90 Clark (n.d. but 1937), pp 154/5.
91 Habakkuk (1962), p 156.
92 Redman (1972), p 59.
93 Clark (n.d. but 1937), pp 55/6.
94 Article, *The Railway Engineer*, Vol. VIII, No.10, October 1887, pp 304-309.
95 Redman (1972), p 21.
96 Articles of Apprenticeship, Sixth day of October 1823, Library of the Institution of Mechanical Engineers, Phillimore Collection.
97 Letter, George Stephenson to William Salvin, Liverpool, 25th June 1827, Durham County Record Office, D/Sa/C/139.2.
98 Letter, Robert Stephenson to Thomas Richardson, Dieppe, July 11th 1833, quoted in Jeaffreson (1864), p 182.
99 Article, *Biographer and Review*, Vol. III, No.40, September 1900, Copy in the Northumberland County Record Office, NRO765.
100 Nasmyths Gaskell & Co. to G. Grundy, 10th November 1838, Nasmyth Gaskell collection, Letter Book 3, p 240.
101 Letter, Nasmyths Gaskell & Co. to S. Wardle, 28 June 1838, Nasmyth Gaskell collection, Letter Book 2, p 253.
102 Letter Robert Stephenson to Joseph Pease, Weedon, 7 Sept:1836, Pease-Stephenson Collection, D/PS/2/55.
103 *Royal Commission on Trades Unions, Tenth Report*, Parliamentary Papers, 1868, Vol. 39, Q.19,201, quoted in Cantrell (1984), p 239.
104 Smiles (Ed.) (1883), p 227.
105 Thomas (1964), p 103.
106 Drummond (1995), p 42.
107 Clarke, (n.d. but 1979), p 29.
108 Clark (n.d. but 1937), p 175.
109 Clark (n.d. but 1937), Chapters 8, 9 and 10.
110 Charlesworth, Gilbert, Randall, Southall and Wrigley (Eds) (1996), Section B 1850-1900, p 61.
111 Taksa (1998), p 7.
112 Jefferys (1945), p 58.
113 Indentures for John Simpson, December 8th 1827, Crow Collection, *op cit* (25); for Michael Hobson, May 15th 1830, McDowell Collection of the Stephenson Locomotive Society, p 14; and for Edward Bates, October 25th 1830, Crow Collection.
114 Indenture, Robert Millward, 15 March 1839, Science Museum Library.
115 Smiles (Ed.) (1883), pp 227/8.
116 Clarke (n.d. but 1979), p 29.
117 Orcutt (1902), pp 551-554 & 703-710.
118 Brown (1995), pp 136-143.
119 Royal Commission on Children's Employment in Mines and Manufactories, Parliamentary Papers 380/1/2, 1842, p.558.
120 Drummond (1995), p 63.
121 Royal Commission on Children's Employment in Mines and Manufactories, Parliamentary Papers 380/1/2, 1842, p.720.
122 Martin (1974), p 45.
123 *The Blyth and Bedlington Literary Supplement*, 1839, quoted in Martin (1974), p 38.
124 Drummond (1995), p 64.
125 Lane (1980), p 166.
126 Millar (1976), p 48.
127 Report of speech by R. Brown at New Years' Eve dinner for the boiler-makers of Robert Stephenson & Co. at the Garrick's Head, Newcastle on Tyne, *Newcastle Courant*, January 1st 1870.
128 Report, *Newcastle Chronicle*, September 6th 1845.
129 Hills and Patrick (1982), p 20.
130 Reports, *Newcastle Chronicle*, 1st January 1869 and 1870.
131 Taksa (1998).
132 Article, 'Workers in Steel and Iron, 1', Messrs. R. Stephenson & Co. Ltd., Newcastle Upon Tyne, *The Shipping World*, March 1 1888, pp 338/9.
133 Thomas (1964), p 84.
134 Hills and Patrick (1982), p 19.

135 Drummond (1989), p 12.
136 Letter G.R. Stephenson to George Crow, Newcastle upon Tyne, July 13th 1860, Northumberland County Record Office, NRO793/5.
137 Letter, G.F. Duncombe, for the Board of Trade, to Messrs. R. Stephenson & Co., London, 5th April 1856, Northumberland County Record Office, NRO793/4.
138 Article, *Newcastle Courant*, 20th February 1836.
139 Article, *Newcastle Courant*, 19th August 1853.
140 Mewburn (1876); also report, *The Times*, October 21st 1859.
141 Edwin Kitson Clark, quoted in Clark, (n.d. but 1937), p 170.
142 Printed notice, 'Articles, Rules & Regulations of the Millwrights' Benevolent Benefit Society, instituted at Newcastle Upon Tyne, February 19th 1825… To be held at the House of Mr. Heron, The Sign of the Cock Inn, Head of the Side, Newcastle Upon Tyne'.
143 *Fifth Report from Select Committee Respecting Artisans* etc., Parliamentary Papers, 1824, Vol. 5, pp 566-569.
144 Letter, Michael Longridge to George Stephenson, Bedlington Iron Works, 5 Mar 1825, Crow Collection, *op cit* (25).
145 Rowe (1971), p 41.
146 *Newcastle Courant*, 24th September and October 1st 1836.
147 Copy letter, Nasmyths Gaskell & Co. to Braithwaite Milner & Co., 2nd November 1838, Nasmyth Gaskell Collection, Letter Book 3, p 262.
148 *Royal Commission on Trade Unions, Tenth Report*, Parliamentary Papers, 1868, Vol. 39, Q.19,112, quoted in Cantrell (1984), p 241.
149 *Royal Commission, ibid*, Q.19,234, p 237.
150 Printed pamphlet: 'An Address Delivered to the Workmen Employed at the Bedlington Iron Works, Upon the 21st June 1839', Michael Longridge, Newcastle, 1839, Newcastle Public Library, Ref. L042, Local Tracts D34, Political No.7.
151 Letter, Michael Longridge to William Longridge, Bedlington, July 14th 1839, privately held.
152 Mather (1974), pp 115-140; Also Jenkins (1980), *passim*.
153 Charlesworth, Gilbert, Randall, Southall and Wrigley (Eds) (1996), Chapter 7, The General Strike of 1842, pp 51-58.
154 Mather (1974), pp 133-135.
155 Jefferys (1945), pp 26/7; Also Burgess (1975), pp 11 & 16.
156 Charlesworth, Gilbert, Randall, Southall, and Wrigley, (Eds) (1996), Chapter 9, 'Lock-Outs and National Bargaining in the Enginering Industry 1852', pp 72/3.
157 *The Times*, 20th December 1851.
158 *The Times*, for example letter, 14th January 1852. See also Byrom (2017), pp 164-69.
159 *Royal Commission*, 1868, Q.19,134, quoted in Cantrell (1984), p 243.
160 Smiles (Ed.) (1883), p 311.
161 Charlesworth, Gilbert, Randall, Southall, and Wrigley, (Eds), 1996.
162 Rowe (1971), p 91. Also, Article, Local Industrial Sketches, *Northern Echo*, January 3rd 1873.
163 *Engineering*, Vol. XI, March 10 1871, p.172; also, Clarke (n.d. but 1979), pp 21/2.
164 McCord (1971), pp 100-168.
165 Presidential Address of George Robert Stephenson, *Proceedings of the Institution of Civil Engineers*, Session 1875-76, Vol. 44, Part II, January 11 1876, pp 14-15.
166 Article, *Engineering*, Vol. XII, December 29th 1871, p 424.
167 Wigham (1973), p 12.
168 Clarke (n.d. but 1979), pp 29/30.
169 Stephenson Presidential Address, *op cit* (165).
170 Davis (1979), p 44.
171 Notebook prepared by 'AI', senior clerk of the Neilson Co. Author's collection.
172 Clark (n.d. but 1937), pp 146/7.
173 Drummond (1989), p 13.
174 Charlesworth, Gilbert, Randall, Southall and Wrigley (Eds) (1996), pp 76/7.
175 McKinlay and Zeitlin (1989), pp 33-47.
176 Clarke (n.d. but 1979), pp 58/9.
177 Thomas (1964), p 155.
178 Lane (1980), p 165.
179 Clark (n.d. but 1937), p 177.
180 Clarke, (n.d. but 1979), p 59.
181 Charlesworth, Gilbert, Randall, Southall and Wrigley (Eds) (1996), p 77.
182 McKinlay and Zeitlin (1989), p 36.
183 Clark (n.d. but 1937), pp 155/6.
184 Charlesworth, Gilbert, Randall, Southall and Wrigley (Eds) (1996), p 62.
185 Clark (n.d. but 1937), p 168.
186 McKinlay and Zeitlin (1989).

Chapter 7

Strategy and Administration

Introduction

The growth of the manufacturing firms engaged in the locomotive industry through the nineteenth century introduced new kinds of decision making, both for the profitable execution of their day-to-day business and the strategies for their long-term development. The proprietors' motivations in running their businesses, and their competence in managing the complex requirements of large, integrated operations governed their strategies on expansion, investment, diversification, incorporation, amalgamations and even survival. Their managing competence required leadership attributes and delegation of authority, as well as sufficient information on which to base their decisions.

The entrepreneurial motivation and combined expertise of the earliest proprietors gave them the confidence and ability to raise sufficient capital to commence locomotive production. However, in the following generations as further capital was needed for replacement, upgrading and expansion, proprietors displayed a more cautious interpretation of market development, derived from experience of its uncertain evolution. This was combined with a growing sense of loyalty towards their longer-serving workforce of managers, craftsmen and clerks. These wider capital commitments depended upon continuing entrepreneurial flair or acceptance of the need for incorporation and the transfer of strategic decision-making to a wider directorate.

The size, diversification and integrated structure of the firms called for the development of management information and accounts, which the proprietors and their senior managers could use to monitor profitability and guide them in their tactical policies. The development of management and financial accounts and information

Fig. 7.1 4-4-0ST *PLUTO* built in 1866 by Avonside Engine Co. of Bristol for the South Devon Railway (as taken in to Great Western Railway stock from 1876 as No. 2123). *R.H. Bleasdale Collection – National Railway Museum.*

systems made extraordinary progress through the century, particularly in serving the integrated requirements of the heavy manufacturing industry. They provided the potential for firms to make good use of assets, minimise costs of raw materials and finished components, deal with receipts and payment pressures, and contain their working capital requirements. The introduction of allowances for overhead costs and other general charges, and of 'cost-centre' monitoring, provided the potential for sounder decisions on workshop profitability and wider production costing.

Financial accounting and depreciation policies, both through balance sheet and profit and loss statements, developed slowly through the century, with diverse definitions and practices. Following the Joint Stock Companies Act of 1856, and until the Companies Act of 1900, there was no compulsory financial reporting or auditing for firms with limited liability, and thus no formalisation of financial accounts.[1] England had the most permissive commercial law in the whole of Europe as far as general manufacturing companies were concerned.[2] Whilst it was in their own best interests for firms to prepare accounts in sufficient detail, and with sufficient discipline, to reflect their true financial position, there would have been a particular requirement to demonstrate competence, and stimulate confidence, when seeking investment capital from their backers.

Against the background of an unpredictable market, manufacturers' re-investment decisions related to expansion and modernisation of manufacturing capacity and capital equipment, diversification into second factories, at home and abroad, and backwards integration through supplier acquisition. The different interpretations of the market, and the varied entrepreneurial inclinations of the proprietors, caused a divergence in strategic policies, which ranged between major investment for the 'progressive' specialised locomotive factories and the maintenance of 'craft'-based general manufacturing sites, whose output included locomotives. The raising of sufficient capital for expansion received a significant boost with the formation of some limited companies from the 1860s. With the move towards incorporation tempered by the proprietors' desire to retain control of their firms, however, the industry's conversion to private and public companies was spread throughout the remainder of the century.

To accommodate the severe market fluctuations, the manufacturers were faced with difficult medium-term strategic decisions to optimise their production capacity and maintain satisfactory levels of employment for their men and capital equipment. Whilst much of the capacity variation was absorbed through the recruitment and lay-off of un-skilled machinists, each demand peak raised consideration of investment for further capacity. Options with each downturn in the market, however, ranged between complete withdrawal from locomotive manufacture, and partial withdrawal through diversification into alternative markets. The correct interpretation of the locomotive and other heavy engineering markets allowed manufacturers to sustain their production capacity through diversification. The causes of company failures, and the absence of company amalgamations, provides further insight into the continued proliferation of locomotive firms at the end of the century.

Cost and Management Accounting

The foundations of cost accounting had been well established since the 18th century by large companies such as Boulton & Watt and Josiah Wedgewood,[3] although 'total' cost accounting did not become an established management tool until the last quarter of the 19th century.[4] Lack of progress in cost accounting had been due to the absence of keen competition, allowing prices which produced generous profit margins, whilst secrecy hampered the diffusion of

Fig. 7.2
Product of an early public limited company. A 2-4-0 built in 1879 by Sharp, Stewart & Co. Ltd. (SS 2861/79) for the Java State Railways. *Author Collection.*

ideas and techniques in the metal producing industries from the 18th century.[5]

Cost accounting methods and management information systems were therefore developed in accordance with the perceived requirements of individual firms, and similar cost-accounting practices in the coal, iron and steel industries suggest a measure of skill transfer through recruitment of experienced cost-accounting clerks.[6] These practices provided management information systems, evolving in tandem with financial accounting. Industry in the 19th century was thus well versed in integrating cost with financial accounting, but not until the 1870s was it recognised in Britain as a text-book subject that would lead to a more consistent approach. This compared badly with French practice, where the discipline had become well developed, both in text and in practice, from the 1820s.[7] There is no explanation for this delay, but it could well reflect a more general antipathy towards formal learning in Britain.

The locomotive sector, as part of the heavy manufacturing industry, was in the forefront of the development of cost accounting in Britain. The accounts, including prevailing raw material and bought-in component charges, were essential management information on which were based not only tender quotations, but the monitoring of locomotive production costs. The early development of cost-centre accounting allowed an increased understanding of the cost-effectiveness of each production process, but there was a wide diversity in its use and interpretation which was reflected in company profitability.

The first locomotive manufacturers inherited the accounting practices of previous manufacturing industry. Robert Stephenson & Co.'s first accounts were overseen by its partner, Michael Longridge, with practices transferred from the Bedlington Iron Works. The Stephenson Company's first ledger reveals that cost and financial accounts were fully integrated.[8] The ledger is a compendium of folio entries in the company's materials purchase book, wages book, Goods (sales) account, Trade (maintenance and manufacturing overheads) account, Stock (capital goods) account, and Cash Received (income) account.

Raw materials and sub-contracted components for each order were recorded by weight and unit price. The men's time may also have been set down against each order in the wages book, as off-site wage rates and hours allocated (installing stationary engines for example) were separately recorded in the ledger. The detail allowed the company to predict the likely production, delivery and installation costs for each quotation. Multiple items, such as track-work and wagon components were quoted by weight, including machining and finishing costs. Multiple component machines, such as locomotives, were quoted a single ex-factory or delivered price. Separate accounts were kept for the firm's foundry and stable, including purchase, goods and trade books, with each credit matched as a debit in the main works accounts. This early cost-centre accounting allowed the proprietors to identify the foundry's losses, which prompted their decision to close it in 1832 (Chapter 5).

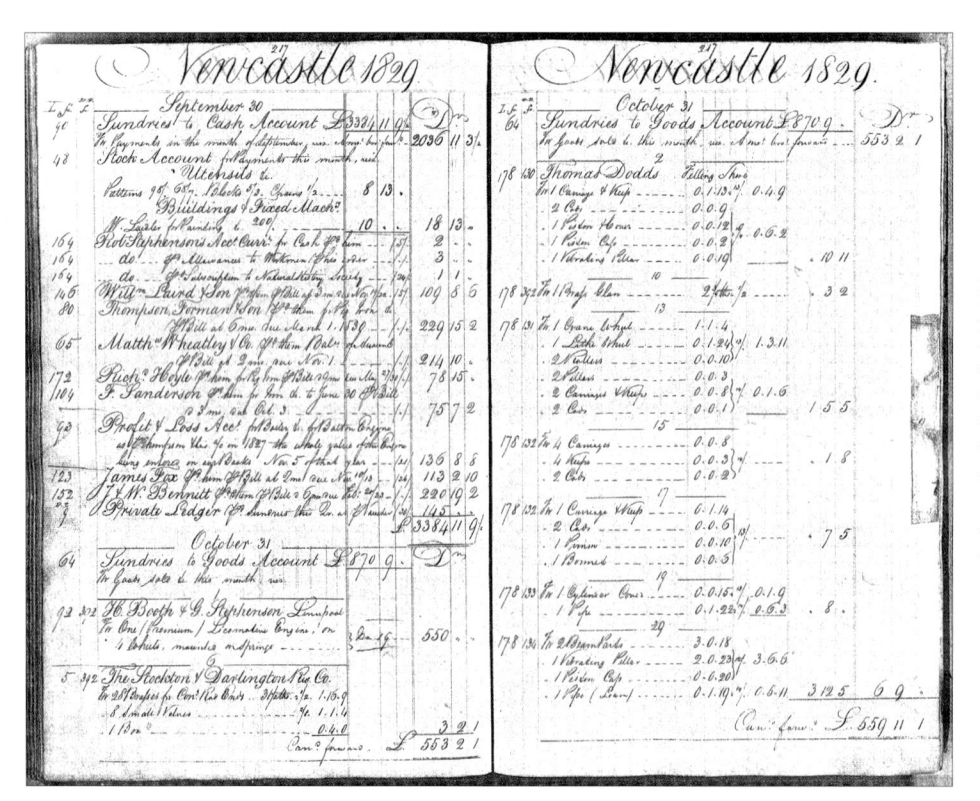

Figure 7.3
Robert Stephenson & Co. ledger (1823-1831) folio 217, showing end of month cash statements for September 1829 and sales account entries for October 1829 including the entry for the 'Premium' No. 19 locomotive (*ROCKET*). *Robert Stephenson & Co. collection, ROB/4/1.*

CHAPTER 7 – STRATEGY AND ADMINISTRATION

Fig. 7.4
Robert Stephenson & Co. ledger 1823-1831, folio 220 showing end of month cash statements for October 1829, including workmen's wages, materials and incidental expenses.
Robert Stephenson & Co. collection, ROB/4/1.

Whilst the ledger does not confirm that overhead costs were analysed to form a percentage on-cost for quotations, the allocation of overheads is confirmed in a surviving Cost and Profit Account Book for 1834/5.[9] This shows that the Stephenson Company was making full use of the cost information at its disposal. Costs were separately recorded for each locomotive.

Fig. 7.5
Example of Locomotive Costings in 1835 by R. Stephenson & Co. Entry for the *RAPID* Locomotive (RS 109/1835) built for the Newcastle & Carlisle Railway.
Cost & Profit Account Book, Bidder Papers, Science Museum Library, (Arch:Bidd 27/8).

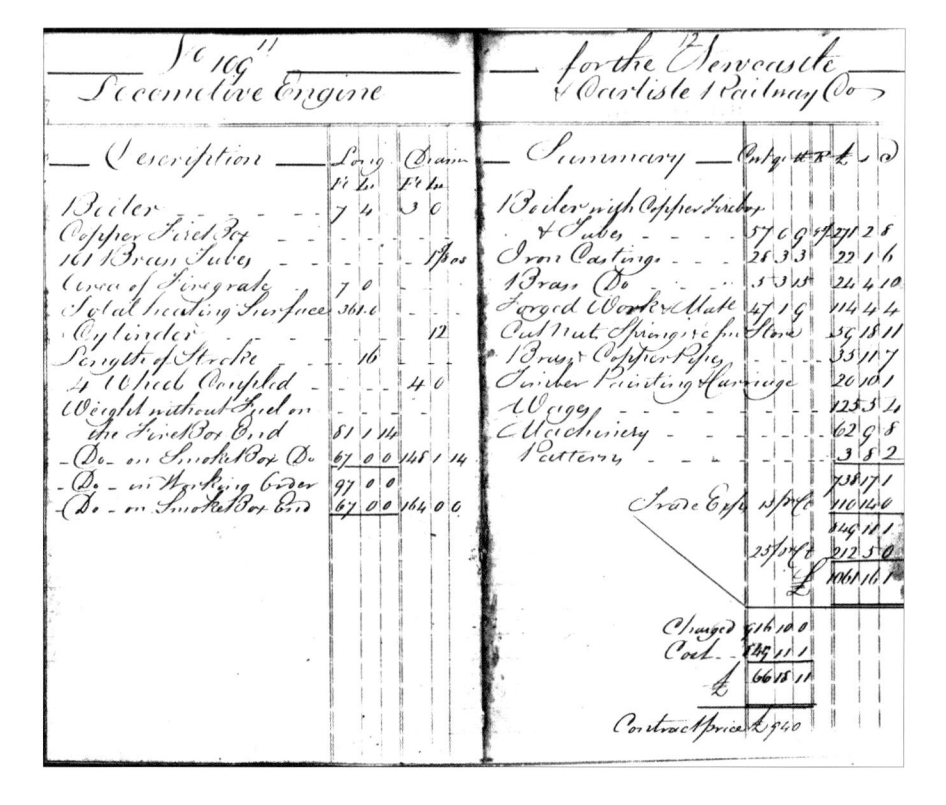

163

This example includes the apportionment of overhead costs through the 15% on-charge. The further 25% charge may have been for comparative purposes, perhaps reflecting a target profit margin that was being considered as the market began to expand. The 2.5% difference between the charged and contract prices may have been a working capital allowance due to bills of exchange credit delays.

Cost accounts were made available by the manufacturers to customers needing to check the accuracy of locomotive charges. Daniel Gooch, Locomotive Superintendent of the Great Western Railway, reported to I.K. Brunel, the railway's engineer in 1839: 'I have gone carefully through the whole of Hawthorn's accounts, and find everything charged very moderately…'[10] The engineer, T.E. Harrison, similarly reported that Hawthorns had 'placed the whole of the Books at my command' and 'I must say [amounts] are down in a most clear and satisfactory manner.'[11]

By 1839, Kitson Laird & Co. had adopted job numbers in its wages book to allow ready abstraction of cost data for each job, as well as for making up the wages.[12] Time, to the nearest quarter-hour, was related to each man's wage rate to determine the costs for each job. Piece-work was introduced in 1839, and the wages book recorded additional wages paid for productivity improvements, in addition to component weights and machinery and fitting times. Charles Tayleur & Co. employed an 'Abstract Book' from 1844, in which the material and wages costs of each locomotive batch were recorded, and which reflected the value of bulk purchasing of raw materials in reducing locomotive costs.[13] The grouping of wage costs for each activity similarly reflected an understanding of the benefits of batch production. Beyer Peacock & Co. set up similar detailed cost analyses in a 'Cost of Work' book from 1855, showing profits or losses on each locomotive batch after deduction of commission.[14]

That cost accounting in manufacturing companies had developed, and been fully integrated with financial accounting by about 1860, was demonstrated by the recollections of a chartered accountant, Thomas Plumpton. He had been instructed in cost accounting in the early 1860s by a professional accountant:[15]

> I had the advantage of a thorough training in a large engineering concern, manufacturing locomotive and marine engines, boilers, and every kind of machinery, where the Cost Accounts were so interwoven with the Commercial Accounts as to form an integral part of the whole on the system known as the Italian System [double-entry accounting], which until recent years was so universally adopted …

An expanded form of cost-centre accounting was being practised by R.&W. Hawthorn during the 1860s.[16] Its profit and loss management summary included separate accounts for its forge, brass foundry and machine shop, with internal debits being traced through from those shops to the erecting shop and included in its profitability assessment. That assessment was further divided between locomotives and marine engines is shown in Fig. 7.7:

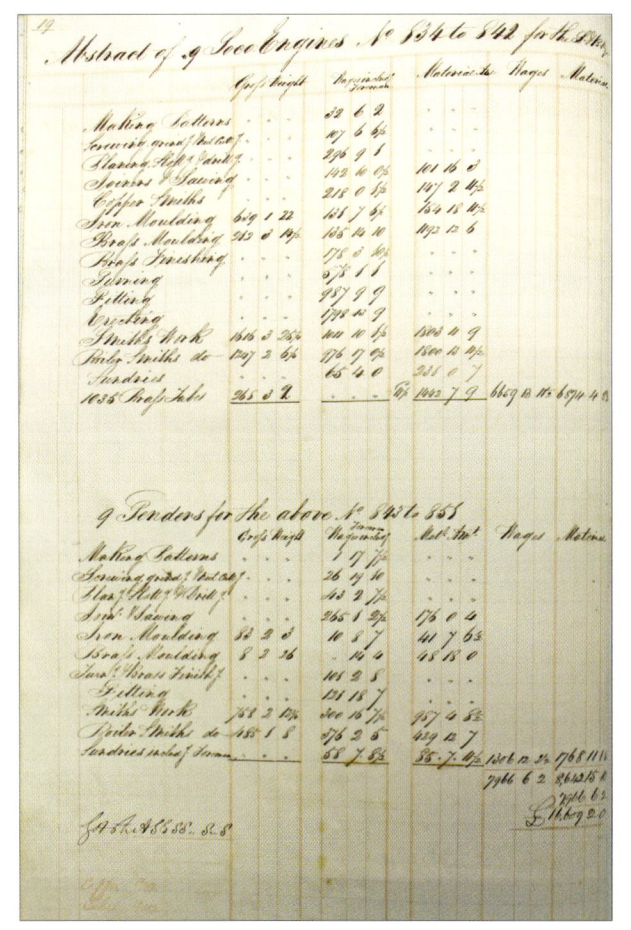

Fig. 7.6 Example of Materials and Wages cost allocation for a batch of nine 4-2-0 locomotives built by Charles Tayleur Co. in 1846/7 (VF rotation Nos. 254-262) for the London & Birmingham Railway (Delivered to the London & North Western Railway, Nos. 159-167). *Abstract Book 1844-1870, Merseyside Museum, Vulcan Foundry collection (B/VF/5/6/1), p.19.*

	1864 £	1865 £	1866 £	1867 £	1868 £
Locomotive Engines & Tenders	3680	4443	9949	-1284	-1572
Marine & Stationary Engines	9737	3154	1726	2035	676
[other] Orders	3568	2949	2915	2388	3965
Brass foundry	400	408	1506	1162	757
Forge	-247	1272	1049	337	110
Tools	1648	2321	1902	163	-556
Discounts & Charges	-2591	-1369	604	-1455	-2734
Total (Profit)	16196	13180	19654	3349	645

Fig. 7.7 Example of Cost Centre Profit and Loss Accounting. R.&W. Hawthorn (Engine Works) 1864-1868. *Clarke, (n.d. but 1979), p16.*

Management accounting was generally well established by the 1870s. For example, regular cost analyses became a feature of the Vulcan Foundry's management control systems.[17] Wage comparisons, between similar locomotives made for different railways, were used to identify high cost examples, allowing corrective action to be taken, and batch production costs to be monitored. Its quarterly analyses of workshop expenses were more frequent than the half-yearly analyses that had generally been undertaken in the coal, iron and steel industries in the second half of the 19th century.[18] During the last third of the century, six-year summaries were compiled to provide longer-term trend analyses.

Close monitoring of raw material charges and other costs allowed manufacturers to respond better to the external factors affecting their profitability. The Vulcan Foundry's 'Abstract Book' recorded price variations for raw materials and sub-contracted components, used both in submitting quotations and in determining the profitability of each locomotive batch.[19] It reveals, for example, that between 1871 and 1873 there was a 50% increase in coal prices with a consequential affect on iron prices, no doubt arising from the economic effects of the Franco-Prussian War, and compounded by the affects of the 1871 'nine-hour' dispute. The Hawthorn Company's Chairman, Benjamin Browne, noted that 'we had therefore to work off a very heavy order book, taken at low prices, at greatly increased costs and heavy loss.'[20] Weekly meetings of the partners delivered the tactical decisions necessary to minimise the firm's losses.

Not all firms maintained good management accounts, and some failed to identify loss-making areas and take decisions to rectify problems. The failure of Andrew Barclay & Son in 1874 was attributable to poor management and lack of financial control.[21] Its reconstitution with new backers the following year was dependent upon new financial controls being introduced. Day books were started up, setting out details of each job undertaken and its cost.

The level of cost detail also varied considerably between manufacturers. By the late 1880s, Neilson & Co.'s' cost accounting records were detailed, with materials and wages booked to each batch of components.[22] In 1889, however, the Hawthorn Leslie Company was only recording broad estimates of costs because of the lack of clerical staff. William Cross, the Company's Managing Director, complained about the lack of accurate records which prevented the successful continuation of piece-work and suggested that 'one, if not two, clerks employed in such work [keeping accurate cost sheets] would repay their cost over and over again…'[23]

By 1886, Robert Stephenson & Co. Ltd.'s detailed cost analyses used a printed ledger with headed columns.[24] Data was transferred from workshop folios, summarising material costs, wages, trade expenses, other expenses and commission payments (where applicable). Cost totals were matched with invoice figures to determine profits or losses. Larger engines and components had separate analyses, whilst the many smaller items were grouped into half-yearly summaries by customers. Half-yearly reports were prepared for its three businesses, 'Engine Works', 'Boiler Shop' and 'Shipyard', which were separately accounted for, work done for one being charged by another, for example a marine engine charged to the shipyard.

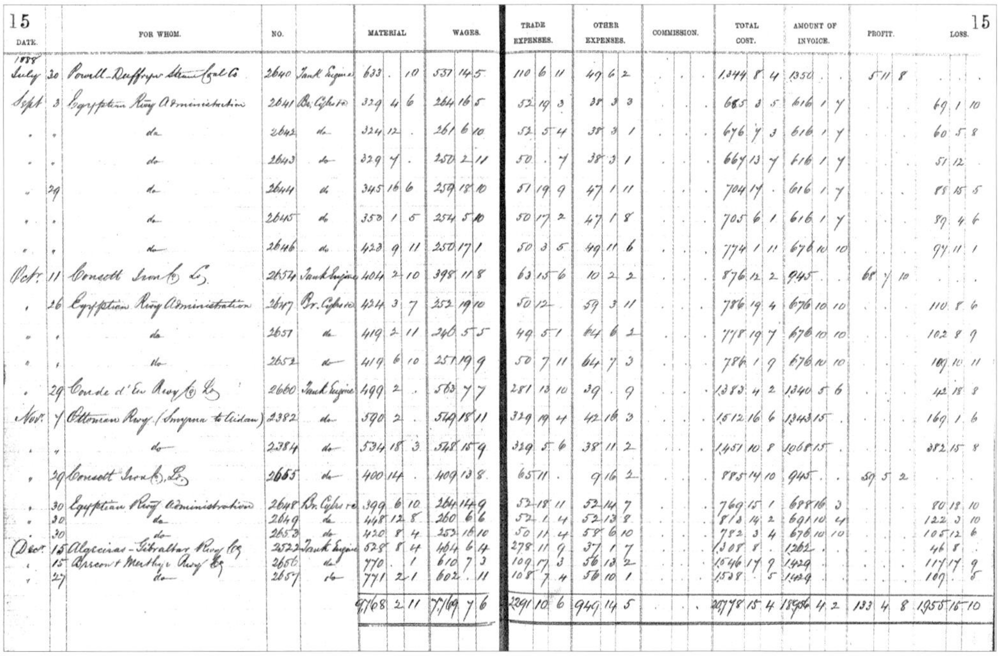

15 Date.		For Whom.	No.		Material.	Wages.	Trade Expenses.	Other Expenses.	Commission.	Total Cost.	Amount of Invoice.	Profit.	Loss. 15
1888 July	30	Powell-Duffryn Steam Coal Co	2640	Tank Engine	633 . 10	537 14 5	110 6 11	49 6 2	. . .	1,344 8 4	1350 . .	5 11 8	. . .
Sept	3	Egyptian Rwy Administration	2641	Br. Cylrs &c	329 4 6	264 16 5	52 19 3	38 3 3	. . .	685 3 5	616 1 7		69 1 10
"	"	do	2642	do	324 12 .	261 6 10	52 5 4	38 3 1	. . .	676 7 3	616 1 7		60 5 8
"	"	do	2643	do	329 7 .	250 2 11	50 . 7	38 3 1	. . .	667 13 7	616 1 7		51 12 .
"	29	do	2644	do	345 16 6	259 18 10	51 19 9	47 1 11	. . .	704 17 .	616 1 7		88 15 5
"	"	do	2645	do	353 1 5	254 5 10	50 17 2	47 1 8	. . .	705 6 1	616 1 7		89 4 6
"	"	do	2646	do	423 9 11	250 17 1	50 3 5	49 11 6	. . .	774 1 11	676 10 10		97 11 1
Octr	11	Consett Iron Co. Ld.	2654	Tank Engine	404 2 10	398 11 8	63 15 6	10 2 2	. . .	876 12 2	945 . .	68 7 10	. . .
"	26	Egyptian Rwy Administration	2647	Br. Cylrs &c	424 3 7	252 19 10	50 12 .	59 3 11	. . .	786 19 4	676 10 10		110 8 6
"	"	do	2651	do	419 2 11	246 5 5	49 5 1	64 6 2	. . .	778 19 7	676 10 10		102 8 9
"	"	do	2652	do	419 6 10	251 19 9	50 7 11	64 7 3	. . .	786 1 9	676 10 10		109 10 11
"	29	Conde d'Eu Rwy Co. Ld.	2660	Tank Engine	499 2 .	563 7 7	281 13 10	39 . 9	. . .	1383 4 2	1340 5 6		42 18 8
Novr	7	Ottoman Rwy (Smyrna to Aidan)	2382	do	590 2 .	549 18 11	329 19 4	42 16 3	. . .	1512 16 6	1343 15 .		169 1 6
"	"	do	2384	do	534 18 3	548 15 9	329 5 6	38 11 2	. . .	1451 10 8	1068 15 .		382 15 8
"	29	Consett Iron Co. Ld.	2655	do	400 14 .	409 13 8	65 11 .	9 16 2	. . .	885 14 10	945 . .	59 5 2	
"	30	Egyptian Rwy Administration	2648	Br. Cylrs &c	399 6 10	264 14 9	52 18 11	52 14 7	. . .	769 15 1	688 16 3		80 18 10
"	30	do	2649	do	448 12 8	260 6 6	52 1 4	52 13 8	. . .	813 14 2	691 10 4		122 3 10
"	30	do	2653	do	420 8 4	252 16 10	50 11 4	58 6 10	. . .	782 3 4	676 10 10		105 12 6
Decr	15	Algeciras-Gibraltar Rwy Co	2522	Tank Engine	528 8 4	464 6 4	278 11 9	37 1 7	. . .	1308 8 .	1262 . .		46 8 .
"	15	Brecon & Merthyr Rwy Co	2656	do	770 . 1	610 7 3	109 17 3	56 13 2	. . .	1546 17 9	1429 . .		117 17 9
"	27	do	2657	do	771 2 1	602 . 11	108 7 4	56 10 1	. . .	1538 . 5	1429 . .		109 . 5
					9168 2 11	7769 7 6	2291 10 6	949 14 5	. . .	20778 15 4	18956 4 2	133 4 8	1955 15 10

Fig. 7.8
Robert Stephenson & Co. 'Cost Analysis' Ledger for 1886-1901 showing folio 15 covering the period July to December 1888. This analysis highlighted the severe losses being incurred on the Company's locomotive business at that time for works Nos. 2640-2660. *Robert Stephenson & Co. collection (ROB/4/2).*

The detail to which most British companies recorded their manufacturing costs was followed, to a limited extent, by the Globe Locomotive Works in Boston, USA, which by 1851 assembled unit cost data, including labour, materials and some overhead charges.[25] This was, however, in marked contrast to the Baldwin Locomotive Works, which kept no cost accounting records until the mid-1870s recession forced upon it the necessity of knowing and controlling costs. Until then Baldwins merely charged what the market would bear for locomotives and, so long as the company was in profit, considered cost accounting a drag on its business.[26] After 1878, however, the trend away from standard locomotive designs in America increased the importance of cost accounting, and Baldwin's cost controls grew increasingly sophisticated and detailed.

Although the locomotive industry was generally advanced in its use of cost and management accounts, the wide divergence in their interpretation, and in the decisions based upon them, seems to confirm that secrecy hampered the diffusion of accounting techniques, and accounting knowledge depended upon recruitment of experienced cost-accounting clerks. Although proprietors were used to receiving cost-centre based information, they would not necessarily use the data to improve their capacity to make managerial decisions and take appropriate action to remedy cost over-runs. In spite of good accounting procedures, for example, the Stephenson Company did not act to stem its major losses in the 1880s which eventually led to its failure.

Receipts, Payments and Working Capital

The necessity to be alert to cash-flow problems and the adequacy of working capital was constantly addressed by the proprietors and their senior managers. Tactical judgement was required with the timing of both payments and receipts. When markets were buoyant advanced payments with orders were sought, but at times when credit was limited considerable ingenuity was necessary to prevent short-falls of funds. The larger, progressive manufacturers undoubtedly benefited from more frequent receipt of payments, and their greater purchasing power allowed them to delay raw material payments. The 'craft' manufacturers, however, were less well placed because of longer periods between payments for their small-batch products, and their propensity to deal with 'doubtful debt' customers.

Until the 1860s, payments were in the form of bills, the credit for which was redeemable after a specified time and which sometimes incurred interest. Manufacturers' working capital therefore made provision for the redeemable periods, both for supplier payments and customer receipts. Cash flow projections were occasionally undertaken, particularly when trading conditions were tight and problems anticipated. Robert Stephenson & Co.'s head clerk, for example, prepared statements for the partners to assist their tactical decision making. A statement prepared for early 1830, when the company was embarking on its first multiple locomotive orders, showed expected liabilities for loan interest, bills payable, tradesmen, wages, salaries and a £500 dividend. To meet these liabilities, the credit entries showed bills and cash, 'Capital from Partners' (loan capital), and book debts. A shortfall of £2,246 was shown to be 'wanted'.[27]

Insufficient working capital frequently led to cash flow problems, particularly with large locomotive orders. As early as 1832, R.&W. Hawthorn, when manufacturing six locomotives for the Stockton & Darlington Railway, had to write to its customer that 'our pecuniary source…is rather low' and asked for 'a remittance either in Cash or Bills on London for One Thousand on account…'[28] The Stephenson Company also withstood long periods of cash shortage, which were met by borrowings from its partners, repayable with interest. In July 1833, the company faced up to an accumulation of bad debts and wrote off nearly £1,000 from the balance sheet.[29]

With the buoyant market of 1835-38, however, the Stephenson Company sought advance payments when orders were placed. Its head clerk wrote in 1837:[30]

> We have been very well off for some time past in money matters and at present have a Balance in hand in cash & Bills of nearly £5000. We are moreover looking for large remittances….. and may 'ere long look for the confirmation of a contract upon which we are to receive an advance of £4000…. Now all this will not only make us quite easy, but I anticipate will enable us to pay off our borrowed Capital which will of course be very desirable as we shall thereby save the Interest. Upon these advances we pay no Interest.

Towards the end of the 1843 recession, during which advance payments had ceased, the Stephenson Co.'s cash position had reduced, the head clerk reporting that: 'We have paid off all the money we had upon the advance account & therefore are as it were entirely dependent upon our resources.'[31] The situation was soon reversed with the extraordinary demand for locomotives from 1844, and advanced payments were again sought. By June 1846, the Company had £10,000 in hand through advanced payments.[32]

Smaller manufacturers, such as Jones & Potts, were particularly aware of money markets and the relative merits of cash and bill payments. Even during the boom year of 1846, Arthur Potts wrote to his partner about a customer payment by 'bank bill for £500 and you may draw upon them at 2 months for the remainder adding Intt [interest],

2 months only, I should do this if I was you at once… The money market is in an awful state. Would it not be better to have cash for the whole…'.[33]

The 1847-50 recession caused severe liquidity problems for the manufacturers due to the poor financial position of their railway customers, leading to the failure of some companies. Potts' correspondence reveals some of the extraordinary payment problems that arose, and as early as August 1847 he wrote that 'money is indeed very tight here [London].'[34] In the following month he wrote:[35]

> …I thought I had better get this money matter settled at once – London is in a fearful state people do not know who to trust… Mr. Rankine [customer] informed me that he had sent to the works a 6 months promisory note for the amount of our account. Should he have done so please return it as any Bill at 6 months however well Backed is worth in London no more than waste paper.

By the summer of 1848, the company's cash flow was very poor, and Potts was obliged to press for payment from several railways, including the Caledonian, whose Edinburgh headquarters he visited several times. After a frustrating day he wrote that he had:[36]

> …waited upon the Caledonian Board to day for fully 2 hours and we have not succeeded in procuring an interview we have been put off until tomorrow If we are not seen tomorrow D---n em we will pitch into them I'm in a d---l of a rage at them.

Fig. 7.9 Example of a locomotive supplied at a time of severe cash-flow problems. A 2-2-2 built by Jones & Potts of Newton-le-Willows in 1848 for the Scottish Central Railway, later absorbed into the Caledonian Railway as No. 322. *Author's collection.*

By 1849, some railways were unable to pay their creditors. The Scottish Central Railway had 'no money or Bonds' and could only pay Jones & Potts by 'selling 5 engines to the Aberdeen Railway and on receipt of this money they will pay us our debt –'[37] Even the London & North Western Railway found difficulty meeting its payments, and having extracted a promise to pay, Potts reflected '…there is so many slips between the cup & the lips that I shall not believe I shall get it until I have got it.'[38]

Negotiations over discount arrangements for bills of exchange could provide welcome additional revenue. When the London & Blackwall Railway paid £3,000 for its locomotives with a three month bill, Jones & Potts sought to find the best exchange terms from the discount banks. Potts wrote to Jones:[39]

> I wish you to… ask them [Messrs. Parr & Co.] what they will discount the Blackwall Cos. Bill… Curries have offered to discount it at £5 per cent – without commission Dont mention this to Parr as if they charge the ¼ per cent and £5 besides I shall get Curries to discount it –

A favourite form of extending credit for railways with cash problems was to make payment in their own debenture bonds which could be subsequently exchanged, albeit being discounted each time. A payment dispute with the Eastern Counties Railway saw Jones & Potts offered payment in this form, the railway's Secretary, Roney, suggesting to Potts that he should make a proposition to the Directors:[40]

> That provided the Company pays us the Int[erest] we will take the Debenture Bonds – (which Roney promises to dispose of for us at £98 for each £100 –) this interest will more than cover the Loss – by the Bonds being at a discount of £2 per £100… Should they accept my offer of taking their Bonds, (provided they allow us this Intt.) we shall be a gainer of £100 or more and we shall get the money within one week –

If the shares and bonds could not be immediately exchanged, they were divided out as a dividend between the partners and were shown as a double entry in the manufacturers' balance sheets. Beyer Peacock & Co., for example, occasionally received preference share payments both for home and London-based overseas railways from the 1850s, which were shown in its balance sheets.[41]

Following the failure of the Overend Gurney Discount Bank in 1866, the British banking system was radically changed. Credit was sought from, and approved by, commercial banks, and direct payments were made by cheques rather than bills. The manufacturers were as much to benefit from the new arrangements as their customers. Some came under considerable financial strain following the 1871 strike and the sharp increase in commodity prices in the early 1870s. R.&W. Hawthorn was obliged to borrow heavily from its bankers to maintain sufficient working capital, and, by 1876, its overdrafts had risen to nearly £44,000.[42]

Whilst the larger main line railways were usually credit-worthy, other main line and industrial customers were not always so, and there was often concern regarding their ability to pay on time, or at all. Delays were costly for the manufacturers but, in a tight market, there was a reluctance to drive away customers. Awaiting an outstanding payment from the Alexandra Dock Co. in 1890, the Hawthorn Leslie company feared that '[if we] bully them they won't come to us for any more locos…'[43] Part exchange with old locomotives was occasionally accepted, usually requiring an independent valuation, if only to establish a scrap value. Hawthorn Leslie & Co. Ltd., for example, sold a small tank locomotive in 1896 for £700 plus an old locomotive valued at £200.[44] In 1897, Benjamin Browne warned that a quarter of Hawthorn Leslie's 'doubtful locomotive customers' would in the end have to be written off as bad debts.

Some manufacturers went into liquidation in the worst recessions and down-turns in the locomotive market during the century, quite possibly through poor judgement and financial naivety by their proprietors. Conversely, the majority of manufacturers survived the worst recessions, which is indicative not only of the effectiveness of their diversification policies, but also of close attention to payment timing and adequacy of working capital.

Financial Accounting

Although there was no statutory requirement to prepare financial accounts during the century, manufacturers prepared them, both for their own internal knowledge and as the means to satisfy their backers of a continuing return on their investments. Decisions taken on investments, and the awarding of dividends, depended on accurate balance

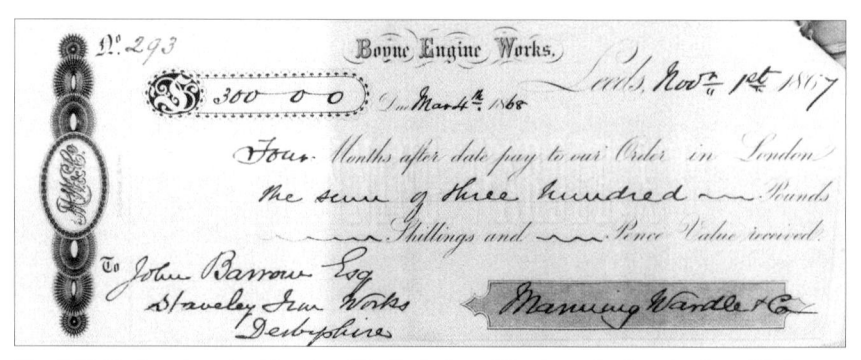

Fig. 7.10 Draft written by Manning Wardle & Co. of Leeds to the Staveley Iron Works on November 1st 1867 due to be credited four months later. Payment for material supply in respect of Works No. (MW 238/67) for the Manchester & Milford Railway.
National Archives, RAIL 456/24.

Fig. 7.11 A manufacturer with ongoing 'doubtful debt' problems. R.&W. Hawthorn Leslie & Co. 4-4-0 locomotive built in 1884 (HL 2000/84) for the North Eastern Railway (No.1496).
R.H. Bleasdale Collection – National Railway Museum.

sheets and profit and loss statements, the preparation for which required supporting ledgers and periodic valuations. Manufacturers were equivocal in applying depreciation policies, which had become more widely practised after 1800, particularly in the textile industries.[45] After 1830, significant developments in the treatment of depreciation occurred in the railway industry, and some, at least, of the locomotive manufacturers introduced it from that time. Their treatment of depreciation indicates how they developed their capital investment programmes, whilst accommodating fluctuating incomes from the cyclical trading patterns.

There were model Articles of Association shown in the 1856 Act, re-iterated in the 1862 Companies Act, which may have encouraged companies to prepare comparable accounts, but there was no requirement to adopt a depreciation policy. The biggest variant in accounting practice during the century was asset depreciation,[46] and 'secret reserve' accounting was prevalent before 1900.[47] Secret reserve accounting was an expedient used by

Fig. 7.12 0-6-0ST built in 1875 by Nasmyth Wilson of Patricroft (NW 173/75) for the Rhymney Railway (No. 38). *Peter Wardle collection.*

firms to depreciate quickly, or by under-valuing assets during prosperous trading years, and making little or no allowance during less prosperous years. It was a common practice to reverse depreciation by altering asset values to cover up trading losses. Exaggeration of liabilities through undisclosed transfers not entered on balance sheets was a further method of creating secret reserves as a hedge against poor trading years.[48] There was no debate on the subject of undisclosed reserves until 1895, and in 1899 the '*Accountant*' was reporting the belief among Manchester accountants that secret reserve accounting was widespread among manufacturing companies.[49]

The best set of surviving financial accounts for a locomotive manufacturer is a discontinuous set for Robert Stephenson & Co. between 1824 and 1855.[50] As with its cost accounts, its balance sheet and profit and loss accounts were initiated under the experienced supervision of its partner, Michael Longridge. The balance sheets provided the partners with a generally good presentation of assets and liabilities at the end of each calendar year. When working capital was insufficient, loan capital, mostly provided by the partners themselves, was shown as a liability, usually at 5% interest. Asset valuation was maintained in a 'stock account' which, by inference, appears to have been an inventory of assets with current values as marked down by depreciation, to which new items were added when acquired, and removed if sold or scrapped. As asset value increased with new buildings and equipment, the partners' capital was increased correspondingly in the balance sheet.

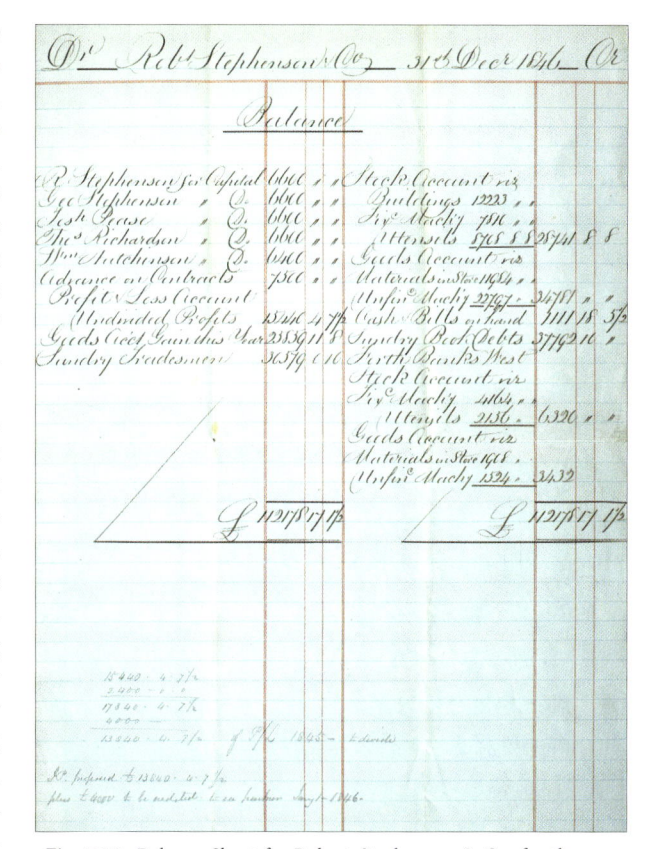

Fig. 7.13 Balance Sheet for Robert Stephenson & Co. for the year ending 31st December 1846. A remarkably healthy statement following the 'mania' for orders in that year. *Pease-Stephenson Collection, Durham County Record Office, Darlington Public Library (D/PS/2/26).*

The statements confirm that depreciation, separately recorded for 'Buildings', 'Fixed Machinery' and 'Utensils', was undertaken throughout those years. In 1829, a 15% (by calculation) depreciation was shown, being a 'catching up' figure to represent reduction in asset value, 'diminished value if sold', over the firm's first six years of operation. A 5% figure was put in for 1830, whilst the 1831 figure was 'about 5 p.cent' for buildings and fixed machinery and 'near 2½ p.cent' for utensils. 'Nothing was shown during the loss-making years, 1833-36, suggesting that secret reserve accounting was being practised, with a 10% depreciation in 1838/39 representing a measure of 'catching up'. 'However, from 1840 an annual 5% depreciation was adopted, regardless of profit or loss levels, and thus without recourse to reserve accounting.

The Company's Head Clerk explained the depreciation entry in the 1845 balance sheet:[51]

> I may state that the amount shewn as the value of <u>Buildings Fixed Machinery</u> &c is the <u>net</u> amount after deducting 5 per cent from the former years stock for depreciation – This deduction is regularly made every year – The amount of '<u>Utensils</u>' viz Smiths' Fire-places, vices, moveable tools such as Files, Chisels &c &c appears heavier than last year partly in consequence of additions to the stock & partly because the Inventory has been very carefully made out so as to include everything on the premises without the least omission, which has not been done for a year or two past. The valuation is according to Mr. Hutchinson's Estimate of the selling price of the various materials.

In the absence of further balance sheets, with the exception of 1855 which makes no reference to depreciation, it is not possible to confirm the continuation of this policy.

In contrast, Beyer Peacock & Co. made no allowance for depreciation until as late as the 1870s, after nearly twenty years of manufacture.[52] The firm's accountants tried to introduce depreciation in 1869 and again in 1872, but for unexplained reasons, the partners would not adopt the practice. Capital assets remained on the balance sheet at their full purchase value, or scrap price, credited to the books. From 1878, however, after the death of Charles Beyer, an annual 5% depreciation was introduced, with 10% for the shorter-life shop boilers. Even then, the half-yearly returns still included the original equipment value up to 1888, when Henry Robertson died, after which annual depreciation figures were reported.

The few surviving profit and loss accounts for Robert Stephenson & Co. in the 1830s/1840s show them to be adjuncts to balance sheets rather than trading records, with summaries of financial movements over twelve month periods.[53]

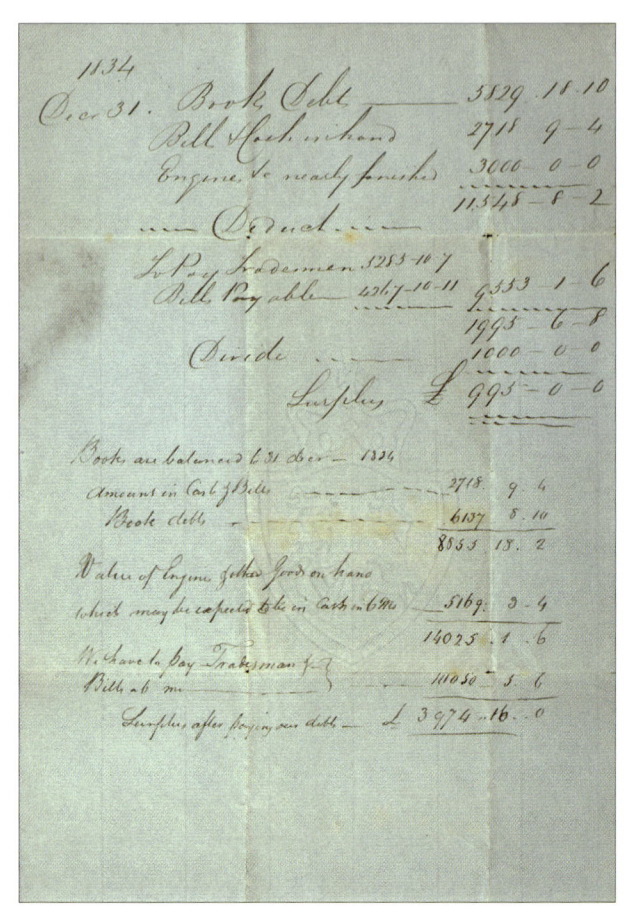

Fig. 7.14 Profit & Loss statement for Robert Stephenson & Co. for 1834. *Pease-Stephenson Collection, Durham County Record Office, Darlington Public Library, D/PS/2.*

Entries related to liabilities, such as increases in capital holdings, dividend payments, asset depreciation and interest on loan capital and borrowings, which were balanced by credit movements. Commencing with the balance from the previous account, entries were made for the net profit from the 'Goods Account' (manufacturing profit) and 'undivided profits' (net profits after dividend payments). The true summaries of operating profits were encompassed in the 'Goods Accounts' and other 'profit centre' accounts, few of which have survived. For the year ending December 1834 (Fig. 7.14), debit entries represented opening stock value, total amounts paid for materials and wages and a derived surplus for the year-end. The two credit entries show the value of 'sales' and the closing valuation of the stock.

R.&W. Hawthorn's surviving profit and loss accounts for 1864-68 show them to be more comprehensive by those years and, being an annual statement rather than a ledger, were more akin to latter-day profit and loss statements.[54] Net profit (loss) entries were provided for locomotives and tenders, marine and stationary engines, and other orders (duplicates and a wide variety of general manufactured items), with separate profit or loss entries for the brass foundry, forge and 'tools' (machine shop), as well as entries for discounts and charges.

Although manufacturers developed their financial accounts in the absence of any accounting conventions, the surviving examples demonstrate an impressive level of detail. From these it is clear that from the outset, they maintained a good understanding of their financial position and could demonstrate to their backers the current state of their business and the returns that were being made. Although there is no evidence of business plans or any form of projections being made based on these fundamental statements, it is likely that they provided financial 'comfort' for raising capital and loans for investment and expansion and, from the 1860s, for negotiations leading to incorporation.

Diversification

Throughout the century, proprietors were faced with uncertain business prospects arising from the unpredictable nature of the manufacturing markets, and the locomotive market in particular. All manufacturers, to a greater or lesser degree, hedged against market fluctuations through diversification. In the earlier, craft-based, manufacturing years, this policy worked well and allowed most companies to remain in business. In later years, and based on this early experience, proprietors sought to predict the long-term trend in locomotive demand, whilst re-investing to meet a sustainable and competitive proportion of that market.

Their judgement included divergent expectations of medium-term economic recessions, and their strategic decision-making therefore included a wide range of diversification strategies to accommodate the periodic decline in locomotive orders. In this way they sought to maintain sufficient work to retain their skilled craftsmen, and maintain sufficient income to meet their loan and other short-term financial commitments. The choices, which largely fell on the ingenuity of the craft manufacturers to pursue were:

- to withdraw from the locomotive market temporarily, or on a permanent basis, and revert to, or diversify into, alternative markets,

- to continue to pursue locomotive orders, whilst seeking to use up excess capacity through alternative markets,

- to manufacture locomotives for 'stock' and later sale, in order to pursue batch production and sales benefits when the demand returned,

- to manufacture locomotive components, such as replacement boilers and wheelsets, for locomotive refurbishments, or undertake locomotive refurbishment programmes themselves.

Whilst many firms halted locomotive production during recessions, thirty-nine factories withdrew permanently from locomotive work during the course of the 19th century and concentrated on alternative markets.[55] There is no evidence to indicate how withdrawal decisions were made, but an analysis of their timing provides some indications of the reasoning.

Eleven firms withdrew in the 1839-1843 period when the first serious downturn in orders was experienced. They ranged from Summers Groves & Day, which built just six locomotives before concentrating on marine engineering,[56] to Kirtley & Co., which had promoted itself widely with published drawings of its locomotive designs, but then opted to revert to colliery work.[57] It would seem that the uncertainty of the locomotive market in the early 1840s caused these companies to abandon locomotive manufacture and revert to the markets that they knew better.

A further ten firms withdrew in the post-1847 depression, ranging from Hick Hargreaves & Co., which switched production fully to the textile and colliery industries, to W. J. & J. Garforth, a general manufacturer which had been sub-contracted to make locomotives at the height of the 'mania' order boom. The 'feast or famine' nature of the locomotive market, and the failure of one of the largest manufacturers, Bury Curtis & Kennedy of Liverpool, may well have influenced several firms, including Hick Hargreaves.

The remaining withdrawals, in the second half of the century, cannot be related to poor or uncertain market conditions, suggesting that the proprietors rejected the investment requirements for competitive locomotive production in favour of their alternative markets. Both the Canada Works in Birkenhead and the Teesside Engine Works, for example, concentrated largely on bridge-building as their main activity, allowing them to preserve their craft manufacturing base, and avoiding investment in capital equipment for competitive locomotive production.[58] The recession of 1848-1851 prompted Robert Stephenson & Co. to re-deploy part of its capacity to the manufacture of marine engines, as well as revert to industrial engine manufacture, although evidence towards the policy decision is lacking.[59] Although the locomotive market recovered in the 1850s, the suspicion about the effects of a further recession maintained the perception of the need for a broad market. The Stephenson Company further diversified into the manufacture of steam-plough machinery for John Fowler of Leeds, until his own works began production in 1862.[60]

its production from main-line to industrial locomotives, for which the characteristics of standard designs with interchangeable components were more closely allied to its main agricultural equipment market.[62]

Beyer Peacock established its factory at Gorton in the 1850s, chiefly for 'The manufacture of locomotive engines' but, together with several other manufacturers, it also made a range of machine tools, to supplement its business, particularly at times of low locomotive demand.[63] In its early years, Beyer Peacock supplied machine tools for maintenance workshops for railways in Sweden, Austria and Spain.

Beyer Peacock & Co., Kitson & Co., Henry Hughes & Co. and several other, smaller, locomotive manufacturers diversified into the extensive urban steam tram market from the 1870s, which helped to offset the downturn in locomotive demand in both the late 1870s and the late 1880s.

Fig. 7.15 Entry in catalogue of Steam Cultivating Machinery, 1862, showing John Fowler's patent windlass equipment sub-contracted to Robert Stephenson & Co.
The Museum of English Rural Life, University of Reading.

Ingenuity was required by some manufacturers to maintain work for their workshops. For R.&W. Hawthorn, 1857 was 'a time of very great industrial depression… all kinds of orders had to be picked up… a machine for cutting tobacco, some flour grinding machinery and even cooling apparatus for Allsopp's Brewery…'[61]

Industrial locomotive manufacture was usually closely allied to mining or iron-works machinery, its prosperity often rising and falling according to the health of those industries. John Fowler & Co. decided in 1875 to switch

Fig. 7.17 Steam tram built by Kitson & Co. in 1885 for the Dudley, Sedgley & Wolverhampton Tramways. *The Locomotive Railway Carriage & Wagon Review, Vol.41 (1935), p.367.*

The need for diversification to offset the irregularity of orders was also experienced by overseas manufacturers. The French Fives-Lille Company, for example, diversified into hydraulic lifting and handling equipment and eventually into armaments.[64] American manufacturers were similarly obliged to diversify, alternative markets including steam fire-engines, marine engines and bridge-building.[65] The Baldwin Locomotive Works, however, limited its diversification to other forms of motive power, notably in the industrial and 'street-car' sectors, in order to maximise the use of 'standard' components.[66]

During recessions, a few firms took the risk of making main-line locomotives, of their 'standard' designs, for 'stock' and subsequent sale, whilst others avoided this policy, aware of the dangers of excessive working capital

Fig. 7.16 Specialised manufacture for the industrial locomotive market. John Fowler & Co. 1876-built 0-4-0ST for the Nitrate Railway's fleet in Chile (No. 5). *The Locomotive Railway Carriage and Wagon Review, Vol. 38 (1932), pp.86/7.*

requirements risking their viability. Nasmyths Gaskell & Co., for example, made three locomotives for stock in 1840 but, in a poor market, was obliged to sell them for much below their expected price.[67] From the 1850s, stock locomotives became less acceptable to the larger railways, who increasingly favoured their own specifications. Some had to be sold at lower prices, in order for the manufacturers to cut their losses.

E.B.Wilson's policy of standardised locomotive production in the 1850s, pursuing the cost benefits of batch production, encouraged excessive stock production which placed severe financial strains on the company. In the last three years of its trading, 1855-57, the firm built seventy-nine locomotives for which it had no immediate customers.[68] Wilson fell out with his non-executive partners, Pollard and Turner, almost certainly over this policy, the resulting litigation forcing the winding up of the company.

Fig. 7.18 A company undertaking locomotive building for stock. E.B. Wilson & Co. of Leeds. 0-6-0 tender locomotive (EBW 387/54) built for the North Eastern Railway in 1854 (No. 231).
R.H. Bleasdale collection – National Railway Museum.

Although the larger industrial locomotive manufacturers, such as the Hunslet Engine Co., Hudswell Clarke & Co. and Manning Wardle & Co., could better anticipate subsequent orders for their standard designs to risk making locomotives for stock, main-line manufacturers generally resisted such a move, perhaps in the knowledge of what had happened to E.B.Wilson & Co. Between 1886 and 1898, however, Hawthorn Leslie maintained production at its Forth Banks works by making industrial locomotives for stock, eighty-eight out of their total of 227 being built without orders. This policy put financial strains on its whole operation, and differing opinions within the Board of Directors were evident when, in 1890, the Chairman, Benjamin Browne, retorted: 'No doubt it is the least disgusting part of the locomotive trade but it is not much good.'[69]

Although William Cross, the Works Managing Director, feared the Board would consider 'a proposition… as to the desirability of altogether closing the works,' it decided instead to modernise them to reduce production costs.[70] The Company returned to manufacturing for stock in the early 1890s recession, however, a policy which again forced it to consider closing the Forth Banks works and writing off its investment. In 1897, Browne saw the loss as being 'so heavy a drag on the Company' and argued that 'the value of the Company's property would be materially increased if the locomotive business was abandoned.'[71] The works were rescued from closure by the extraordinary demand for locomotives from 1898.

There was a substantial locomotive refurbishment market, particularly during economic recessions, which provided a cheaper alternative to new locomotive orders for both railways and industrial customers. Because of the small order sizes and sporadic timing, it was more attractive to the 'craft' than the 'progressive' manufacturers. There is no record, for example, of the large batch-production companies, such as Neilson & Co. and Dübs & Co., undertaking this work, whereas Robert Stephenson & Co. Ltd. and R.&W. Hawthorn Leslie & Co. Ltd. frequently competed for it in the 1880s and 1890s.

The Stephenson Company's refurbishment market replaced its profitable 'duplicates' market, which declined from the 1850s as component reliability improved. The increase in the Company's refurbishment work, compared with its reducing locomotive market after 1880, is illustrated in Fig. 7.20.[72] Whilst the refurbishment work was largely for the smaller British railway and industrial companies, replacement boilers were also made for overseas customers.

Fig. 7.19 A company building locomotives for 'stock'. Hawthorn Leslie & Co. An 0-6-0ST built in 1886 (HL 2147/86) for the Alexander (Newport & South Wales) Docks & Railway (No. 17). *The Locomotive Railway Carriage & Wagon Review, Vol. 33 (1927), pp.154/5.*

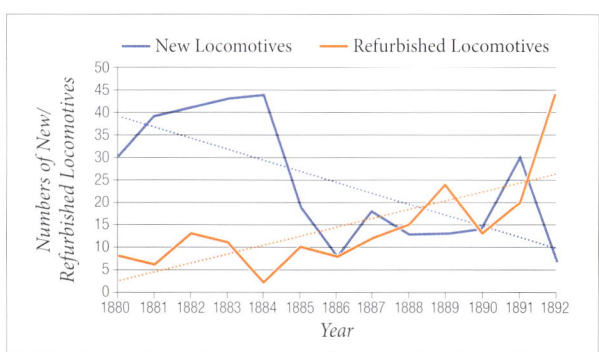

Fig. 7.20 Comparison of new locomotive manufacture to refurbishment locomotives by R. Stephenson & Co. between 1880 & 1892. *Analysis of Engines Delivered Books, Robert Stephenson & Co. Collection ROB/2/3.*

Hawthorn Leslie made modest profits on its locomotive refurbishment business (as well as its general manufacturing business and marine work) which offset its continuing locomotive manufacturing losses.[73]

Investment and Company Status

Partnership enterprises benefited from the combined expertise of both executive and non-executive partners. Their ability to interpret the potential of the locomotive market was built on their previous manufacturing experience, and their entrepreneurial flair had a capacity to persuade potential backers that attractive returns on investment could be made from locomotive factories. The ability of the early proprietors to take the decisions necessary for the development of their manufacturing businesses is demonstrated by the fact that there are no recorded examples of consultants being brought in, as was necessary, for example, for the iron industry.[74] Proprietors could easily demonstrate their capacity to supervise the preparation of cost and financial accounts and to monitor the profitability of their enterprises, as well as take the tactical decisions necessary to maintain that profitability.

The growth of the locomotive industry from the 1840s brought with it the potential for expansion, on existing or second sites, the further capital requirements for which benefited from the potential for incorporation from the 1860s. The decisions made by the proprietors in pursuing this potential were diverse and reflected a wide degree of willingness to share their enterprises with non-executive directors. The diversity was regardless of the size of firms or their depth of involvement in locomotive work; early incorporation of some small firms contrasted to the status of other, larger firms, which remained as partnerships throughout the century. There was a growing divergence within the industry from the 1860s, through the decisions of the second generation of proprietors, between the 'progressive' manufacturers, which had the confidence to pursue increased locomotive production, and the more cautious approach of the 'craft' manufacturers, which maintained a broad market base.

The will to remain independent, let alone to survive, was very strong. The reluctance of some proprietors to incorporate, caused by their desire to remain in absolute control of their firms, applied equally to the potential for amalgamations. There were no amalgamations until the creation of the North British Locomotive Company in 1903, even though there would have been economies of scale benefits. Yet, in all too many cases, the will to survive was insufficient on its own to prevent company failures.

Although partners usually approached their ventures with entrepreneurial as well as technical flair, partnerships frequently changed according to personal circumstances, and the partners' ability to attract loan capital and sustain and increase their financial holding. The Airedale Foundry in Leeds is an example, where, in 1863, Kitson & Co. became the fourth partnership to own the works in twenty-four years. Where a proprietor retired or felt unable to raise sufficient capital for further investment, he would sell out to other proprietors with adequate financial backing, with the assurance of continuity of work for all, or most of the factory personnel.

Other firms, however, were much more stable. Charles Beyer and Richard Peacock, of Beyer Peacock & Co., for example, retained their involvement and shareholding until they died, and then passed on their interests to other members of their family.[75]

> MR. RALPH COULTHARD, begs to intimate to his Patrons and the Public generally, that he has RETIRED from the Business which he has hitherto carried on at Gateshead-upon-Tyne, as an Engine Builder, &c., and takes this opportunity of thanking them for the past favours.
>
> The Business will henceforth be carried on upon the same premises by MESSRS. BLACK, HAWTHORN, & CO., and, from his personal knowledge, he can with confidence recommend his Successors to their favourable consideration.
>
> MESSRS. BLACK, HAWTHORN, & CO., beg respectfully to refer their Friends and the Public to the above Notice, and to inform them that they have purchased the Plant, Premises, and Good-Will, and made arrangements for carrying on the Business so long successfully conducted by Mr. R. Coulthard, at Gateshead, and to assure them that it shall be their endeavour to execute all orders entrusted to them in such a manner as to merit and ensure a continuance of the support which has been so liberally bestowed upon their predecessors.
>
> LOCOMOTIVE, MARINE, AND STATIONARY ENGINE WORKS, Gateshead-on-Tyne, 12th July, 1865.

Fig. 7.21 An example of a factory sale to new proprietors. Newspaper advertisement announcing the transfer of the Quarry Field Ironworks in Gateshead from Ralph Coulthard & Co. to Black Hawthorn & Co. in 1865. *Gateshead Observer, 22nd July 1865.*

The raising of investment capital today requires rigorous assessment through business plans, requiring considered predictions of expenditure, income and cash-flow. Although in setting up and expanding their sites, and renewing capital equipment, 19th century proprietors usually required to attract equity or loan capital, there is no evidence of even basic business plans being prepared to demonstrate what sort of returns they expected. The capital appears to have been attracted largely through the technical capabilities of the proprietors, their powers of persuasion and, where they were already in business, their 'track record'.

No consistent pattern of capital funding can be discerned. The Manchester-based Birley family business, for example, provided the Nasmyth brothers and Gaskell with loan capital to build and equip the Bridgewater Foundry in 1836.[76] The most obvious reason for this investment was their own perception that the future looked healthy for the manufacturing industry in the 1830s, and that James Nasmyth had shown business acumen with his first small workshop in Manchester. Nasmyth who, together with his fellow partners, was under thirty years of age, was described as having 'excellent sense and experience', possessing 'genius in a variety of ways', and uniting 'force of character' with 'sharp hard cleverness'.[77] Although the factory's early years were profitable, for unknown reasons the Birleys withdrew their investment after just two years, but, again, Nasmyth had no difficulty in attracting replacement capital. It is most likely that other early manufacturers also relied heavily on personal reputations to raise their requisite capital.

The risks associated with workshop investment were brought home to investors in 1847, when E.B. Wilson & Co. built perhaps the largest and best equipped workshops thus far provided for locomotive manufacture. Wilson announced that 'no expense has been spared to obtain the newest machinery of the day' when he 'opened' them with a major dinner in December 1847, but it was just as the country's recession was commencing and locomotive orders had dropped substantially.[78]

Provision of additional capital for expansion of manufacturing capacity and new equipment, was addressed in different ways by the proprietors. Whilst some firms sought loan capital, others acquired additional equity, such as Sharp Brothers who in 1852 brought in Charles Stewart to provide additional capital as well as expertise. Other firms, such as Beyer Peacock & Co., brought in non-executive partners. When it was formed in 1854, both Beyer and Peacock held £10,000 in shares, funded from their own resources, including loans.[79] The third, non-executive, partner, Henry Robertson, also invested £10,000 having been convinced of the credentials of Beyer and Peacock by the contractor, Thomas Brassey. The Partnership Memorandum determined that further investment up to £90,000 would be met out of profits, thus delaying the first dividends, but avoiding any dependency on external investment or loans.

Partnerships could raise significant sums of capital for major expansion programmes. The establishment of purpose-designed locomotive factories by E.B. Wilson and Beyer Peacock, encouraged other 'progressive' firms to follow suit. Walter Neilson and Henry Dübs, of Neilson & Co., and later James Reid of Dübs & Co., commanded particular respect in Glasgow banking circles that enabled them to raise sufficient capital to fund their several expansion programmes. In America, similarly, the Baldwin Locomotive Works had fourteen different partnerships before becoming an incorporated company in 1909, each able to encourage sufficient capital for the factory's several expansion programmes.[80]

Fig. 7.23 Product of a company able to raise capital through demonstrable managerial and technical expertise – Dübs & Co. of Glasgow. A 4-4-0 passenger locomotive built in 1881 (D 1560/81) for the Intercolonial Railway of Canada (No. 4). *Author's collection.*

However, as the size of factories and the capital value of their equipment increased from the 1860s, it became increasingly difficult to stimulate sufficient partnership equity, and other firms became limited companies in order to raise the requisite capital. Other firms, which remained as largely craft-based manufacturing organisations, found it increasingly difficult to raise capital. In 1870, the new proprietors of R.&W. Hawthorn acquired the firm's Newcastle premises and goodwill for £60,000 but achieved this only through substantial loans from family and business contacts, and a £25,000 mortgage from William Hawthorn.[81] Again, there is no evidence of a business plan, but the Chairman, Benjamin Browne, observed that 'the Company was strengthened' by the 'extraordinary ability' of the Company Secretary, J.H. Ridley, 'to make accurate forecasts.'[82]

Some proprietorial firms maintained their family status, but the succession from one generation to the next was not always accomplished. Pioneering attributes were not necessarily passed on or, if they were, they proved unequal to the changing business environment. Some proprietors put more emphasis on their sons acquiring engineering and production experience than they did on business matters, which was largely to be gained when they were promoted to senior positions. John Hawthorn Kitson, for example, served an apprenticeship at Kitsons' Airedale Foundry,

Fig. 7.22 A factory established with financial input from a non-executive director. Beyer Peacock & Co. A backdrop to a 2-4-0 passenger locomotive built in 1859 (BP 127/59) for the Swedish Government Railway, No. 8 *STOCKHOLM*.
The Locomotive Magazine, Vol. 54 (1948), p.56.

and was promoted through several departments before becoming a partner.[83] In the next generation, Edwin Kitson Clark, was also apprenticed at the Airedale Foundry and was promoted through several senior posts, before his partnership appointment in the twentieth century.[84]

Several companies considered establishing a second manufacturing site. As early as 1830, the threat from the Lancashire manufacturers of textile, mining and marine equipment, notably Edward Bury & Co., to diversify into locomotive manufacture following the opening of the Liverpool & Manchester Railway, prompted Robert Stephenson to consider establishing a second factory in Liverpool.[85] Without a business plan, but stressing the geographical disadvantage of the Newcastle factory, he discussed the proposal with his fellow partners of Robert Stephenson & Co. They were opposed to his proposal so soon after the difficult financial years of the 1820s, when closure of the Newcastle factory had been considered. Their views were summed up by Michael Longridge:[86]

> The establishment you contemplate at Liverpool appears to me fraught with injury to Forth Street: the chief employment of which will be transferred to your new establishment. How then are you ever to pay the Dividends and recover the Capital?

By March 1831, Longridge was sufficiently convinced about the competition to concede: 'It is certain the demand for Engines will be much greater than we can supply at Newcastle, and if we can keep that place fully employed, we shall in a few years receive back our Money at any rate.'[87] The partners compromised by allowing Robert Stephenson to enter another partnership in order to establish the new factory, but he was obliged to pledge that:[88]

> Should I become connected with another Manufactory for building Engines in Lancashire or elsewhere I have no objections to bind myself to devote an equal share of my time and attention to the existing establishment at Newcastle. I will also pledge myself <u>not to hold a larger interest in any other factory</u>, than I have in Forth Street and <u>to divide the Locomotive Engine orders equally.</u>

George and Robert Stephenson formed a partnership with Charles Tayleur, a Liverpool businessman and Director of the Liverpool & Manchester and Grand Junction Railways, and his son, also Charles Tayleur. The Stephenson company partners agreed 'That the firm at Liverpool shall be Charles Tayleur, Junr. & Co. or any other Firm not embracing the name of 'Stephenson' so as to distinguish it entirely from the Newcastle House.'[89] Locomotive orders were equally divided between Newcastle and the Vulcan Foundry, which began production in 1834,[90] but by the end of 1835, the agreement ended and the Stephensons withdrew from the partnership.[91] The cause of their withdrawal cannot be confirmed, but it probably arose from the disagreements over early locomotive specifications, from which the Tayleurs concluded that their investment would be better protected without the Stephensons.[92]

In the peak market conditions from 1844, the Stephenson Company found itself with a severe capacity constraint and lengthening delivery times. With the experience of previous demand cycles, it opted to take a short (seven-year) lease on a site half a mile from its main premises. The 'West' factory required significant new capital as observed by the firm's partner, Edward Pease:[93]

> the increased number of Engines you are laying yourselves out for completing next year, will no doubt keep your capital near its tension yet as you have some small advance from purchasers I cannot think our capital will cause account below respectable – If a small number of the contemplated RWays only be sanctioned by the ensuing Parliament, what a number of Loco factorys [sic] may be expected to spring up & many a sorry engine be made!

R. & W. Hawthorn was similarly inundated with orders, and relieved the problem by acquiring and adapting a works at Leith as a locomotive erecting shop, to which components made at its Newcastle works could be shipped and erected for delivery to its several Scottish customers.[94] By 1850, the downturn in locomotive demand made the plant superfluous, and the Hawthorn family sold two thirds of its holding to local land and marine engineering interests.[95] A new company, Hawthorns & Co., enlarged the site into a locomotive factory which, for twenty-two years years manufactured industrial locomotives. Shortly after the sale of his Newcastle factory, William Hawthorn also sold his interests in the Leith Works, after which it concentrated on its main marine engineering work.[96]

Some manufacturers believed there would be financial

Fig. 7.24 The earliest second factory for locomotive manufacture. Invoice form for Charles Tayleur & Co.'s Vulcan Foundry showing its new factory opened in 1834. *Author's collection.*

advantages through acquiring their main suppliers of iron, although there is no evidence to indicate whether those advantages were realised. In 1854, the Kitson family acquired Whetham's Forge in Holbeck (Leeds), which they enlarged and renamed the Monk Bridge Iron Works.[97] The ironworks supplied Kitsons' Airedale Foundry with iron plate, bars and sections, as well as forgings. The Monk Bridge Company became a large supplier of iron, principally for 'railway material', and diversified into steel production from the 1860s.

At around the same time, Neilson & Co. acquired the Summerlee Ironworks in Coatbridge, although it is not clear for how long. It was unlikely to have acquired it prior to its move to Springburn in 1862, and its first attempt to sell the site in 1873 suggests that its involvement was of short duration.[98] It is possible that the switch to steel would have required substantial investment, which Neilsons were, perhaps, unwilling to make; a limited liability company hoping to acquire the site in 1873 was seeking a capital of £400,000. In 1872, Andrew Barclay Sons & Co. set up the North British Iron Hematite Co., but this was a financial disaster which brought both firms into sequestration within two years.[99]

Overseas Ventures

As the overseas locomotive market expanded, some manufacturers considered setting up subordinate factories in other countries. The earliest was by the Fairbairn family, who 'issued prospectuses of an establishment' at Malines in Belgium in 1839: 'Mr. William Fairbairn is to superintend the heavy department for engines, locomotives, &c, and Mr. Peter Fairbain that for spinning machinery.'[100] Malines was the centre of the new Belgian State railway system, suggesting that locomotives were to have formed a large part of the business. The factory would have competed with Cockerill's large works near Liège, but there is no further record of the Fairbairn proposal.

Also in 1839, Robert Stephenson was said to have agreed with the French engineer, Paulin Talabot, to set up a locomotive factory 'somewhere in France'.[101] There is no evidence that Stephenson proceeded with this idea, indeed, the division of the 1844 locomotive order for the Chemin der fer Marseille à Avignon (Marseilles-Avignon Railway) between the Newcastle factory and the Benet Company would seem to confirm this. Following its acquisition of the Leith Works to accommodate the large locomotive demand in the mid-1840s, further erecting facilities were found by R.&W. Hawthorn through association with the German firm of Lindheim, whose factory was at Ullesdorf in Silesia.[102] Although components for three locomotives for Silesia were sent to Ullesdorf, there is no evidence of a lasting association.

Two overseas factories were built with British capital, neither being associated with a British-based works. In 1842, the Sotteville Works in Rouen were built and operated by William Buddicom, to supply locomotives for the Paris-Rouen and later French railways.[103] The second was established in Montreal in 1853, to make locomotives for the Grand Trunk Railway, by the two Kinmond brothers who had withdrawn from the Kinmond Hutton & Steel partnership of Dundee at the time of low locomotive demand in 1850. Despite a promising start, the venture failed in 1857.[104]

Immediately after establishing the Gorton Foundry in 1855, Beyer Peacock & Co. considered a design, build and management contract for an engineering works in Vienna, probably responding to an approach from Austrian interests.[105] The company proposed to supply all designs and drawings for the works for a charge of 5% of its outlay, and to supervise its construction for 2½%, as well as supply and install all capital equipment. Beyer Peacock recruited Henry Dübs to manage the Vienna Works for five years, for which it was to receive an unspecified percentage of the turnover. Beyer Peacock was to invest £10,000 in the venture, a substantial sum so soon after establishing the Gorton Foundry, and nominate a director to the Vienna Board, but there is no evidence that they proceeded, and Dübs only remained at Gorton for six months, before taking up his position with Neilson & Co.[106]

When the contracting partnership, Peto Brassey & Betts, won the contract to build and equip the Canadian Grand Trunk Railway, including the Victoria Bridge near Montreal, the three proprietors established their own works. Although they had the option of setting up a factory in Canada, they established the Canada Works in Birkenhead, alongside a deep-water berth.[107] With an urgent requirement, their decision was probably influenced by Brassey's familiarity with Birkenhead, and his ability to build the works, recruit the craftsmen and have it up and running within six months. The works were also available for the manufacture of locomotives for other world markets.

The majority of locomotives built for early Canadian railroads were made by American manufacturers. In the thirty years from 1860, only 200 were exported from Great Britain to Canada, the worst market in the Empire for the British manufacturers.[108] Against this background, Dübs & Co. established, in the mid-1880s, at Kingston, Ontario, the only example of a British locomotive manufacturer opening a subordinate overseas factory.[109] Production lasted until 1900, but it has yet to be determined the arguments for its establishment or the reasons for its termination after the short period.

The lack of diversification into overseas plants is difficult to explain, but it is likely that, with the exception of Canada, the need was perceived not to be required because of the dominance of London in the locomotive market. Also, most manufacturers had experienced the cyclical locomotive market changes from the 1830s onwards, which persuaded them to maintain interests in non-locomotive markets. As these were often of a local or regional nature, for example mining equipment for a local coalfield, the British-based manufacturers may have felt less secure in an overseas market with which they were less familiar.

Incorporation

From the 1860s, the expansion of locomotive works and re-investment in new machinery began to exceed the ability of some proprietors to raise sufficient capital which was perceived to be too great a risk. Following the passing of the 1856 Joint Stock Companies Act, and the consolidating Company Act of 1862, 'private' or 'public' limited liability status opened the way for the introduction of new capital. The Vulcan Foundry Co. was incorporated as a private company in 1864, whilst Sharp Stewart & Co. and the Avonside Engine Co. were the first manufacturers to be floated as public companies, also in 1864. The Yorkshire Engine Co. was started up with public company status in 1865.

Fig. 7.25 One of the first locomotives built by a public limited company, the Yorkshire Engine Co. Ltd. of Sheffield. Delivered in 1867 (YE 2/67) for the Great Northern Railway (No. 268).
The Locomotive Magazine, Vol. 60 (1954), p.10.

Similar limited liability opportunities opened up in Europe, to provide for a significant expansion of locomotive manufacturing capacity. In France, the parent, Schaken Caillet et Cie. partnership was incorporated as the Compagnie de Fives-Lille in 1865, allowing it to increase its capital. Three years later, it became the first of three public locomotive manufacturing 'Sociétés Anonymes', the further injection of capital providing for an increase in production to one locomotive per week.[110] In Germany, similarly, several locomotive manufacturers were incorporated as 'Aktien-Gesellschafts' after 1870, and the increase in capital provided for the major expansion of the country's locomotive industry.[111]

No British firms were incorporated in the depression years of the 1870s, but as confidence returned in the 1880s/1890s, eleven manufacturers became limited companies.[112] These included 'progressive' companies, such as Beyer Peacock & Co., 'craft' companies, such as Hawthorn Leslie & Co., which became a public company in 1886 after only a year as a private company, and industrial locomotive manufacturers, such as Fletcher Jennings and Dick Kerr & Co. Hawthorns' new status, and its amalgamation with the Andrew Leslie shipyard, prompted Robert Stephenson & Co. to acquire its own shipyard the following year, prior to which it also became a private limited company.

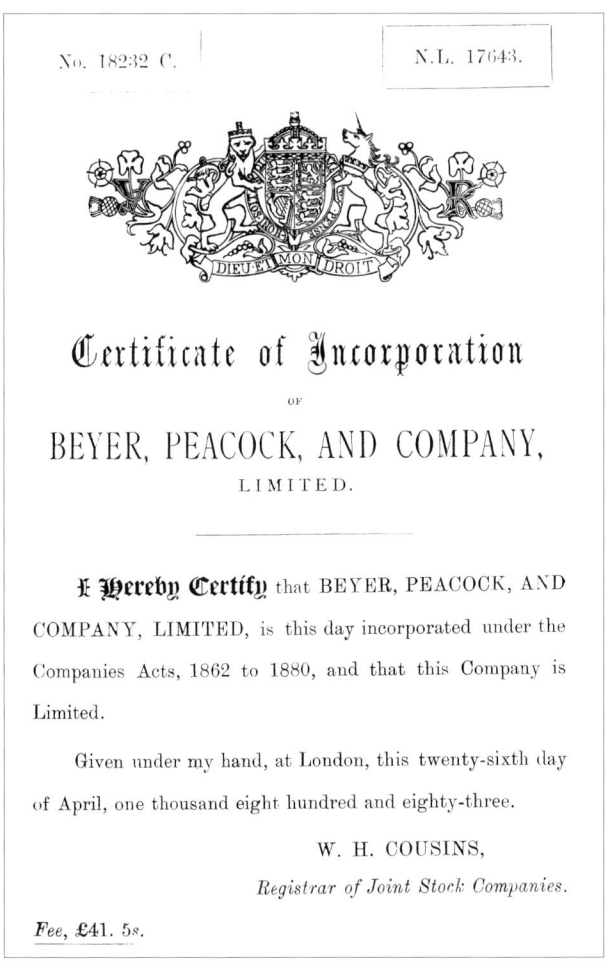

Fig. 7.26 Certificate of Incorporation for Beyer Peacock & Co. Ltd., taken out in April 1883. *Museum of Science & Industry, Manchester.*

In practice, public company status was not always pursued vigorously, and the sharing of strategic decision-making responsibilities with a Board that included non-executive directors caused discontent amongst the longer-serving executive directors. Only a quarter of the Hawthorn Leslie shares were taken up at first, for example, the remainder being retained by the original Board members. The Company Secretary felt that quotation on the London Stock Exchange was not possible because 'our Company is to all intents and purposes a Private one as our Directors hold more than 5/6ths of the Capital and it is not our wish to give the concern a public character for some years to come.'[113] Shortly afterwards, the Company's chairman, Benjamin Browne, criticised his fellow directors when writing about the responsibilities of public companies:[114]

> The principle that nothing should be left permanently to one director is, to my mind, a vital point in any Public Company… Being a Director of a Public Company involves some extra trouble, but this is inevitable and should have been thought of before… No man likes to be told he must not act independently but it is inevitable not only that he be told so but that he should accept the fact and act accordingly…

The extraordinary market events of 1897-1900 also stimulated further incorporation of public companies, the largest of which were Kitson & Co. and, after thirty-three years as a private company, the Vulcan Foundry Co.

Amalgamation
No amalgamations took place between any of the locomotive manufacturers during the 19th century, even though pooling of capital and goodwill would have had clear advantages, and public company amalgamations had been well understood since the first railway mergers in the 1830s. Firms such as Robert Stephenson & Co. and R.&W. Hawthorn, on adjacent sites in Newcastle, and the Hunslet Engine Co., Hudswell Clarke, and Manning Wardle on adjacent sites in Leeds, could have amalgamated with economies of scale advantages. Hawthorn's merger with the Andrew Leslie shipbuilding company in 1885 confirmed the principle of amalgamation in the manufacturing industry, albeit towards horizontal integration of related activities, rather than production economy within one of them.

The lack of interest in amalgamation is difficult to explain in the absence of firm evidence, but it seems to have been due to the tenacious desire of the partners themselves to remain in control of their firms. Even after incorporation, companies were usually dominated by the senior directors, for whom amalgamation would have represented loss of influence or control. Mistrust could also rule out any moves in this direction. There had been a long-standing antagonism between the Stephenson and Hawthorn families, for example, which kept their firms in competition. The row over patent infringement in 1844 deepened the mistrust, which continued after the demise of the original protagonists. The Stephenson Company, which had been remarkably profitable,[115] could have acquired the

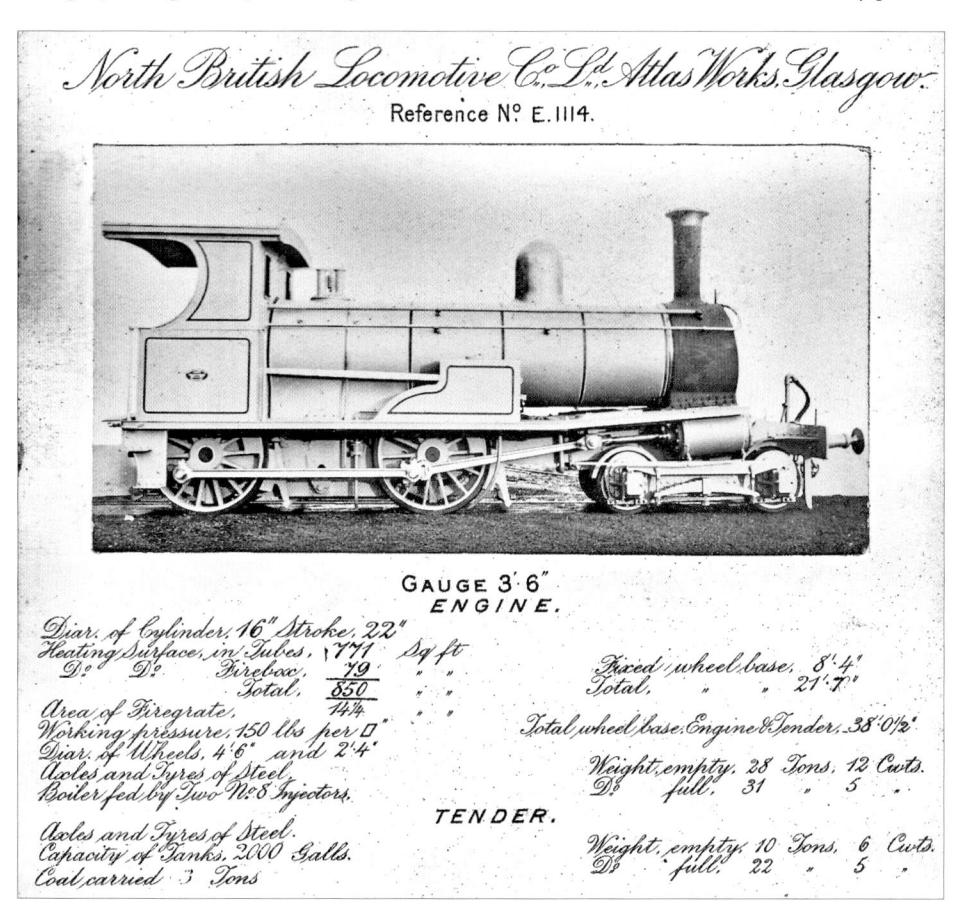

Fig. 7.27
The first example of amalgamation – formation of the North British Locomotive Co. from 1903. A 4-4-0 locomotive built to the designs of Sharp Stewart & Co. Ltd., used as a standard design offer for future customers. *Author's collection.*

Hawthorn Company in 1870, when it had been available 'for some time' and 'the place might probably be bought cheap.'[116] It had an excellent opportunity to combine the two sites and, albeit with major investment, create a progressive locomotive manufacturing site. The subsequent purchase of the Hawthorn site by other interests, and its integration with the Leslie shipyard, may have caused the Stephenson Company proprietors to regret the lost opportunity.

A proprietor, such as Andrew Barclay, was tenacious in his desire to remain independent, even though it was clear that he was unable to maintain financial independence and attract sufficient funds for investment at the Caledonia Works in Kilmarnock. Such tenacity could even have far-reaching effects for the progressive manufacturers, and occasional disagreements amongst partners could threaten the survival of a company, such as occurred with E.B.Wilson & Co. Far from considering amalgamations, Walter Neilson, the charismatic proprietor of the Glasgow firm that bore his name, had major disagreements, firstly with Henry Dübs and latterly with James Reid, both leading to the establishment of large competitors in the same city. Neilson later claimed that 'Dübs made himself so excessively disagreeable and having offered to give up his partnership and leave the works upon my paying him a certain sum of money… he left.'[117] Neilson chose to retire in 1884 and invited his fellow partner, James Reid, to buy out his share, but there was a deep disagreement over the valuation, 'I considered I was being robbed by Mr. Reid of a sum equal to £20,000.'[118]

As if to prove a point to James Reid, Neilson convinced Glasgow's banking community, by virtue of his experience and strength of character, to provide loan capital to establish a new factory, the Clyde Locomotive Works. His claim that his personal reputation would retain a sizeable portfolio of customers proved to be exaggerated, and Neilson's career ended in 1888 when his firm ceased trading through lack of orders. The works were acquired for about a third of their cost by Sharp Stewart & Co. Ltd., which, as a public company, raised additional capital to move from its cramped central Manchester site and re-locate to the failed Clyde Locomotive Works.

The American and European locomotive industries also resisted amalgamations. Serious consideration was given to it in America after 1892, but the prolonged recession in the decade saw a serious downturn in its domestic locomotive market. In such an unsettled situation, the proprietors could not agree on terms for any mergers but agreed instead to face the recession by co-operating under a freshly constituted American Locomotive Manufacturers Association.[119] This collusion was seen as a successful response to narrowing profit margins and mounting capital costs.

Only as the locomotive market regained strength from 1898, and through the initiative of a financier and industrialist, Joseph Hoadley, was there confidence for a major amalgamation, in 1901, of the ten smaller manufacturers to form the American Locomotive Company (ALCo).[120] This extraordinary move, to achieve rough parity with Baldwin's production capacity, recognised both the economies of scale of large batch production, and the negotiating strength of large suppliers in pursuing the standardisation of components to achieve large batch orders. The lesson was quickly learned by the three main Glasgow firms (Neilson Reid & Co., Dübs & Co. and Sharp Stewart & Co. Ltd.), already Britain's most progressive main-line locomotive firms, which amalgamated in 1903 to form the North British Locomotive Co. Ltd.[121]

Failures
There were thirty-three failures of locomotive manufacturing companies during the course of the 19th century, including two sites which failed on two occasions.[122] Twenty-two of the failures, such as that of the Worcester Engine Co. in 1872, saw the ending of locomotive production altogether.

In the absence of company papers, the circumstances of most failures are unknown, although their timing goes some way towards explaining them. Some occurred during recessions and times of low locomotive demand, such as Mather Dixon in 1843; whilst the liquidation of Bury Curtis & Kennedy in 1851 was the largest of all the 19th century failures. The majority occurred at other times, however, suggesting that proprietors failed to take the strategic decisions necessary to maintain their competitiveness, or the tactical decisions necessary to deal with raw material price movements and working capital requirements.

Tenacity alone was insufficient in keeping a business solvent, as demonstrated by Andrew Barclay, the largest employer in Kilmarnock in 1871. He faced a liquidity crisis after a re-investment programme at the Caledonia Works coincided with a downturn in demand and prices for mining machinery. Fearing a collapse of his business, he set up his five sons in a separate business, Barclays & Co., at the Riverbank Works in the town, to which most locomotive orders were then directed.[123] In spite of many endeavours, Barclay could not recover financial independence and profitability. His business was sequestrated in 1874, re-instated the following year through new securities, but sequestrated again in 1882. Its re-instatement in 1886 was through the control of the firm's creditors, under whom it continued trading before incorporation as a limited company in 1892. That Barclay's autocratic style of direction was partly at fault for the firm's difficulties was born out with the firm's move into profit after 1894, in which year he had stood down as Managing Director, after forty-seven years in charge of the Caledonia Works.

The failure of Robert Stephenson & Co. Ltd. in 1899,

Fig. 7.28
Example of a failed locomotive manufacturer – the Worcester Engine Co. An 1868-built 0-6-0T for the Taff Vale Railway (No. 90).
R. H. Bleasdale Collection – National Railway Museum.

at a time of extraordinary demand for locomotives, and when prices were high, was the outcome of its long policy of inadequate investment. As early as 1883, Sir Joseph Pease (1828-1903), who had inherited his partnership shares from his father, Joseph Pease (1799-1872), observed that 'the management seems all asleep and wants waking up.'[124] Following the Company's incorporation in 1886, it appointed a new General Manager, G.H.Garrett. In spite of his rigorous cost analysis, showing that some activities were unprofitable, including locomotive manufacture, Garrett, who died in 1889, appears to have taken insufficient action to stem the losses.

Garrett's replacement from 1891, John Walker, similarly took insufficient action to rectify the losses on locomotive production.[125] Between 1886 and 1900, most locomotives were made at a loss, including the forty tank locomotives for the Midland Railway in 1899, when the locomotive market was high and the prevailing prices should have earned a good profit. The Newcastle site could no longer sustain multiple locomotive manufacture down to the costs of other manufacturers, and its output was no longer acceptable to any but a handful of loyal customers. Between 1876 and its failure in 1899, the site accumulated losses of £580,000.[126] A new public company was formed that year to acquire the Stephenson Company's assets and goodwill, and to provide the capital for the building of a new factory in Darlington, which allowed the Stephenson Company name to be carried forward into the twentieth century.

Fig. 7.29
Failure of the first locomotive manufacturing business. The Robert Stephenson & Co. Ltd. site in South Street, Newcastle at the end of the nineteenth century, just prior to its closure.
The Stephenson Locomotive Society, McDowell collection, Tyne & Wear Record Office.

Notes – Chapter 7

1. Lee (1978), pp 235-261, and Arnold (1996), pp 40-54.
2. Cottrell (1980), p 41.
3. Edwards and Newell (1991, pp 35-57, quoting from Roll (1930), pp 244-252. Also McKendrick (1970).
4. Pollard (1965), and Solomons (1968). Also Garner (1954), Chapter 2.
5. Solomons (1968). Also Edwards and Newell (1991), p 53.
6. Boyns and Edwards (1997), pp 1-29.
7. Boyns, Edwards and Nikitin (1996), pp 15-20.
8. Ledger, 1823-1831, Robert Stephenson & Co. Collection, ROB/1/1. Also Bailey (1984), p 19.
9. Robert Stephenson & Co., Cost & Profit Account Book, Bidder collection, Arch:Bidd 27/8.
10. Letter, Daniel Gooch to I.K. Brunel, August 1839, quoted in Clarke (n.d. but 1979), p 8.
11. Report by T.E. Harrison, August 1839, quoted in Clarke (n.d. but 1979), p 8.
12. Kitson Laird & Co., Wages Book, May 18th 1839 to March 7th 1840. (Kitson Collection, The Stephenson Locomotive Society).
13. C. Tayleur & Co., Abstract Book 1844-1870, Vulcan Foundry Collection, 1970/37/6/1, p 3.
14. Cost of Work Book, Beyer Peacock Collection, quoted in Hills and Patrick (1982), p 38.
15. T. Plumpton, 'Manufacturing Costs', *Accountant*, Vol. 20, 1894, p 990, quoted by Boyns and Edwards (1997), p 11.
16. Clarke (n.d. but 1979), p 16.
17. Vulcan Foundry Cost Book, 1870-1939, Vulcan Foundry collection, B/VF/5/6/3, *passim*.
18. Boyns and Edwards (1997), p 12.
19. C. Tayleur & Co., Abstract Book 1844-1870, Vulcan Foundry Collection, B/VF/5/6/1, p 87
20. Quoted in Clarke (n.d. but 1979), p 23.
21. Moss and Hume (1977), p 72.
22. Cost and Weight Books, Neilson Reid & Co. collection, UGD 10/4.
23. Clarke (n.d. but 1979), pp 55/6.
24. Cost Analysis Ledger, 1886-1901, Robert Stephenson & Co. Collection, ROB/4/2.
25. Brown (1995), p 121.
26. *ibid*.
27. 'Finance Statement' for R. Stephenson & Co. to 30 Jun 1830, Pease-Stephenson collection, D/PS/2.
28. Clarke (n.d. but 1979), p 7.
29. Partners Minute Book, Robert Stephenson & Co. collection, ROB/1/1, p 37, 7 Mo 3 1833.
30. Letter, Edw. J. Cook (for R. Stephenson & Co.) to Edward Pease, Newcastle, 22 Decr 1837, Pease-Stephenson Collection, D/PS/2/38.
31. Letter, Edw. J. Cook, for R. Stephenson & Co., to Edward Pease, Newcastle, 11 Decr.1843, Pease-Stephenson Collection, D/PS/2/44.
32. Estimate of Finance Statement for R. Stephenson & Co. for 3 months ending June 12 1846, Pease-Stephenson Collection, D/PS/2.
33. Letter, Arthur Potts to John Jones, Chester, March 28th 1846, Jones & Potts collection, IMS 248/1.
34. Letter A. Potts to J. Jones, London, August 16th 1847, Jones & Potts collection, IMS 249/1.
35. Letter A. Potts to J. Jones, London, Sepr. 18th 1847, Jones & Potts collection, IMS 249/3.
36. Letter A. Potts to J. Jones, Edinburgh, July 3rd 1848, Jones & Potts collection, IMS 250/3.
37. Letter A. Potts to J. Jones, Perth, Augt 7th 1849, Jones & Potts collection, IMS 252/4.
38. Letter A. Potts to J. Jones, London, n.d. but postmark SP (September) 26.1848, Jones & Potts collection, IMS 250/5.
39. Letter A. Potts to J. Jones, London, Jany 2 1849, Jones & Potts Collection, IMS 251/1.
40. Letter A. Potts to J. Jones, London, Jany 2 1849, Jones & Potts Collection, IMS 251/2.
41. Hills and Patrick (1982), p 35.
42. Clarke (n.d. but 1979), p 23.
43. Clarke (n.d. but 1979), p 56.
44. *ibid*.
45. Edwards and Newell (1991), p 52.
46. Marriner (1980), pp 219.
47. Lee (1975), p 33.
48. Arnold (1996), Section VI, pp 45-49.
49. *Accountant*, 1895, pp 75-6, cited by Arnold (1996), p 47.
50. Pease-Stephenson Collection, D/PS/2.
51. Letter W.H. Budden to Edward Pease, Newcastle-upon-Tyne, March 12th 1846, Pease-Stephenson Collection, D/PS/2/47.
52. Hills and Patrick (1982), p 70.
53. R. Stephenson & Co. profit & loss accounts, Pease-Stephenson Collection, D/PS/2.
54. Clarke (n.d. but 1979), p 16.
55. Analysis of Appendix.
56. W.H.Wright and S.H.P. Higgins, 'Summers Groves and Day', *Journal of the Stephenson Locomotive Society*, October 1952, pp 263-264.
57. Templeton (1841), pp 1-41.
58. Millar (1976), p 78; also Lowe (1975), p 192.
59. For example the Starbuck papers, Tyne & Wear Record Office, File 131. Also Pease-Stephenson Collection, D/PS/2/72.
60. Fowler's illustrated catalogue for 1858, shown in Lane (1980), p 22.
61. Note by Wigham Richardson, quote in Clarke (n.d. but 1979), p 9.
62. Lane (1980), p 146.
63. Beyer Peacock Memorandum of Partnership, cited

CHAPTER 7 – STRATEGY AND ADMINISTRATION

63 in Hills and Patrick (1982), p 39.
64 François Crouzet, 'When the Railways Were Built', in Marriner (1980), p 114.
65 Brown (1995), p 48.
66 Brown (1995), pp 83-85.
67 Nasmyths Gaskell & Co., Letter Book 3, p 54, and Sales Book A, pp 158/182, Nasmyths Gaskell collection.
68 Redman (1972), pp 17-19.
69 Clarke (n.d. but 1979), pp 54.
70 Clarke (n.d. but 1979), p 55.
71 Clarke (n.d. but 1979), p 57.
72 Analysis of Engines Delivered Books, Robert Stephenson & Co. collection, ROB/2/3.
73 Clarke (n.d. but 1979), pp 24/5.
74 Boyns and Edwards (1997), p 35, refer to three consultants for the Consett Iron Co. in the 1860s.
75 Hills and Patrick (1982), pp 53/73.
76 Cantrell (1984), p 20.
77 K.M. Kyell (Ed), *Memoirs of Leonard Horner*, 1890, Vol 1, p 356, & Vol 2, pp 3/106; and M.J. Shaen, (Ed), *Memorials of Two Sisters, Susanna & Catherine Winkworth*, 1908, p 133; cited in Cantrell (1984), p 21.
78 Speech by E.B. Wilson at the opening of the Railway Foundry extension, December 1847, cited in Redman (1972), p 10.
79 Hills and Patrick (1982), pp 13/14.
80 Brown (1995), pp 96/7.
81 Clarke (n.d. but 1979), pp 15-21.
82 *ibid*.
83 Clark (n.d. but 1937), p 177.
84 Clark (n.d. but 1937), Chapter VIII, pp 139-170.
85 Bailey (1984), Section 10.2, pp 150-159.
86 Letter, M. Longridge to R. Stephenson, B[edlington] I[ron] W[orks], 13 Dec.1830, Pease-Stephenson collection, D/PS/2/34.
87 Letter, M. Longridge to Thos. Richardson, Bedlington Iron Works, 24 Mar 1831, Pease-Stephenson collection, D/PS/2/62.
88 Letter R. Stephenson to fellow partners of R. Stephenson & Co., Newcastle, 7 March 1831, Robert Stephenson & Co. collection, ROB/5.
89 R. Stephenson & Co. Minute Book, entry for 27: day of June 1831, Robert Stephenson & Co. collection, ROB/1/1.
90 Gooch (1972), p 18.
91 Bailey (1984), p 157.
92 Minutes of the Grand Junction Railway Board, 4th November 1835, National Archives, Rail 220/1.
93 Letter E. Pease to E.J. Cooke (for R. Stephenson & Co.), Darlington, 11 Mo 29. 44, Institution of Mechanical Engineers, Crow Collection.
94 Clarke (n.d. but 1979), p 10.
95 Vamplew (1972), p 327.
96 Vamplew (1972), p 332.
97 Clark (n.d. but 1937), pp 70/176.
98 *Engineering*, Vol. XVI, October 24th 1873.
99 Moss and Hume (1977), p 72.
100 Report cited by Pollard and Holmes (1968), p 321.
101 Letter M. Longridge to W. S. Longridge, Brussels, 19 Sept 1839, in private collection.
102 Clarke (n.d. but 1979), p 10.
103 Payen (1986), pp 123-130.
104 Vamplew (1972), pp 324-327.
105 Draft Agreement, Manchester, Decr.19th 1856, Beyer Peacock collection, MS0001/255.
106 Letter, C. Beyer to H. Roberston, Gorton Foundry, March 5th 1857, Beyer Peacock collection, MS0001/256.
107 Millar (1976), pp 44-46.
108 Legget (1973), p 50.
109 Saul, in Aldcroft (Ed) (1968), p 201, quoting O.S.A.Lavallee and R.R.Brown, 'Locomotives of the Canadian Pacific Railroad Company', *Railway and Locomotive Historical Society Bulletin*, Vol. LXXXIII (1951), p 9, and R.R. Brown, 'British and Foreign Locomotives in Canada and Newfoundland', *Railway and Locomotive Historical Society Bulletin*, Vol. XLIII (1937), pp 6-23.
110 François Crouzet, 'When the Railways Were Built', in Marriner (1980), p 106.
111 Analysis of data contained in Slezak (1962).
112 Analysis of Appendix.
113 Statement by Thomas Ridley, *Newcastle Journal*, quoted in Clarke (n.d. but 1979, p 49.
114 *ibid*.
115 Kirby (1984), p 43.
116 Communication, Thomas Hodgkin to Benjamin Browne, quoted in Clarke (n.d. but 1979), p 19.
117 Walter Neilson, autobiographical notes, Neilson Reid & Co. collection, UGD10.5/1.
118 *ibid*.
119 Brown (1995), p 53.
120 *ibid*.
121 *A History of the North British Locomotive Co Ltd*, 50th Anniversary publication, Glasgow, 1953, pp 44/5.
122 Analysis of Appendix.
123 Moss and Hume (1977), pp 70-75.
124 Kirby (1984), p 79.
125 Warren (1923), pp 416/7.
126 Kirby (1984), p 79.

Chapter 8
Conclusions

It is difficult to conclude that British industry's reluctance through the 19th century to embrace mass production and re-organise into 'managerial' enterprises, can be applied to the heavy manufacturing industry.[1] Firms were dominated by demand-led markets for the small batch locomotive industry, requiring vertically-integrated production facilities. It follows that only multiple product manufacturing firms, that controlled their markets, could reasonably be criticised if they failed to embrace structural changes, or could have sustained a supply-led market in order to benefit from larger batch production. Greater standardisation of components would have permitted economies of scale, but to achieve the production economies of the American and German industries by the end of the century, British industry would have required fewer factories, each producing more units of fewer designs.

The industry's failure to achieve these production levels was due to its inability to control the locomotive market. In the progressive early years until the 1850s, the industry had moved strongly towards standardised designs and batch production of components. Thereafter, however, it became increasingly dominated by the railway market which it then served largely as a contract manufacturing industry, with only the growing industrial locomotive sector able to sustain a production-led market. The factors that led to this regression to contract manufacturing were three-fold, namely the determination of each firm to survive, the introduction and expansion of British railway company workshops, and the isolating effects of the London-based overseas market.

The culture in heavy manufacturing firms of preserving a broad market base to provide continuity of work for skilled personnel and capital equipment had been present since the pre-railway era. To overcome market fluctuations these firms sought continuity of work through diversification, not just of product, but of market, a welcome new form

Fig. 8.1 Vulcan Foundry-built 4-4-0 (VF 1162/86) for the Waterford & Limerick Railway No. 12, *Earl of Bessborough*. *Peter Wardle collection.*

of which was the manufacture of locomotives. The culture was continued as manufacturers experienced the sharply fluctuating locomotive demand during the 19th century, which was closely related to cyclical movements in the domestic British economy and capital export market, as well as to overseas economic and political events. The ability of manufacturers to survive the low-points in these cycles through diversification into other markets explains the continued existence of so many firms in the locomotive market. This became more significant later in the century as the very characteristics enabling their survival, a broad market base using a high proportion of skilled labour, increasingly constrained firms' ability to compete with more progressive companies that specialised in locomotive production.

The rate of survival among British locomotive manufacturers contrasts with the American industry, which developed to exploit the rapid growth of its domestic locomotive market. This it achieved largely through specialisation in locomotive production but, without a broad manufacturing base, the industry was more vulnerable to market fluctuations, and several firms failed, particularly in the 1850s. With their greater vulnerability to market trends, and faced with a greater shortage of skilled labour, the sixteen or so surviving American manufacturers at the end of the century accommodated the fluctuations through layoffs and recruitment of un-skilled labour. They also increasingly developed overseas markets during the worst recessions in the American domestic market. The four or five largest manufacturers, plus the dozen smaller firms, which fulfilled the requirements of the large domestic American market, provided the opportunities for production economies denied to the British industry, and gave it the strength to sustain a production-led market.

The British locomotive manufacturers adapted well to the introduction and development of the tendering and contracting practice required by the 'transparency' of competition for public railways. Their marketing became solely representational, as each sought inclusion and retention on railway tender lists from which obtaining orders was determined by price quotation and delivery times. Their inability to expand quickly enough to meet the extraordinary demand of the domestic railways in the mid-1840s, however, produced delivery dates which were quite unacceptable, and prompted the larger railways to diversify into locomotive manufacture.

The start up and development of the British railway workshops from the 1840s was the greatest influence on the evolution of the independent locomotive industry from that time. The increasing loss of much of the domestic main-line market was compounded by the conversion of the remainder from a supplier-led to a customer-led, contract market. The initiative for technological innovation passed to the railway development teams, and the opportunity for design innovation was increasingly restricted to smaller railways which did not have a design capability.

Fig. 8.2 An example of a smaller railway locomotive market. Sharp Stewart & Co. Ltd. 1877-built 0-4-2 (SS 2716/1877) for the West Lancashire Railway No. 1 *EDWD HOLDEN*. *Peter Wardle Collection.*

The proliferation of railway workshops led to a proliferation of design teams, in turn leading to a proliferation of designs. The absence of any co-operative dialogue on designs, and the 'Empire-building propensities' of the locomotive superintendents,[2] saw the standardisation of components within, but not between, railway companies. These multiple standards were of more consequence to the independent manufacturers than the proliferation of designs, as the resulting small batch orders offered few production economies unlike the systemisation of production pursued by the American manufacturers.

The proliferation of domestic railway companies, encouraged by Parliament from the 1870s through its suppression of mergers, helped to maintain the large number of railway-owned locomotive workshops. The independent manufacturers were called upon to supplement the main-line locomotive stock with relatively small batches when demand was beyond the capacity of these workshops. There was an acceptance by the railway companies that prices for these small batches would be higher than would have been the case if the independent industry alone had provided all their fleet requirements. This was acceptable to the railways because greater use was made of their workshops, which were primarily maintenance factories.

Until the 1850s, the manufacturers' foreign markets, particularly in Europe, were mostly conducted through commission agents who were of much importance in dealing with the large risks then associated with exporting.[3] The agents negotiated prices, payment terms, delivery and proving arrangements, as well as providing expertise in credit transactions and shipping. They were also well experienced in providing intelligence about market opportunities and the activities of their British and foreign competitors. From the 1850s, however, their use diminished as the locomotive market in the major European economic centres moved in favour of the developing continental capital markets.

The change in emphasis of British capital exports for railways, from portfolio to foreign direct investments, including government 'guarantee' lines, in the developing countries of Europe, the Empire and South America, opened up the locomotive market to many new opportunities. Over 40% of Britain's overseas investments between 1865 and 1914 were for railway projects.[4] However, there was a close relationship between British foreign direct investment and the locomotive industry. This illustrates the close liaison between the manufacturers, on the one hand, and the railway companies and government agencies on the other, focused on London.

Overseas railways funded through foreign direct investment, and colonial governments mostly had offices

Fig. 8.3 A main line company locomotive built by an independent manufacturer. William Yates-designed 0-6-0, built by the Vulcan Foundry in 1880 (VF 880/80) for the Lancashire & Yorkshire Railway (No. 575). *Peter Wardle Collection*.

or representation in the City of London, providing the manufacturers with a much easier way to sell locomotives than their former reliance on commission agents. Market intelligence was maintained by firms' London managers, or their representative agents, in the close confines of the London business houses, and the market became the exclusive preserve of the British industry. Manufacturers' marketing efforts were focused on obtaining and retaining places on the tender lists of the railways and their consulting engineers.

The growth of the commercial, guaranteed and government railways was accompanied by the growth of the London-based consulting engineering firms representing those railways.[5] Their role in specifying their principals' locomotive requirements increasingly developed into overall locomotive design work, thus further removing the manufacturers' discretion to innovate.

With a few notable exceptions, the manufacturers were mostly restricted to detailing design work prior to manufacture, and, as with domestic railway designs, there was no agreed standardisation for components between overseas railways. Each manufacturer was free to adapt as many of its templates and gauges as the design would allow. This practice, which was closest to that of the American domestic railroad market, did allow a measure of standardisation for each manufacturer. However, the general lack of standardisation between manufacturers was the cause of considerable concern, particularly in larger economic regions such as India, whose railways acquired locomotives from several British firms.

The convenience of the London market saw the manufacturers largely withdraw from other foreign markets, in which they faced competition from American and continental industries which benefited from tariff impositions on imported locomotives. Without the spur of competition from those industries, however, a growing diversity developed, both technical and commercial, between the London-based market and the rest of the world. The convenience and lack of competition from non-British manufacturers led to a collective market assurance in which wider marketing through international exhibitions was largely felt to be unnecessary and, as with the domestic market, the sole emphasis was on maintaining the right to inclusion on tender lists.

As the world market opened up from the 1880s, in geographical areas outside the dominance of the British, German and American industries, the British firms were faced with direct competition from their counterparts. This revealed both the extent of the diversity in the market and the reduced ability of the British firms to compete. The British locomotive industry, used to the substantial 'rigid' designs perpetuated by consulting engineers, became less

Fig. 8.4 A locomotive built for export to an overseas Government obtained through the London market. This 2-6-2T was constructed in 1888 for the Imperial Railways of Japan (No. 102) by Nasmyth Wilson & Co. Ltd. (NW 336/88). *Peter Wardle collection*.

able to influence even the principles of some designs, and was less able to provide locomotives of the more 'flexible' American type, and price comparisons frequently favoured the standardised products of the American and German industries.

Subsequent organisational reforms did nothing to alleviate these problems. The Locomotive Manufacturers Association, born out of the perceived threat to the industry's existence from British railway workshops, became the vehicle for protective cartel arrangements from the 1880s. This served to accommodate the inefficiencies of the smaller, craft-based firms, the consequent raising of prices being indicative of an industry that had got out of touch with wider railway developments after years of dependency on the London market.

The British locomotive industry had to develop strategies for growth and re-investment in accordance with the reducing proportion of the domestic main-line market and the uncertain and fluctuating overseas market. The marked divergence in the manufacturers' strategies resulted from these uncertainties and was also affected by the degree of confidence felt by proprietors when it came to re-investment and expansion, and by their strong will to survive. 'Progressive' firms were prepared to risk investments to develop their businesses as specialist locomotive manufacturers whose increasing emphasis on batch production would require expanding capital equipment programmes and employment of unskilled labour. The more cautious approach by the 'craft' firms saw them risk less capital for investment as they relied on a broad market base of small batch orders requiring a higher proportion of skilled labour.

Until the 1850s, the industry controlled the locomotive market and evolved new decision-making practice as it developed policies in marketing, selling, technology, design, manufacturing, employment and administration. It had begun with an 'intensive burst' of technological development, between 1828 and 1830,[6] that had made possible the rapid expansion of railway networks. The locomotive then developed incrementally, in terms of thermodynamics and material technology as well as design, to fulfil the railways' requirements for improving economy, power and speed. Firms, such as Nasmyths Gaskell & Co., Sharp Stewart & Co. and E.B. Wilson & Co., emulated the earlier progressive role of Robert Stephenson & Co. in pursuing standard design strategies, and demonstrated extraordinary progress in manufacturing development. The industry pioneered new capital equipment and production processes which reduced manufacturing time and cost for increasingly standardised and interchangeable components, as well as reducing the requirement for craft labour.

However, as the industry's influence over its market declined from the 1850s, it largely lost discretion for technology and design improvements with the transformation in both domestic and overseas sectors from supplier-led to customer-led markets. This rôle was taken on by railway workshops and consulting engineers, and, by the 1890s, as empirical advancement gave way to scientific progress much initiative had passed to foreign railways, manufacturers and technical institutions. The changing markets, with limited scope for standardisation and the resulting proliferation of designs, reduced considerably the opportunity for manufacturing procedures to evolve towards the 'American system' of manufacture.

The main-line locomotive manufacturers were too small and diverse to counter the rise of the large railway workshops, and too dependent upon the consulting engineers to counter the fragmented development of the London overseas market. They had no option but to accept the largely contract manufacturing role, which would continue to be subject to the market fluctuations. The progressive manufacturers accommodated the new market requirement by pursuing specialised locomotive production.[7] They introduced 'Systematized, but not standardized' batch control procedures, as far as the market would allow, which did much to reduce production time and cost.[8] The 'craft' manufacturers opted to retain their traditional methods of manufacture, through their broad, heavy manufacturing market base, to accommodate market fluctuations. By the end of the century, therefore, there was a wide diversity of skills, capital equipment and production procedures, although the true costs of manufacture were partly concealed by the cartel pricing agreements of the ten firms forming the Locomotive Manufacturers Association.

The loss of market control, which resulted in this diversity, therefore prevented the industry from continuing its progress towards greater standardisation and batch production and, ultimately, an 'American system' of production. Failure to pursue greater economies of scale and scope therefore has some relevance in regard to the locomotive industry, in which too many firms continued in business chasing small batch orders.[9] Had the industry not been subjected to these market changes, it is likely that it would have evolved in a similar manner to the locomotive industry in the United States. More firms would have had a sufficiently large market base to encourage further investment and specialisation in locomotive production, although the survival rate may have been lower, with some firms unable to protect themselves from market recessions. These companies would, in turn, have been of sufficient size to have played a more influential role in the overseas market by offering a higher degree of standardisation and design in the manner successfully pursued by the German locomotive industry.

The plausibility of this argument is strengthened by reference to the industrial locomotive sector which had quite

Fig. 8.5
A standard industrial tank locomotive of the late 1880s. Hunslet Engine Co. 0-6-0T built in 1887 (HE 415/1887) for colliery use, PONTYPRIDD.
Industrial Locomotive Society, Negative No. 29789.

different characteristics from the main-line sector, and in which manufacturers retained discretion for specification and design. Although there was no scope for technological development, industrial locomotives followed main-line practice with new materials and increasing performance specifications, to provide more power without an increase in weight or dimensions. Manufacturers of industrial locomotives, both specialist firms and the larger firms also engaged in main-line production, took full advantage of their discretionary strength to maintain and develop fleets of standard designs. Their vigorous marketing and selling adopted the practices of the light manufacturing sector, using catalogues, trade fairs and selling agents. Production, similarly, used mostly standardised components, whilst accommodating the variations of track and profile gauge required by the customers.

The early locomotive firms achieved a major adaptation of their skill-base, transforming a 'craft culture' to a 'factory culture' among their workers. Discretionary responsibilities for design, selection of materials and work administration was passed to the specialist managers and foremen. The shortage of craftsmen during the century, accentuated at times of high locomotive demand, was alleviated by the introduction of self-acting machine tools and the employment of un-skilled labour for repetitive machining. The majority of tasks, however, continued to require the presence of 'time-served' journeymen boiler-makers, forge-men, foundry-men, fitters and erectors.

The dependence on these craft skills, and the ongoing shortage of craftsmen, meant that manufacturers were obliged to maintain skilled workers in employment as far as possible. There is, however, no evidence of a policy to make machine tools 'firm-specific' in order to deter free labour movement and suppress wage claims, as was the case with the railway companies in their workshops.[10] Rather, fluctuations in overall labour requirements were absorbed through the engagement and dismissal of un-skilled men to carry out routine machining tasks. This provided an essential employment cushion allowing firms to provide continuity of work for their craftsmen. The higher proportion of un-skilled men employed by the progressive manufacturers, following their programmes of investment into more types of advanced machine tools, gave them greater facility to reduce their total work-force when demand was low. The 'craft' manufacturers, however, were less able to reduce their work-force at such times and were obliged to retain a higher proportion of craftsmen in order to maintain their breadth of production.

The locomotive industry was generally successful in its tactical decision-making, although the failure of several firms through the century confirms that adequate provision was not always made for sufficient working capital or to deal with market changes. From its outset the locomotive industry was run by, and was dependent upon, managing partners assisted by specialist managers.[11] The latter were largely selected on merit rather than nepotism, and usually invested in the firms that employed them. Employment of specialist and general managers was increasingly adopted by all firms in the sector, particularly from the 1860s as the early proprietors were ageing and faced with expanding businesses and more demanding decision-making. A form of functional line management evolved. Head foremen and head clerks were delegated responsibilities for employment, production, procurement, sales and marketing, cost and financial accounting. Workshop foremen were delegated full responsibility for hiring and firing, discipline and production control, which they achieved without the necessity for sub-contracted 'piece-mastering'.

The quality of business information available to manufacturing proprietors and senior managers was generally good during the century. Most proprietors were well versed in credit arrangements, debt recovery and, within the limits of small batch production, in raw material cost limitation. Although cost and management accounting procedures were well developed, however, the use of this information in making tactical decisions was somewhat variable. Several firms used the information effectively, with separate management information for each of their main workshops, making them effectively 'cost-centres'. Other firms paid less attention to this detail and were less aware of cost escalation.

The view that factory employers secured both ideological and cultural authority over their workforce, largely related to volume industries and did not apply to the locomotive industry.[12] With most locomotive factories being located in urban areas, paternalism was generally limited to works' events rather than fulfilling a deeper community involvement. However, even this modest action was seen by the proprietors as an important means of fostering a 'factory culture' which helped to maintain craft employment levels. Although paternalism was not as pervasive as in the railway workshop towns, such as Crewe,[13] the depth and early date of the paternalistic endeavours at Bedlington Iron Works mark it out as being the progenitor for the independent industry.

The manufacturers were obliged to confront major industrial relations issues and were in the forefront of some of the major industrial disputes during the century. Urban manufacturing craftsmen held real bargaining power, but there was no accommodation between the trades unions and the manufacturers over the erosion of craft skills and the employment of un-skilled labour.[14] The issue was a running sore throughout the century, re-surfacing with each claim for improved hours and wages. Even after the long-running 1897 dispute, the divisions within the Employers' Federation left the issue un-resolved to be carried into the next century.

It is apparent that throughout the century most proprietors were well served with management information and accounts, and had good knowledge of wages, component and material costs, as well as overhead cost assessments from which their direct cost manufacturing base was determined. The use made of this information in making tactical decisions was, however, somewhat uneven. Some proprietors responded well in making tactical decisions, including negotiations to minimise cash flow problems arising from sudden rises in raw material costs and bad debts. Equally, they took advantage of increasing demand through advance payments to provide periods of healthy cash flow to reduce their loan capital. Other firms did not keep adequate cost accounts and allowed costs to escalate.

The proprietors were also generally well served with financial accounts and knowledge about their company performances and profitability, with depreciation allowance well developed from the industry's earliest years. In spite of this information, the proprietors were generally poor at taking strategic decisions upon which the long-term future of their companies depended. Such caution could lead to conservatism with damaging long-term consequences. In particular, several firms did not respond to the general reduction in prices in the 'Great Depression' from 1873, believing that it was to be temporary and that strategic decisions could wait until prices returned to their former levels.

Accountancy in its wider sense was used only minimally to guide businessmen in their decisions, and where it was used the guidance was often unreliable.[15] Decisions relating to site expansion, development of new sites at home and abroad, re-investment in capital equipment and diversification into and out of locomotive manufacture, were based on caution born from the uncertain market trends, and without the discipline of business plans.

The proprietors in the first 30 years of the railway era showed extraordinary entrepreneurial flair in committing equity and loan capital to develop manufacturing concerns which employed advanced capital equipment and production control procedures. In spite of the limited evidence to suggest how proprietors made their strategic decisions, it is evident that, at the end of the 19th century, the locomotive industry as a whole remained dominated by individualism and lacked the capability for making strategic business plans for long-term growth and profitability. With the exception of the progressive manufacturers, the tenacious desire to survive was not matched by an understanding of the developing economics of heavy manufacturing. There were still too many craft firms offering non-standard, small-batch, high-cost products manufactured as part of wider heavy manufacturing activities.

This is best illustrated by comparing the number of British and overseas firms involved in locomotive manufacture. In the USA, the 40 manufacturers making locomotives in 1854 had dropped to 16 by 1877.[16] Five large concerns and eleven smaller firms provided not only for the whole North American market, but for a growing export market as well. In Britain, even with the loss of much of the domestic market to the railway workshops, there were 27 large and medium size companies (and 16 smaller companies) making locomotives in 1877, all with varying degrees of involvement in other manufacturing activities.

In the last quarter of the century, the number of American firms remained at about 16, with the five largest companies investing heavily to meet the extensive growth in both domestic and foreign locomotive markets.[17] In Germany by the end of the century, there were eight

'Aktien-Gesellschafts' and 12 other large manufacturers making locomotives for the whole domestic market, and a growing foreign market.[18] The French locomotive industry had three 'Société Anonyme' and six other manufacturers, also with large domestic and foreign markets.[19] Britain, however, still retained 26 large and medium manufacturers (and 7 smaller ones) by 1900, 13 being private limited companies, 8 public limited companies and 12 partnerships. The production capacity of 230 locomotives a year of the largest manufacturer, Neilson Reid & Co., was much less than Baldwins' production of 1100 a year.[20]

The development of the locomotive industry was largely dictated by the nature of its market rather than any limitations in enterprise of manufacturers. This runs counter to the theme that British industry did not pursue 'managerial capitalism' quickly enough.[21] The success of some of the larger partnerships which relied upon specialist managers demonstrated that many of the benefits of separating ownership from management could be achieved without the wholesale separation inherent in managerial capitalism. 'Partnership capitalism' served the locomotive industries of Britain, America and Germany well throughout the 19th century, proving to be a system that could accommodate generational transition through external recruitment and internal promotion, whilst attracting increasing managerial specialisation.

Partnership capitalism was not a cause of the industry's failure to maintain control of the locomotive market from the 1850s. Progressive companies, such as the Vulcan Foundry Co. Ltd. and the Yorkshire Engine Co. Ltd., which were themselves early examples of managerial enterprises, were just as influential in the conduct of the market as the partnership enterprises, and also just as vulnerable to its limitations.

The incorporations in the locomotive industry did not constitute full 'managerial capitalism'.[22] Indeed, the private limited companies were a means to attract additional capital without the partners having to give up full control of their companies.[23] The take-up of public limited company status was rather higher for the locomotive industry, there being eight such firms by the end of the century who used the status to raise capital for significant investment programmes.[24] Although limited companies were amongst the most progressive locomotive firms, employing career managers with ever-greater experience and expertise, the largest and most profitable British firms, Neilson Reid & Co. and Dübs & Co., both remained partnerships throughout the 19th century. Only when they amalgamated in 1903 with Sharp Stewart & Co. Ltd. (by then a public company of long-standing) to form the North British Locomotive Company Ltd. did the combined enterprise become a public company.

Similarly, in America, successful companies remained as partnership enterprises until such time that their re-capitalisation requirements made incorporation desirable. Notably, the most progressive locomotive manufacturer, the Baldwin Locomotive Works, remained a partnership enterprise through to 1909, and did not convert to a public corporation until 1911.[25] The Baldwin and North British incorporations were to provide substantial capital for investment. This was perceived to be necessary as a defensive response to the major amalgamation and incorporation in 1901 of the American Locomotive Company (ALCo), the motivation for which was to pursue greater economies of scale in production.

Non-executive partnerships employing general and specialist managers were not necessarily going to provide a level of expertise that would guarantee a firm's long-term prosperity, or even survival. There were examples of hereditary partnerships amongst the locomotive firms, some of which succeeded and some of which failed. It cannot be said that 'gentrification' contributed to Britain's

Fig. 8.6
Product of a managerial enterprise. Yorkshire Engine Co. Ltd. 1869 - built 0-6-4T (YE 108/69) for the Russian Poti-Tiflisskaya ZH.D (Poti-Tiflis Railway), (No. 61). *The Locomotive Magazine, Vol 25 (1919), p.15.*

relative economic decline,[26] although the first locomotive company to employ a General Manager, Robert Stephenson & Co., which continued as a 'craft' enterprise after his appointment in 1862, nevertheless declined and failed at the end of the century.[27]

These comparisons between successful and failed companies illustrate the importance of combining the talents of individuals with experience, capability, vision and entrepreneurial drive, whether they be partners or directors, to ensure the long-term prosperity of enterprises.

The presence of general and specialist managers could not alter the dominance of main-line railways and the London-based overseas market, which resulted in locomotive design remaining in the hands of railway locomotive superintendents and consulting engineers. Only a significant increase in the size of the manufacturing firms, through substantial investment or through amalgamations, would have provided economies of scale sufficient to have encouraged a return of some market control to the industry, emulating the influence of the Baldwin works in America.

The industry, which had benefited substantially from Britain's large foreign direct investments, became insular through its monopoly of the London-based overseas locomotive market. At the end of the century, as it became subjected to increasing competition from foreign manufacturers, the industry was shown to have too many firms, manufacturing locomotives of too many designs with less economy than their competitors. The progressive firms, however, with their advanced managerial organisations, carried into the 20th century the entrepreneurship that had been present across the very complex industry, and that had played such an important part in the 19th century British economy.

The criticism that partners tried to maintain an assured income at the expense of investment for long-term growth is also difficult to sustain. Failure to invest there certainly was in some firms, but this reflected the divergent interpretations of market growth between the progressive and craft firms, the latter being more cautious in investing for batch production. This was more a reflection of poor strategic decision-making than a deliberate policy of financial benefit to the partners, a number of whom lost substantial sums through the poor performance of their firms in the last quarter of the century.

From the commencement of the railway era there were many non-executive partners who oversaw the financial well-being of their firms, and who directed the managing partners towards corporate strategies in much the same way that would have been achieved by latter-day managerial enterprises. The perceived success of this form of enterprise gave no cause for change nor perception of resistance to managerial capitalism.[28] The incorporation of public locomotive companies was not undertaken as a means of hiring in specialist managers, but to gain access to new capital sources for investment.

It is most likely that the 'will to survive' in some partnerships was so strong that it actively discouraged mergers that could have consolidated markets and produced economies of scale and scope. This will to survive was perhaps a form of cultural restraint believed to be one of the main 'institutional rigidities' of the British economy.[29] There was, however, an equally strong desire for American manufacturers to remain independent, and mergers in that country did not occur either until the formation of ALCo in 1901. Undoubtedly, more could have been done in this direction by the British industry, which lost an opportunity to consolidate through the medium of the Locomotive Manufacturers Association. The Association lacked strength of purpose in its early years and the very formula agreed by the ten member firms, by which the progressive manufacturers diluted the value of their production economies to support the craft firms, would, in another era, have led to mergers.

The lack of consideration of business mergers, which would have provided larger capital concentrations and economies of scale, suggests that survival and proprietorial pride was always more important than profit. It can only be speculated that if Walter Neilson had applied as much energy in pursuing the growth of his Glasgow factory to three times its end-of-century size as he did in spreading the potential market between three factories by falling out with his partners, Britain might have had at least one factory better able to compete with the growing American and German competition.

Even though the three leading progressive firms, Neilson & Co. Dübs & Co. and Sharp Stewart & Co. Ltd., had clearly demonstrated that investment for higher production brought down unit costs, the risk of further capital commitment was too great for some firms, even with limited liability protection. That ten manufacturers protected themselves through the cartel pooling arrangement of the Locomotive Manufacturers Association, suggests that this was a further disincentive to amalgamations, even though maintenance of high production costs was threatening the export market by the end of the century.

It is thus apparent that the proprietors' decisions were more dependent upon their behavioural characteristics and personal persuasions than on clear business objectives.[30] When the Hawthorn Company was acquired by new partners in 1870, it was fully their intention:[31]

> to confine ourselves to marine engines, but the North Eastern Railway Co. were buying new locomotives very freely and encouraged us to go on with the trade, and, what with the old reputation of the firm and orders being so easy

to get, we yielded to temptation and decided to go on with this business also.

The Stephenson Company was similarly motivated during the loss-making years prior to its failure in 1899. The lack of decisive action revealed an extraordinary inertia born of the contemporary opinion that 1873 to 1896 represented a period of 'Great Depression' in trade and industry, and that by taking a long-term view, the company would return to profitability. It clearly took the major market upturn from 1897, and the Stephenson Company's inability even to make a profit on the large order it did receive, to convince the proprietors to wind up the business, and make a fresh start on a new site.

Long-term profit was perceived by some locomotive manufacturers to be less important than short-term survival and loyalty to their workforce.[32] Indeed, it is also apparent that the strategic decisions of some proprietors were more dependent upon their personal persuasions than on clear business objectives.[33] The Pease family, for example, non-executive proprietors of the Stephenson company throughout the 19th century, allowed it to continue trading in spite of its accumulating losses, partly for 'non-entrepreneurial' reasons.[34]

Resistance to change, perhaps the primary non-entrepreneurial influence on proprietors, can best be summed up as being sentiment for their firms and loyalty to their long-serving craftsmen. This resistance has been shown in the American context to reflect proprietors' social status and pecuniary awards, as well as their unthinking continuation of operational routines embedded in old corporate cultures.[35] Some proprietors perceived sentiment to be of greater importance than the more radical alternative strategies of closure or merger, which the more productive use of capital might otherwise have suggested. Perhaps the best reflection of the feelings that determined the survivability of some companies, in spite of uncertain profitability, was expressed by Benjamin Browne, the Chairman of Hawthorn Leslie & Co. Ltd.:[36]

> ...when the North Eastern Railway ceased to order locomotives regularly there were strong grounds for saying that it would have paid us to give up the locomotive trade altogether as far as mere money is concerned, but there was a widespread feeling of unwillingness to abandon an old and celebrated business: we also had a body of particularly high-class and loyal workmen whom we did not want to turn adrift; and, of course, we always hoped that something would turn up sooner or later.

Browne's views confirm how, even when all the management information may have strongly suggested a contrary action, decision-making was influenced by considerations other than those of profit maximisation.[37] Personal agendas, such as the individualism pursued by Walter Neilson, and the sentiment and loyalty, pursued by Benjamin Browne, could override the search for a return on capital investment that may otherwise have dictated expansion or closure. These agendas serve to emphasise that the industry's development was determined not just by business judgement but also by personal persuasions.

There is scope to carry forward the conclusions reached in this book, through more detailed business studies into individual progressive and craft locomotive companies. This will provide further evidence to confirm the motivations and varying levels of vision and entrepreneurship that were present in the industry. There is also considerable scope to follow through the conclusions of this book to determine how the locomotive industry evolved in the 20th century to face the growing challenges of foreign competition, with its greater emphasis on increased output, standardisation and economies of scale and scope. It would need to focus on the industry's corporate development to highlight the weaknesses of the continued proliferation of firms, and to ascertain how quickly it responded to its challenges through a conscious movement towards managerial enterprise. Such enquiry would include the extent to which the industry's lack of experience in technological and design development became a contributory cause in Britain's growing reliance on overseas motive power technology.

In concluding, therefore, it is emphasised that the locomotive industry was central to Britain's extraordinary contribution to technology and business development in the early years of the 'industrial revolution'. Partnership enterprise made possible the co-ordination of technical and business talent, and entrepreneurial drive, which developed this sector of the heavy manufacturing industry from small craft-based activities to large vertically-integrated factories. The industry would undoubtedly have continued its progress towards fewer specialist manufacturers producing larger numbers of standardised locomotives, with greater economies of scale, but for the radical changes to its markets. The industry failed to prevent these changes and, in accommodating the resulting proliferation of orders and designs, diversified between 'progressive' firms, which pursued greater production economy, and 'craft' firms, which retained a broad manufacturing base.

Many of the products of the nineteenth century locomotive manufacturers survive. Over 400 examples of locomotives may be seen in museums and heritage collections in both Great Britain and in many other countries of the world. These survivors are an on-going testament to the work of the industry and the longevity of their products.

BUILT IN BRITAIN

Fig. 8.7 19th century survivors in Australia (1). The oldest working locomotive in Australia: Neilson & Co, 1865-built 3 ft 6 in gauge 0-4-2 for the Queensland Government Railways (A10-class No.6) as displayed in The Workshops Rail Museum, Ipswich, Queensland. *Author.*

Fig. 8.8 19th century survivors in Australia (2). Robert Stephenson & Co. 1864-built standard-gauge 'long-boiler' 0-6-0 (RS 1542/1864) for the New South Wales Railway (No. E18) as displayed in the Rail Museum, Thirlmere, New South Wales. *Author.*

Notes – Chapter 8

1. Chandler (1990), Part III, 'Great Britain: Personal Capitalism', pp 235-294.
2. Kirby (1988).
3. Chapman (1992), pp 129-166.
4. Cottrell (1975), Fig 1, p 14; also, Edelstein (1982), p 37.
5. Saul (1970), pp 146-150.
6. Kirby (1991), pp 25-42.
7. Saul (1968), Table 1, p 192.
8. Scranton (1997), p 99.
9. Chandler (1990), pp 235-294.
10. Drummond (1997), pp 32/3, Note 22.
11. Wilson (1995), p 27; also Pollard (1965), pp 174/185.
12. Joyce (1980), p 92.
13. Drummond (1995), pp 186-208.
14. Southall (1996), p 61.
15. Pollard (1965), p 245.
16. White Jr. (1968), p 19.
17. Brown (1995), p 31.
18. Slezak (1962), pp 6-9.
19. Slezak (1962), pp 10-12.
20. Brown (1995), p 195.
21. Chandler (1990), pp 235-294.
22. Chandler (1990), p 240.
23. Payne (1967), p 520.
24. Cottrell (1980), pp 39-45.
25. Brown (1995), pp 216/220.
26. As discussed by Wiener (1981); with a counter view by Rubinstein (1993), pp 25-44.
27. Roper (n.d. but 1992), *passim*. Also, Kirby (1984), p 79.
28. Lazonick (1991), pp 25-27, 45-49.
29. Elbaum and Lazonick (1986), pp 1-15.
30. Simon (1992), quoted by Boyns and Edwards (1997), p 29.
31. Clarke (n.d. but 1979), p 21.
32. Boyns and Edwards (1995), pp 30/31.
33. Simon (1986), p 223.
34. Kirby (1984), p 115.
35. Marx (1976), p 19. Also Churella (1995), p 196.
36. Clarke (n.d. but 1979), p 25.
37. Boyns and Edwards (1995). Also Simon (1986).

Appendix
List of Independent Locomotive Workshops

Introduction

An attempt has been made to list each of the locomotive building workshops and companies in the 19th century. They are shown in the order of location and workshop title, then broken down further by the name of the partnership or company by which they traded. Dates of the changes are shown, wherever known. Where the trading names changed, the reasons for the change are noted. Company failures, or sale of assets are further noted, as are diversification into other manufacturing sectors when locomotive building ceased.

The leading archives of firms are shown. There is a wide disparity of archive collection and availability, ranging from a complete lack of surviving papers through to comprehensive collections of papers, drawings and photographic collections. Any student pursuing researches into particular workshops are advised to consult the archive administrators beforehand to establish whether the surviving papers they are seeking are included in particular archive collections. Isolated papers concerning workshops and their firms are often included in other archival collections, and no attempt has been made here to identify such individual items.

A number of the workshops and their firms have been the subject of histories, ranging from short monographs or papers through to comprehensive histories. These are listed wherever known, and it is to be hoped that further histories into the industry will be undertaken in future years.

Town	Locomotive Works	Start up Year	Period of Locomotive Manufacture	Name of Firm
Aberdeen (Aberdeenshire)	York Place Ironworks	1840	1845-1853	William Simpson & Co.
Airdrie (Lanarkshire)	Airdrie Engine Works	1790	1864-1890	Dick & Stevenson
	Standard Works	1860	1869-1913	Airdrie Iron Co.
	Victoria Engine Works	1866	1894-1912	Gibb & Hogg Ltd
Arbroath (Forfarshire)	Dens Iron Works	1854	1872-1877	Alexander Shanks & Son
Bedlington (Northumberland)	Bedlington Locomotive Works	1837	1837-1853	R.B. Longridge & Co.
Birkenhead (Cheshire)	Canada Works	1853	1854-*c*1857	Peto, Brassey, Betts and Jackson
			*c*1857-*c*1864	Brassey, Jackson, Betts & Co.
			*c*1864-*c*1866	Peto, Brassey & Betts
			*c*1866-1874	Thomas Brassey & Co.
Bolton (Lancashire)	Hope Foundry	1824	1840-1841	Thompson & Cole
	Soho Ironworks	1833	1833-1842	Benjamin Hick & Sons
			1842-1855	Benjamin Hick & Son
	Union Foundry	*c*1830	1830-1833	Rothwell Hick & Rothwell
			1833-1864	Rothwell & Co.
Bristol (Gloucestershire)	Atlas Engine Works	1864	1864-1880	Fox Walker & Co.
			1881-1914	Peckett & Sons
	Avonside Ironworks	1837	1840-1841	Henry Stothert & Co.
			1841-1856	Stothert Slaughter & Co.
			1856-1864	Slaughter Grüning & Co.
			1864-1879	Avonside Engine Co. Ltd.
			1879-1881	Avonside Engine Co. Ltd.
			1881-1905	Avonside Engine Co. Ltd.
Burton-on-Trent (Staffordshire)	Thornewill & Warham Iron Foundry	1732	1847-1868	Thornewill & Warham
			1868-1890	Thornewill & Warham
Bury (Lancashire)	Vulcan Works	*c*1838	1838-1854	Richard Walker & Brother
Carnarvon (Caernarfon) (Carnarvonshire)	Union Works	1854	1869-1892	De Winton
			1892-1901	De Winton Ltd.
Cardiff (Glamorganshire)	Hayes Foundry	1857	1862-1881	Parfitt & Jenkins
Chesterfield (Derbyshire)	Broad Oaks Works	1870	1888-1889	W. Oliver & Co.
			1889-1914	C. Markham & Co.
Chippenham (Wiltshire)	Railway Works	1858	1858-*c*1869	Rowland Brotherhood
Darlington (Co. Durham)	Hope Town Foundry	1832	1835-1860	W. & A. Kitching
	Whessoe Foundry	1860	1860-1885	C. l'Anson & Co.
	Hope Town Works	*c*1838	1838-1869	W. Lister
Drogheda (Co. Louth)	Drogheda Iron Works	1835	1844-1868	Thomas Grendon & Co.

APPENDIX – LIST OF INDEPENDENT LOCOMOTIVE WORKSHOPS

Type of Firm	Reason for Change	Main Archive	Works History
Partnership	Failed	No Known Archive	None Published
Partnership	Failed	No Known Archive	None Published
Partnership	-	No Known Archive	None Published
Partnership	-	No Known Archive	None Published
Partnership	Diversified	No Known Archive	*Scotland's Engineering Heritage*, J. D. Ellis, 2012
Partnership	Failed	No Known Archive	*Bedlington Iron & Engine Works* 1736-1867, Evan Martin, 1974; 'Short Histories of Famous Firms' by E. L. Ahrons, *The Engineer* – 21 Jan 1921.
Partnership	Partner Change	No Known Archive	*Thomas Brassey: Railway Builder & Canada Works Birkenhead,* J. Millar, 1976.
Partnership	Partner Change		
Partnership	Partner Change		
Partnership	Diversified		
Partnership	Diversified	No Known Archive	None Published
Partnership	Partner Change	Bolton Archives (ZHH)	*Hick Hargreaves 100 Years of Engineering Progress 1833-1933*, Bolton, 1933.
Partnership	Diversified		
Partnership	Partner Change	No Known Archive	None Published
Partnership	Failed		
Partnership	Failed	Search Engine - NRM (PECK)	Frank Jux, *Peckett & Sons*, Industrial Locomotive Society, 1988; 'Peckett & Sons – A Bristol Legend', I.C. Coleford, *Railway Bylines*, Summer Special No.2, pp. 40-57.
Partnership	-		
Partnership	Partner Change	Search Engine (GEC, SCIN, PECK6)	C.P. Davis (unpub. dissertation) *Locomotive Building in Bristol, The Avonside Ironworks (1837-1882)*, Bristol University, 1979; 'Avonside Ironworks, Bristol' by John Cattell, Bristol Industrial Archeological Society, Journal 30, 1997
Partnership	Partner Change		
Partnership	Incorporation		
Public Ltd. Co.	Failed		
Public Ltd. Co.	Failed		
Private Ltd. Co.	-		
Partnership	Partner Change	No Known Archive	K. Gallagher, 'Thornewill Family', *The Local History of Burton-on-Trent website*, recovered April, 2017
Partnership	Diversified		
Partnership	Diversified	No Known Archive	None Published
Partnership	Incorporation	No Known Archive	R.A.S. Abbott, 'Chronicles of a Caernarvon Ironworks', *Trans. of Caernarvonshire Historical Society*, Vol. 17, 1956, pp. 86-94; P.M. Hughes *De Winton's of Caernarfon 1854-1892*, 1994; A. Fisher, G.P. Jones, D Fisher, *De Winton of Caernarfon: Engineers of Excellence*, RCL Publications, 2011
Private Ltd. Co.	Failed		
Partnership	Diversified	No Known Archive	None Published
Partnership	Sale of assets	Derbyshire County Record Office	K.G. Wort, M.G. Bennett, *Markham and Company of Chesterfield 1889-1998*, 2005
Partnership	-		
Partnership	Failed	No Known Archive	Sydney A. Leleux, *Brotherhoods, Engineers for Power, Transport & Weapons*, Oakwood Press, 2019
Partnership	Sale of assets	Durham County Record Office (D/Whes: D/Ki); Search Engine - NRM (ROB5, KITC, HACK)	'Short Histories of Famous Firms, 'W.& A. Kitching', E.L. Ahrons, *The Engineer*, October 29th 1920: G.A. Petch, *The Kitchings of Whessoe. The first Hundred Years of Whessoe Ltd., Darlington, 1790-1890*, 1950
Partnership	Diversified		
Partnership	Diversified	No Known Archive	None Published
Partnership	Diversified	No Known Archive	None Published

Town	Locomotive Works	Start up Year	Period of Locomotive Manufacture	Name of Firm
Dundee (Forfarshire)	East/Victoria Foundries	1790	1833-1843	J. Stirling & Co.
			1843-1850	Gourlay Mudie & Co.
	Wallace Foundry		1838-1850	Kinmond Hutton & Steel
			1850-1857	Kinmond & Co.
Falmouth (Cornwall)	Penrhyn Works	1857	1860s	Sara & Burgess
Gateshead (Co. Durham)	Quarry Field Works	1835	1835-1853	J. Coulthard & Son
			1853-1865	R. Coulthard & Co.
			1865-1892	Black Hawthorn & Co.
			1892-1896	Black Hawthorn & Co. Ltd.
			1896-1902	Chapman & Furneaux
Glasgow (Lanarkshire)	Atlas Works (original)	c1839	1839-1851	James M. Rowan & Co.
	Clyde Locomotive Works (Atlas Works [later])	1884	1884-1888	Clyde Locomotive Co. Ltd.
			1888-1903	Sharp Stewart & Co. Ltd
	Cranstonhill Engine Works	1852	1860-1888	A. Chaplin & Co.
	Glasgow Locomotive Works	1863	1864-1903	Dübs & Co.
	Hill Street Foundry	c1831	1831-1841	Murdoch Aitken & Co.
	Hyde Park Street Works	1837	1843-1855	Neilson & Mitchell
			1855-1862	Neilson & Co.
	Hyde Park Works, Springburn	1862	1862-1898	Neilson & Co.
			1898-1903	Neilson Reid & Co.
	Phoenix Iron Works	1797	1840s	Thomas Edington & Sons
	St. Rollox Foundry	c1835	1835-1847	St. Rollox Foundry Co.
	Stark & Fulton Works	c1837	1839-1849	Stark & Fulton
Govan	Helen Street Works	1891	1891-1895	D. Drummond & Son
			1895-1965	Glasgow Railway Engineering Co. Lt
Greenock (Renfrewshire)	Caird Works	1828	1838-1841	Caird & Co.
	Scott Sinclair Works	1825	1847-1849	Scott Sinclair & Co.
Hartlepool (Co. Durham)	Castle Eden Foundry, Castle Eden	1839	1840-1847	Thomas Richardson
	Hartlepool Iron Works, Middleton	1847	1847-1850	T. Richardson & Sons
			1850-1857	T. Richardson & Sons
Kilmarnock (Ayrshire)	Britannia Works	1873	1873-1879	Allen Andrews & Co.
			1879-1881	Andrews, Barr & Co.
			1881-1884	Barr, Morrison & Co.
			1884-1890	Dick Kerr & Co.
			1890-1899	Dick Kerr & Co. Ltd.
			1899-1919	Dick Kerr & Co. Ltd.
	Caledonia Works	1847	1859-1874	A. Barclay
			1874-1874	A. Barclay & Son
			1874-1875	Trustees, A. Barclay & Son
			1875-1882	A. Barclay & Son
			1882-1886	Trustees, A. Barclay & Son

APPENDIX – LIST OF INDEPENDENT LOCOMOTIVE WORKSHOPS

Type of Firm	Reason for Change	Main Archive	Works History
Partnership	Sale of assets	No Known Archive	None Published
Partnership	Diversified		
Partnership	Partner Change	No Known Archive	None Published
Partnership	Failed		
Partnership	Diversified	No Known Archive	None Published
Partnership	Partner Change	No Known Archive	A.C. Baker, *Black Hawthorn & Co.*, Industrial Locomotive Society, 1988
Partnership	Sale of assets		
Partnership	Incorporation		
Private Ltd. Co.	Failed	National Archives (BT 31 & 34)	
Partnership	-		
Partnership	Diversified	No Known Archive	None Published
Private Ltd. Co.	Sale of assets	Search Engine - NRM (NBL/1)	None Published
Public Ltd. Co.	-	Glasgow University archives; Mitchell Library, Glasgow; Search Engine - NRM (NBL)	*A History of the North British Locomotive Company*, 1953; M.Nicholson, M. O'Neill, *Glasgow Locomotive Builder to the World*, 1987
Partnership	Diversified	No Known Archive	None Published
Partnership	-	Glasgow University archives; The Mitchell Library, Glasgow; Search Engine - NRM (NBL)	*A History of the North British Locomotive Company*, 1953; M.Nicholson, M. O'Neill, *Glasgow Locomotive Builder to the World*, 1987
Partnership	Diversified	No Known Archive	None Published
Partnership	Partner Change	Glasgow University archives; The Mitchell Library, Glasgow; Search Engine - NRM (NBL)	*A History of the North British Locomotive Company*, 1953; M.Nicholson, M. O'Neill, *Glasgow Locomotive Builder to the World*, 1987
Partnership	Moved Site		
Partnership	Partner Change		
Partnership	-		
Partnership	Diversified	No Known Archive	None Published
Partnership	Diversified	No Known Archive	None Published
Partnership	Diversified	No Known Archive	None Published
Partnership	Incorporation	No Known Archive	J.E. Chacksfield, *The Drummond Brothers: A Scottish Duo*, 2005
Private Ltd. Co.	-	University of Glasgow (GB 248)	
Partnership	Diversified	No Known Archive	None Published
Partnership	Diversified	No Known Archive	*Scotts of Greenock – An Illustrated History* By William Kane, Vincent P. Gillen, 2012
Partnership	Moved site	No Known Archive	Peter L. Hogg, 'Richies', *A History of Thomas Richardson & Sons and Richardsons, Westgarth & Co. Ltd., 1832-1994*, Hartlepool, 1994.
Partnership	Partner Change		
Partnership	Diversified		
Partnership	Partner Change	Search Engine - NRM (GEC)	R.Wear, 'The Locomotive Builders of Kilmarnock', *Industrial Railway Record*, No. 69, 1977, pp.325-408
Partnership	Failed		
Partnership	Failed		
Partnership	Incorporation		
Private Ltd. Co.	Incorporation		
Public Ltd. Co.	-		
Partnership	Partner Change	Glasgow University Archives	R.Wear, 'The Locomotive Builders of Kilmarnock', *Industrial Railway Record*, No. 69, 1977, pp.325-408
Partnership	Failed		
Partnership	Partner Change		
Partnership	Failed		
Partnership	Partner Change		

Town	Locomotive Works	Start up Year	Period of Locomotive Manufacture	Name of Firm
Kilmarnock (Ayrshire)	Caledonia Works	1847	1886-1892	A. Barclay, Son & Co.
			1899-1972	A. Barclay, Sons & Co. Ltd
	Riverbank Works	1871	1872-1886	Barclays & Co.
			1889-1894	McCulloch, Sons & Kennedy Ltd.
	Townholme Engine Works	1876	1879-1905	Grant Ritchie & Co.
	Vulcan Works	1847	1876-1889	Thos. McCulloch & Sons
			1889-1890	McCulloch, Sons & Kennedy Ltd.
Leeds (Yorkshire)	Airedale Foundry	1835	1838-1838	Todd, Kitson & Laird
			1838-1842	Kitson, Laird & Co.
			1842-1858	Kitson, Thompson & Hewitson
			1858-1863	Kitson & Hewitson
			1863-1899	Kitson & Co.
			1899-1934	Kitson & Co. Ltd.
	Boyne Engine Works	1858	1859-1905	Manning, Wardle & Co.
	Hunslet Engine Works	1864	1865-1902	Hunslet Engine Co.
	Railway Foundry	1838	1839-1844	Shepherd & Todd
			1844-1846	Shepherd & Wilson
			1846-1847	Fenton Craven & Co.
			1847-1858	E.B. Wilson & Co.
			1860-1870	Hudswell & Clarke
			1870-1880	Hudswell, Clarke & Rodgers
			1880-1899	Hudswell, Clarke & Co.
			1900 on	Hudswell, Clarke & Co. Ltd.
	Round Foundry	1795	1812-1826	Fenton, Murray & Wood
			1826-1843	Fenton, Murray & Jackson
	Steam Plough Works	1862	1866-1886	John Fowler & Co.
			1886-1941	John Fowler & Co. (Leeds) Ltd.
Leith (Edinburghshire)	Leith Engine Works	c1835	1846-1850	R. & W. Hawthorn
			1850-c1886	Hawthorns & Co.
Lincoln (Lincolnshire)	Sheaf Iron Works	1857	1866-1889	Ruston Proctor & Co.
			1889-1897	Ruston Proctor & Co. Ltd.
Liverpool (Lancashire)	Bath Street Foundry	1826	1833-1843	J.P. Mather Dixon
	Clarence Foundry	c1828	1830-1842	Edward Bury & Co.
			1842-1851	Bury Curtis & Kennedy
	St. George's Engine Works	1852	1853-1863	John Jones & Sons
	Vauxhall Foundry	1827	1834-1847	George Forrester & Co.

APPENDIX – LIST OF INDEPENDENT LOCOMOTIVE WORKSHOPS

Type of Firm	Reason for Change	Main Archive	Works History
Partnership	Incorporation	Glasgow University Archives	R.Wear, 'The Locomotive Builders of Kilmarnock', *Industrial Railway Record*, No. 69, 1977, pp.325-408
Private Ltd. Co.	-		
Partnership	Failed	No Known Archive	
Private Ltd. Co.	Failed		
Partnership	-	No Known Archive	
Partnership	Incorporation	No Known Archive	
Private Ltd. Co.	Diversified	No Known Archive	
Partnership	Partner Change	National Archives; The Stephenson Locomotive Society Library - Kitson Collection; Search Engine - NRM (KIT); Armley Mills Industrial Museum, Leeds Archives	Short Histories of Famous Firms (XVII), *The Engineer*, Nov. 23 1923; E.K. Clark, *Kitsons of Leeds*, n.d. but 1937; E.F. Clark, 'A Very Special Family Birthday', *The Railway Magazine*, May 1989
Partnership	Partner Change		
Partnership	Partner Change		
Partnership	Partner Change		
Partnership	Incorporation		
Private Ltd. Co.	-		
Partnership	-	National Museums Liverpool (Maritime Archives & Library) (B/VF), Armley Mills Industrial Museum	F.W. Harman, *The Locomotives Built by Manning Wardle & Co.*, (3 Vols.), 1998
Partnership	-	West Yorkshire Archives (WYL 1617 + 4105), Armley Mills Industrial Museum, Leeds	D.H. Townsley, *The Hunslet Engine Works*, Norwich, 1998; L.T.C. Rolt, *A Hunslet Hundred*, 1964
Partnership	Partner Change	I. Mech E. archive (END/10/2); Armley Mills Industrial Museum, Leeds; Search Engine - NRM (HUDS); Staffordshire & Stoke-on-Trent Archive Service (D/240)	R. Redman, *The Railway Foundry Leeds 1839-1969*, 1972
Partnership	Partner Change		
Partnership	Partner Change		
Partnership	Failed		
Partnership	Partner Change		
Partnership	Partner Change		
Partnership	Incorporation		
Private Ltd. Co.	-		
Partnership	Partner Change	IMechE Archives (END/10)	E. Craven, Locomotive Builders of the Past (III), *The Stephenson Locomotive Society Journal*, Vol. 30, pp.411-3; P.M. Thompson, *Matthew Murray and the Firm of Fenton Murray & Co.*, 2015
Partnership	Failed		
Partnership	Incorporation	Museum of English Rural Life, Reading (TR9 FOW)	*The Story of the Steam Plough Works*, Michael R. Lane, 1980
Public Ltd. Co.	-		
Partnership	Sale of assets	Tyne & Wear archives	Hawthorns & Co. and The Leith Engine Works, Robert Humm, *Archive* No. 110, June 2021, pp. 3-20
Partnership	Diversified		
Partnership	Incorporation	Museum of English Rural Life (TR)	R. H. Clarke, *Steam Engine Builders of Lincolnshire*, 2nd Ed. City of Lincoln Local History Society; B. Newman, *One Hundred Years of Good Company, Ruston & Hornsby*, 1957; C.H. Codling, Locomotive Builders of the Past (Part V), Ruston Proctor & Co., *Stephenson Locomotive Society Journal*, Vol.34, pp.66-68
Partnership	Diversified		
Partnership	Failed	No Known Archive	None Published
Partnership	Partner Change	IMechE Archives - Catalogue by C.E. Stretton	E.L.Ahrons, 'Short Histories of Famous Firms' (XVI), *The Engineer* 2/2/1923
Partnership	Failed		
Partnership	Diversified	No Known Archive	None Published
Partnership	Diversified	No Known Archive	None Published

Town	Locomotive Works	Start up Year	Period of Locomotive Manufacture	Name of Firm
Llanelly (Carmarthenshire)	South Wales Foundry	1846	1846-1848	Grylls & Co.
London	Fairfield Works (Bow)	1843	1847-1850	W.B. Adams
	Hatcham Ironworks (New Cross)	1839	1849-1869	George England & Co.
			1869-1871	Fairlie Engine & Steam Carriage Co.
	Holland Street Works (Blackfriars)	1833	1838-1843	G. & J. Rennie
	New Road Works (Bloomsbury)	1818	1829-c1836	Braithwaite & Ericsson
			c1836-1837	Braithwaite Milner & Co.
			1837-1841	Braithwaite & Milner
	Thames Bank Ironworks (Rotherhithe)	?	1848-1849	Christie Adams & Hill
Loughborough (Leicestershire)	Falcon Works	1863	c1865-1877	Henry Hughes & Co.
			1877-1881	The Hughes Locomotive & Tramway Works Ltd.
			1882-1889	Falcon Engine & Car Works Ltd.
			1889-1949	Brush Electrical Engineering Co. L
Manchester (Lancashire)	Atlas Works	1843	1843-1852	Sharp Brothers
			1852-1863	Sharp Stewart & Co.
			1864-1888	Sharp Stewart & Co. Ltd
	Banks Works	c1833	1835-1840	T. Banks & Co.
	Bridgewater Foundry, Patricroft	1836	1839-1850	Nasmyths Gaskell & Co.
			1850-1857	J. Nasmyth & Co.
			1857-1867	Patricroft Iron Works
			1867-1882	Nasmyth Wilson & Co.
			1882-1919	Nasmyth Wilson & Co. Ltd.
	Caledonian Foundry	1806	1831-1839	Galloway, Bowman & Glasgow
	Canal Street Works	1817	1839-1859	W. Fairbairn & Sons
			1859-1863	Fairbairn & Co.
	Garforth Works, Dukinfield		1847-1850	W.J. & J. Garforth

APPENDIX – LIST OF INDEPENDENT LOCOMOTIVE WORKSHOPS

Type of Firm	Reason for Change	Main Archive	Works History
Partnership	Failed	No Known Archive	R.Craig, R.Protheroe Jones & M.V. Symons *The Industrial and Maritime History of Llanelli and Burry Port, 1750 to 2000*
Partnership	Diversified	No Known Archive	H.T. Wood, revised by R. Harrington (2004). *Oxford Dictionary of National Biography*, O.U.P.
Partnership	Sale of assets	No Known Archive	Short Histories of Famous Firms (XII), *The Engineer*, Jul. 15 1921; C.H.Dickson, 'Locomotive Builders of the Past (VI)', *Stephenson Locomotive Society Journal*, Vol. 37, pp.138-143 & 207-213; D. Perrett & O. James, 'The Hatcham Ironworks, New Cross' in *London's Industrial Archaeology No.3*, 1984, Greater London I.A.S.
Partnership	Failed	No Known Archive	
Partnership	Diversified	No Known Archive	E.L. Ahrons, 'Short Histories of Famous Firms' (XI), *The Engineer* – 8/4/1921.
Partnership	Partner Change	London Metropolitan Archives (MS; O)	None Published
Partnership	Partner Change		
Partnership	Failed		
Partnership	Diversified	No Known Archive	None Published
Partnership	Partner Change	No Known Archive	J.H.R. Nixon, *Brush Traction 1865-1965*, Loughborough, 1965. D.F. Hartley, 'Steam Locomotives from Loughborough' in *RCHS Journal*, 241, July 2021, pp 297-310
Partnership	Failed		
Partnership	Sale of assets	Leicestershire, Leicester & Rutland Record Office (DE)	
Public Ltd. Co.	-	Leicestershire, Leicester & Rutland Record Office (DE); I.E.T. Archives (NAEST 11)	
Partnership	Partner Change	Glasgow University archives; Mitchell Library, Glasgow; Search Engine - NRM (NBL)	'Short Histories of Famous Firms', Ernest Leopold Ahrons, *The Engineer* – 24/08/1923; Richard L. Hills, *Life and Inventions of Richard Roberts 1789-1864*, 2002
Partnership	Incorporation		
Public Ltd. Co.	Moved site		
Partnership	Diversified	No Known Archive	None Published
Partnership	Partner Change	Salford City Archive Service (U 268 & 269); Search Engine - NRM (NAS); Lancashire Archives (DDX 260); National Library of Scotland (MS.3241-3)	Samuel Smiles (Ed.), *James Nasmyth An Autobiography*, 1883; *James Nasmyth and the Bridgewater Foundry*, J.A.Cantrell, Manchester 1984
Partnership	Partner Change	Salford City Archive Service (U 49 & U 268 & 269); Search Engine, NRM (NAS); National Library of Scotland (MS.3241-3); I.Mech E. (IMS 99)	
Partnership	Partner Change		
Partnership	Incorporation	Salford City Archive Service (U 49 & U 268 & 269); Search Engine, NRM (NAS); National Library of Scotland (MS.3241-3)	
Private Ltd. Co.	-		
Partnership	Failed	No Known Archive	P.C. Dewhurst and S.H.P. Higgins, 'Locomotive Builders of the Past - II: Galloway, Bowman and Glasgow', *Journal of the Stephenson Locomotive Society*, Vol.XXIX, No. 342.
Partnership	Partner Change	No Known Archive	Sir W. Fairbairn & W. Pole, *The Life of Sir William Fairbairn, Bart*, 1877: *William Fairbairn–The Experimental Engineer*, R. Byrom, RCHS, 2017
Partnership	Diversified		
Partnership	Diversified	No Known Archive	None Published

Town	Locomotive Works	Start up Year	Period of Locomotive Manufacture	Name of Firm
Manchester (Lancashire)	Globe Works	1826	1833-1843	Sharp Roberts & Co.
	Gorton Foundry	1854	1855-1883	Beyer Peacock Co.
			1883-1966	Beyer Peacock Co. Ltd.
	Newton Moor Ironworks, Hyde	1842	1866-1896	Daniel Adamson & Co.
	Soho Iron Works	c1800	1839	Peel Williams & Peel
Middlesbrough (Yorkshire)	Teesside Engine Works	1844	1847-1865	Gilkes Wilson & Co.
			1865-1874	Hopkins Gilkes & Co. Ltd.
			1874-1875	Teesside Iron and Engine Works Co.
Neath (Glamorganshire)	Neath Abbey Ironworks	1792	1829-1880	Neath Abbey Iron Co.
Newcastle-upon-Tyne (Northumberland)	Forth Banks Works	1817	1830-1870	R & W Hawthorn
			1870-1885	R & W Hawthorn
			1885-1886	R & W Hawthorn Leslie & Co. Ltd
			1886-1937	R & W Hawthorn Leslie & Co. Ltd
	Forth Banks West Works	1846	1846-1851	Robert Stephenson & Co.
			1867-1894	J & G. Joicey & Co.
	Forth Street Works	1823	1825-1886	Robert Stephenson & Co.
			1886-1899	Robert Stephenson & Co. Ltd.
			1899-1902	Robert Stephenson & Co. Ltd.
	Elswick Engine Works	1847	1847-1864	W.G. Armstrong & Co.
	Ouseburn Engine Works	1853	1855-1860	Robert Morrison & Co.
Newton-le-Willows (Lancashire)	Viaduct Foundry	1834	1834-1844	Jones Turner & Evans
			1844-1852	Jones & Potts
	Vulcan Foundry	1834	1834-1847	Charles Tayleur & Co.
			1847-1864	Vulcan Foundry Co.
			1864-1897	Vulcan Foundry Co. Ltd.
			1898-1955	Vulcan Foundry Ltd.
Northfleet (Kent)	Northfleet Ironworks	1847	1848	A. Horlock & Co.
Oakengates (Shropshire)	Donnington Wood Works	1802	1862-1880	Lilleshall Co.
			1880-1888	Lilleshall Co. Ltd.
Paisley (Renfrewshire)	Barr & McNab Works	1838	1840	Barr & McNab

APPENDIX – LIST OF INDEPENDENT LOCOMOTIVE WORKSHOPS

Type of Firm	Reason for Change	Main Archive	Works History
Partnership	Moved site	Glasgow University archives; The Mitchell Library, Glasgow; Search Engine - NRM (NBL)	'Short Histories of Famous Firms', Ernest Leopold Ahrons, *The Engineer* – 24/08/1923.
Partnership	Incorporation	Museum of Science & Industry, Manchester archives (YA/1966)	*Beyer Peacock - Locomotive Builders to the World*, R.L. Hills & D. Patrick, Glossop, 1982
Private Ltd. Co.	-		
Partnership	Diversified	Tameside Archives (DD 335), Museum of Science & Industry, Manchester (YMS), Greater Manchester Record Office (B/ADM)	None Published
Partnership	Diversified	Museum of Science & Industry, Manchester archives (YA/2003)	W.H. Wright & S.H.P. Higgins, Locomotive Builders of the Past (IV), *The Stephenson Locomotive Society Journal*, Vol. 32, pp.272-277
Partnership	Incorporation	National Archives (RAIL 667/193)	None Published
Private Ltd. Co.	Incorporation		
Public Ltd. Co.	Diversified		
Partnership	Diversified	Glamorgan archive service (D/D NAI)	'Neath Abbey Ironworks Collection: List of drawings relating to contracts placed in 1792-1882', Glamorgan Archive Service, Swansea
Partnership	Sale of assets	Search Engine - NRM (HL); Tyne & Wear archives (DS - DX); Discovery Museum, Newcastle-upon-Tyne	*Power on Land and Sea*, J. F. Clarke, Newcastle-upon-Tyne, n.d. but 1979
Partnership	Incorporation		
Private Ltd. Co.	Incorporation		
Public Ltd. Co.	-		
Partnership	Moved from site	Search Engine - NRM (ROB)	J.G.H. Warren, *A Century of Locomotive Building 1823-1923*, 1923; M.R. Bailey (Ed) *Robert Stephenson: The Eminent Engineer*, 2003
Partnership	Diversified	No Known Archive	None Published
Partnership	Incorporation	Search Engine - NRM (ROB)	J.G.H. Warren, *A Century of Locomotive Building 1823-1923*, 1923; M.R. Bailey, 'Robert Stephenson & Co, 1823-1829', Trans. Newcomen Society, Vol 50 (1978-79), pp.109-138; M.R. Bailey (Ed) *Robert Stephenson: The Eminent Engineer*, 2003
Partnership	Failed		
Public Ltd. Co.	-		
Partnership	Diversified	Tyne & Wear Archives (DS.VA)	*Biography of W. G. Armstrong*, Peter McKenzie, Newcastle-on-Tyne 1983: *William Armstrong – Magician of the North*, by Henrietta Heald, 2010
Partnership	Diversified	No Known Archive	None Published
Partnership	Partner Change	I Mech E archive (IMS 245 - 252 + MCFW)	'Short Histories of Famous Firms', E.L. Ahrons, *The Engineer* 14/5/1920
Partnership	Failed		
Partnership	Partner Change	Merseyside Maritime Museum (B/VF); Newton-le-Willows Community Library; St Helens Local History and Archives Library; Search Engine - NRM (GEC)	*The Vulcan Locomotive Works 1830-1930*, London, 1930
Partnership	Incorporation		
Private Ltd. Co.	Incorporation		
Public Ltd. Co.	-		
Partnership	Diversified	No Known Archive	None Published
Partnership	Incorporation	Ironbridge Gorge Museum Library & Archives (1986/1 & 1998.320 (DLIL/3)); Staffordshire & Stoke-on-Trent Archives Service (D/641)	"Lilleshall: Economic history', in A History of the County of Shropshire: Volume 11, 1985; Bob Yate, *The Railways and Locomotives of the Lilleshall Company*, 2008
Public Ltd. Co.	Diversified		
Partnership	Diversified	No Known Archive	None Published

Town	Locomotive Works	Start up Year	Period of Locomotive Manufacture	Name of Firm
Paisley (Renfrewshire)	Barr & McNab Works	1838	1840	Barr & McNab
Ripley (Derbyshire)	Butterley Ironworks	1790	1838-1839	Butterley Co.
			1860-1910	Butterley Co. Ltd.
Rochester (Strood, Kent)	Invicta Works	1858	1864-1895	Aveling & Porter
			1895-1937	Aveling & Porter Ltd.
Rotherham (Yorkshire)	Holmes Engine Works	c1840	1849-1867	Dodds & Son
Sheffield (Yorkshire)	Meadow Hall Works	1865	1865-1948	Yorkshire Engine Co. Ltd.
Shildon (Co. Durham)	Phoenix Iron Works (later Soho Works)	1833	1833-1838	Hackworth & Downing
			1838-1839	Thomas Hackworth & Co.
			1839-1840	Hackworth Brothers
			1840-1850	Timothy Hackworth
Southampton (Hampshire)	Millbrook Foundry	1834	1837-1839	Summers Groves & Day
Stafford (Staffordshire)	Castle Engine Works	1875	1876-1887	W.G. Bagnall
			1887-1948	W.G. Bagnall Ltd.
St. Helens (Lancashire)	Providence Works	c1860	1872-1912	E. Borrows & Sons
	Sutton Engine Works	1864	1864-1869	James Cross & Co.
Stockton-on-Tees (Co. Durham)	Fossick & Hackworth's Works	1839	1839-1865	Fossick & Hackworth
			1865-1866	Fossick & Blair
	Teesdale Ironworks	1862	1866-1867	Head Ashby & Co
			1867-1876	Head Wrightson & Co. Ltd.
Stoke-on-Trent (Staffordshire)	California Works	c1875	1880-1891	Hartley & Arnoux Brothers
			1891-1892	Hartley Arnoux & Fanning
			1892-1894	Kerr Stuart & Co.
			1894-1930	Kerr Stuart & Co. Ltd.
Warrington (Lancashire)	Dallam Foundry	c1837	1837-1841	Thomas Kirtley & Co.
Wigan (Lancashire)	Haigh Foundry	1790	1835-1856	Haigh Foundry Co.
	Pagefield Ironworks	1874	1874-1880	J.S. Walker & Brothers
			1880-1888	Walker Brothers (Wigan) Ltd.
Whitehaven (Cumberland)	Lowca Works	1763	1840-1857	Tulk & Ley
			1857-1884	Fletcher Jennings & Co.
			1884-1905	Lowca Engineering Co. Ltd.
Wolverhampton (Staffordshire)	Village Foundry, Covan	1857	1862-1874	J. Smith
Worcester (Worcestershire)	Worcester Engine Works	1865	1865-1872	Worcester Engine Works Co. Ltd.
Wylam (Co. Durham)	Wylam Iron Works	1810	1839-1841	Thompson Brothers

APPENDIX – LIST OF INDEPENDENT LOCOMOTIVE WORKSHOPS

Type of Firm	Reason for Change	Main Archive	Works History
Partnership	Diversified	No Known Archive	None Published
Partnership	Diversified	Derbyshire Record Office (D503)	Philip Riden, *The Butterley Company, 1790-1830*, 1990.
Private Ltd. Co.	-		
Partnership	Incorporation	Museum of English Rural Life, Lincolnshire Archives (AB); Suffolk Record Office (HC30)	J.M. Preston, Aveling & Porter Ltd Rochester, 1987.
Private Ltd. Co.	-		
Partnership	Failed	No Known Archive	S. Snell, *A Story of Railway Pioneers — being an account of the inventions and works of Isaac Dodds and his son Thomas Weatherburn Dodds*, 1921
Public Ltd. Co.	-	Sheffield City Archives (YEC)	*Yorkshire Engine Company, Sheffield's Locomotive Manufacturer*, Tony Vernon, 2006.
Partnership	Partner Change	Search Engine - NRM (HACK)	*Timothy Hackworth and the Locomotive*, Robert Young, London, 1923
Partnership	Partner Change		
Partnership	Partner Change		
Partnership	Failed		
Partnership	Diversified	No Known Archive	W.H. Wright & S.H.P. Higgins, Locomotive Builders of the Past (1), *Stephenson Locomotive Society Journal*, Vol. 28, pp.263-4
Partnership	Incorporation	Staffordshire and Stoke-on-Trent archive service, Staffordshire Record Office (D4338); Search Engine - NRM (WGB)	A. Civil & A.C. Baker, *Bagnalls of Stafford*, 1974; A.C. Baker & T.D. Allen, *Bagnalls of Stafford*, 2007
Private Ltd. Co.	-		
Partnership	-	No Known Archive	None Published
Partnership	Failed	No Known Archive	J.M. Tolson, *The St Helens Railway: Its Rivals and Successors*, 1983
Partnership	Partner Change	Search Engine - NRM (HACK)	None Published
Partnership	Diversified		
Partnership	Incorporation	Teesside Archives (U.HW)	Anon., *A Brief History of Head Wrightson & Co. Ltd. from 1859 to 1952*, 1952.
Private Ltd. Co.	Diversified		
Partnership	Partner Change	Leeds Industrial Museum at Armley Mills	*A Hunslet Hundred*, L T C Rolt, Dawlish, 1964
Partnership	Sale of assets		
Partnership	Incorporation		
Private Ltd. Co.	-		
Partnership	Failed	No Known Archive	None Published
Partnership	Sale of assets	Wigan Archives Service (D/D Hai)	Alan Birch, *The Haigh Ironworks, 1790-1856: A Nobleman's Enterprise During the Industrial Revolution*, 1951
Partnership	Incorporation	Wigan Archives Service (D/DY Pag)	None Published
Private Ltd. Co.	Diversified		
Partnership	Sale of assets	Cumbria Archive & Local Studies Centre, Whitehaven (DBH 23/2)	Anthony Coulls, *The Lowca Legacy*, 2016.
Partnership	Incorporation	Search Engine, NRM (FLET)	
Private Ltd. Co.	-	National Archives (BT 31); Search Engine - NRM (FLET)	
Partnership	Failed	No Known Archive	David H. Tew, 'John Smith of Coven, Engineer, 1827-1879', *Trans. of the Newcomen Society*, Vol.26 (1947-1949), pp.161-168.
Public Ltd. Co.	Failed	No Known Archive	Ian Martin, 'On the outside looking in: a short history of the Worcester Engine Works Co Ltd, 1864–1872', *Journal of the Railway & Canal Historical Society* (238) (July 2020): pp.70–81
Partnership	Diversified	No Known Archive	Philip R.B.Brooks, *Where Railways Were Born: The Story of Wylam and its Railway Pioneers*, 1979/2003

Bibliography

Ahrons, E.L., *The British Steam Railway Locomotive 1825-1925*, London, 1927.

Aldcroft, Derek H., (Ed), *The Development of British Industry and Foreign Competition 1875-1914*, London, 1968.

Alford, B.W.E., 'Entrepreneurship, Business Performance and Industrial Development', *Business History*, Vol. XIX, No. 2, 1977, pp 116-133.

Arnold, A.J., 'Should Historians Trust Late Nineteenth-Century Company Financial Statements?', *Business History*, Vol. 38, No.2, 1996, pp 40-54.

Atkins, Philip, *The Golden Age of Steam Locomotive Building*, Penryn, 1999.

Bailey, Michael R., 'Robert Stephenson & Co. 1823-1829', *Transactions of the Newcomen Society*, Vol. 50, 1978-1979, pp 109-138.

Bailey, Michael R., 'George Stephenson - Locomotive Advocate: The Background to the Rainhill Trials', *Transactions of the Newcomen Society*, Vol. 52, 1980-1981, pp 171-179.

Bailey, Michael R., *Robert Stephenson & Co. 1823-1836*, unpublished MA thesis, University of Newcastle on Tyne, 1984.

Bailey, Michael R., 'Learning Through Replication: The *Planet* Locomotive Project', *Transactions of the Newcomen Society*, Vol. 68, 1996/7, pp 109-136.

Bowen-Cooke, C.J., *British Locomotives*, London, 1893.

Boyns, Trevor, and Edwards, John Richard, 'Accounting Systems and Decision-Making in the mid-Victorian Period: The Case of the Consett Iron Company', *Business History*, Vol. 37, July 1995, pp 28-51.

Boyns, Trevor, and Edwards, John Richard, 'The Construction of Cost Accounting Systems in Britain to 1900: The Case of the Coal, Iron and Steel Industries', *Business History*, Vol. 39, 1997, pp 1-29.

Boyns, T., Edwards, J.R., and Nikitin, M., 'Comptabilité et Revolution Industrielle: Une Comparaison Grand Bretagne/France', in *Comptabilité Contrôle Audit, La Revue de l'Association Française de Comptabilité*, tome 2, Vol. 1, 1996, pp 15-20.

Bradley, Ian, *A History of Machine Tools*, Hemel Hempstead, 1972.

Braverman, H., *Labor and Monopoly Capitalism*, New York, 1974.

Broadberry, S.N., *The Productivity Race: British Manufacturing in International Perspective, 1850-1990*, Cambridge, 1997.

Brooks, Randall C., 'Towards the Perfect Screw Thread: the Making of Precision Screws in the 17th-19th Centuries', *Transactions of the Newcomen Society*, Vol. 64, 1992-93, pp 101-119.

Brown, J., and Rose, M.B., *Entrepreneurship, Networks and Modern Business*, Manchester, 1993.

Brown, John K., *The Baldwin Locomotive Works, 1831-1915*, Baltimore, 1995.

Burawoy, M., *The Politics of Production*, London, 1985.

Burgess, K., *The Origins of British Industrial Relations*, 1975.

Burton, Anthony, *The Railway Empire*, London, 1994; republished, Bradford, 2018.

Cantrell, J.A., *James Nasmyth and the Bridgewater Foundry*, Manchester, 1984.

Cantrell, J.A., 'James Nasmyth and the Steam Hammer', *Transactions of the Newcomen Society*, Vol. 56, 1984-85, pp 133-138.

Casson, Mark, 'Institutional Economics and Business History: A Way Forward?', *Business History*, Vol. 39, No. 4, October 1997, pp 151-154.

Chaloner, W.H., 'New Light on Richard Roberts, textile engineer (1789-1864)', *Transactions of the Newcomen Society*, Vol. XLI, 1968-69, pp 27-44.

Chandler, Alfred D. Jr., *The Visible Hand: The Managerial Revolution*, Harvard, 1977.

Chandler, Alfred D. Jr., and Daems, Herman, *Managerial Hierarchies*, Harvard, 1980.

Chandler, Alfred D. Jr., *Scale and Scope The Dynamics of Industrial Capitalism*, Harvard, 1990.

Channon, Geoffrey, 'A.D. Chandler's 'visible hand' in Transport History - A Review Article', *The Journal of Transport History*, 1981, pp 53-64.

Chapman, S.D., *Merchant Enterprise in Britain: From the Industrial Revolution to World War I*, Cambridge, 1992.

Charlesworth, A., Gilbert, D., Randall, A., Southall, H., and Wrigley, C., (Eds) *An Atlas of Industrial Protest in Britain 1750-1990*, Basingstoke, 1996.

Church, Roy, 'The Family Firm in Industrial Capitalism: International Perspectives on Hypotheses and History', *Business History*, Vol. 35, No. 4, 1993, pp 17-43.

Churella, Albert, 'Corporate Culture and Marketing in the American Railway Locomotive Industry: American Locomotive and Electro-Motive Despond to Dieselization', *Business History Review*, Vol. 69, 1995, pp 191-229.

Clark, Daniel Kinnear, *Railway Machinery: A Treatise on the Mechanical Engineering of Railways*, Glasgow, 1855.

Clark, Edwin Kitson, *Kitsons of Leeds*, London, n.d. but 1937.

Clarke, J.F., *Power on Land and Sea*, Newcastle upon Tyne, n.d. but 1979.

Coase, R.H., 'The Nature of the Firm', *Economica*, Vol. IV, 1937, pp 386-405.

Colburn, Zerah, *Locomotive Engineering and the Mechanism of Railways*, 2 Vols., London, 1871.

Collins, Michael, 'English Bank Lending and the Financial Crisis of the 1870s', *Business History*, Vol. 32, No. 1, 1990, pp 198-224.

Cookson, Gillian, 'Family Firms and Business Networks: Textile Engineering in Yorkshire, 1780-1830', *Business History*, Vol. 39., No. 1, 1997, pp 1-20.

Corley, T.A.B., 'Britain's Overseas Investments in 1914 Revisited', *Business History*, Vol. 36, No. 1, 1994, pp 71-88.

Cottrell, P.L., *British Overseas Investment in the Nineteenth Century, Studies in Economic and Social History*, London, 1975.

Cottrell, P.L., 'Railway Finance and the Crisis of 1866: Contractors' Bills of Exchange, and the Finance Companies', *The Journal of Transport History*, Vol. III, New Series, No. 1, 1975, pp 20-38.

Cottrell, P.L., *Industrial Finance 1830-1914: The Finance and Organisation of English Manufacturing Industry*, London, 1980.

Crouzet, François, 'Essor, Déclin et Renaissance de l'Industrie Française des Locomotives, 1838-1914', *Revue d'Histoire Economique et Sociale*, Vol. 55, 1977, pp 112-209.

Däbritz, Walther, and Metzeltin, E.H. Erich, *Hundert Jahre Hanomag*, Düsseldorf, 1935.

Darbishire, J. Edward, 'Modern Locomotive Design and Construction', *The Railway Engineer*, Vols. IV & V, 1883/4.

Davis, C.P., *Locomotive Building in Bristol, The Avonside Ironworks (1837-1882)*, unpublished BA Dissertation, University of Bristol, 1979.

Devine, P.J., 'The Firm' and 'Corporate Growth', in Devine, P.J., Jones, R.M., Lee, N., and Tyson W.J., (Eds), *An Introduction to Industrial Economics*, London, 1976.

Dickinson, H.W., 'Richard Roberts, his Life and Inventions', *Transactions of the Newcomen Society*, Vol. XXV, 1945-1947, pp 123-137.

Dodsworth, Charles, 'The Low Moor Ironworks Bradford', *Industrial Archaeology*, 1965, pp 122-164.

Drummond, Di, 'Specifically Designed'? Employers' Labour Strategies and Worker Responses in British Railway Workshops, 1838-1914', *Business History*, Vol. 31, No. 2, 1989, pp 8-31.

Drummond, Diane K., *Crewe Railway Town, Company and People 1840-1914*, Aldershot, 1995.

Drummond, Di, 'Technology and the Labour Process: A Preliminary Comparison of British Railway Companies' Approaches to Locomotive Construction Before 1914', *Perspectives on Railway History*, Working Papers in Railway Studies Number One, Institute of Railway Studies, York, 1997, pp 22-34.

Edelstein, Michael, *Overseas Investment in the Age of High Imperialism, The United Kingdom 1850-1914*, New York, 1982.

Edwards, John Richard, and Newell, Edmund, 'The Development of Industrial Cost and Management Accounting Before 1850: A Survey of the Evidence', *Business History*, Vol. 33, No. 1, 1991, pp 35-57.

Elbaum, B., and Lazonick, W., 'An Institutional Perspective on British Decline', in Elbaum, B., and Lazonick, W., (Eds), *The Decline of the British Economy*, Oxford, 1986.

Ericson, Steven J., *The Sound of the Whistle, Railroads and the State in Meiji Japan*, Council on East Asian Studies, Harvard, 1996.

Evans, C., 'Manufacturing Iron in the North-East During the Eighteenth Century: The Case of Bedlington', *Northern History*, Vol. 28 (1992), pp 178-196.

Evans, F.T., 'The Maudslay Touch: Henry Maudslay, Product of the Past and Maker of the Future', *Transactions of the Newcomen Society*, Vol. 66, 1994-95, pp 153-174.

Ewald, Kurt, *125 Jahre Henschel*, Kassel, 1935.

Farnie, D.A., 'The Textile Machine-Making Industry and the World Market, 1870-1960', *Business History*, Vol. 32, No. 4 (1990), pp 150-165.

Fitton, R.S., *The Arkwrights: Spinners of Fortune*, Manchester, 1989.

Fox, Robert (Ed), *Technological Change: Methods and Themes in the History of Technology*, Studies in the History of Science, Technology and Medicine, Harwood Academic, 1996.

Friedman, A., *Industry and Labour*, London, 1977.

Gale, W.K.V., *The Black Country Iron Industry*, London, 1966.

Garner, S.P., *Evolution of Cost Accounting*, Alabama, 1954.

George, K.D., Joll, C., and Lynk, E.L., *Industrial Organisation, Competition, Growth and Structural Change*, London, 1991.

Gilbert, K.R., *The Portsmouth Block-Making Machinery*, London, 1965.

Gilbert, K.R., 'The Control of Machine Tools - A Historical Survey', *Transactions of the Newcomen Society*, Vol. XLIV (1971-72), pp 119-127.

Gordon, D., Edwards, R., and Reich, M., *Segmented Work, Divided Workers*, Cambridge, 1982.

Greenhalgh, Paul, *Ephemeral Vistas, The Expositions Universelles, Great Exhibitions and World's Fairs, 1851-1939*, Studies in Imperialism, Manchester, 1988.

Habakkuk, H.J., *American and British Technology in the Nineteenth Century*, Cambridge, 1962.

Hamilton, S.B., 'Sixty Glorious Years: The Impact of Engineering on Society in the Reign of Queen Victoria', *Transactions of the Newcomen Society*, Vol. XXXI (1957-59), pp 184-187.

Hannah, L., 'Mergers in British Manufacturing Industry, 1880-1918', *Oxford Economic Papers*, Vol XXVI (1974), pp 1-20.

Hannah, L., *The Rise of the Corporate Economy*, 2nd Edition, London, 1983.

Hayward, R.A., *Fairbairns of Manchester*, unpublished MSc dissertation, UMIST, Manchester, 1971.

Hills, R.L., and Patrick, D., *Beyer Peacock Locomotive Builders to the World*, Glossop, 1982.

Jeaffreson, J.C., *The Life of Robert Stephenson, F.R.S.*, 2 Vols., London, 1864.

Jefferys, James B., *The Story of the Engineers*, Amalgamated Engineering Union, London, 1945.

Jenkins, D.T., and Ponting, K.G., *The British Wool Textile Industry 1770-1914*, Pasold Research Fund, 1975.

Jenks, L.H., *The Migration of British Capital to 1875*, London, 1927, re-published 1963.

Jenkins, M., *The General Strike of 1842*, London, 1980.

Jones, Charles, 'Institutional Forms of British Foreign Direct Investment in South America', *Business History*, Vol. 39, No. 2 (1997), pp 21-41.

Jones, Geoffrey, and Rose, Mary B., 'Family Capitalism', *Business History*, Vol. 35, No. 4 (1993), pp 1-16.

Jones, S.R.H., 'Transaction Costs and the Theory of the Firm: The Scope and limitations of the New Institutional Approach', *Business History*, Vol. 39, No. 4 (1997), pp 10-25.

Joyce, P., *Work, Society and Politics: The Culture of the Factory Town in Late Victorian England*, London, 1980.

Kirby, M.W., *Men of Business and Politics, The Rise and Fall of the Quaker Pease Dynasty of North-East England,1700-1943*, London, 1984.

Kirby, M.W., 'Product Proliferation in the British Locomotive Building Industry, 1850-1914: An Engineer's Paradise?', *Business History*, Vol. 30, No. 3 (1988), pp 287-305.

Kirby, M.W., 'Technological Innovation and Structural Division in the UK Locomotive Building Industry, 1850-1914', in Holmes, Colin, and Booth, Alan, (Eds), *Economy and Society: European Industrialisation and Its Social Consequences*, Leicester, 1991.

Lane, Michael R., *The Story of the Steam Plough Works*, London, 1980.

Larkin, Edgar J., and Larkin, John G., *The Railway Workshops of Britain 1823-1986*, Basingstoke, 1988.

Lazonick, W., *Business Organisation and the Myth of the Market Economy*, Cambridge, 1991.

Lee, G.A., 'The Concept of Profit in British Accounting 1760-1900', *Business History Review*, Vol. XLIX (1975), pp 19-33.

Lee, T.A., 'Company Financial Statements, An Essay in Business History 1830-1950', in Marriner, Sheila, (Ed), *Business and Businessmen, Studies in Business, Economic and Accounting History*, Liverpool, 1978.

Legget, Robert F., *Railways of Canada*, Newton Abbot, 1973.

Lewin, H.G., *Early British Railways, &c., 1801-1844*, London, n.d. but 1925.

Lewin, Henry Grote, *The Railway Mania and Its Aftermath: 1845-1852*, London, 1936.

Lindner, Helmut, and Schmalfuß, Jörg, *150 Jahre Borsig Berlin-Tegel*, Berlin, 1987.

Lineham, Wilfred, *A Text-Book of Mechanical Engineering*, London, 1902, pp 313-318.

Lloyd-Jones, R., and LeRoux, A.A., 'Marshall and the Birth and Death of Firms: The Growth and Size Distribution of Firms in the Early Nineteenth-Century Cotton Industry', *Business History*, Vol. XXIV (1982), No. 2, pp 141-155.

Lord, W.M., 'The Development of the Bessemer Process in Lancashire, 1856-1900', *Transactions of the Newcomen Society*, Vol. XXV (1945-1947), pp 163-180.

Lowe, James W., *British Steam Locomotive Builders*, Cambridge, 1975. Later editions 1989 and Barnsley, 2014.

MacLeod, Christine, 'Concepts of Invention and the Patent Controversy in Victorian Britain', in Fox, Robert, (Ed), *Technological Change: Methods and Themes in the History of Technology*, Studies in the History of Science, Technology and Medicine, Harwood Academic, 1996, pp 140/1.

Marriner, S., 'Company Financial Statements as Source Material for Business Historians', *Business History*, Vol. XXII (1980), pp 203-235.

Martin, Evan, *Bedlington Iron & Engine Works 1736-1867*, Newcastle upon Tyne, 1974.

Marx, Thomas G., 'Technological Change and the Theory of the Firm: The American Locomotive Industry, 1920-1955', *Business History Review*, Vol. L, No. 1, 1976, pp 1-24.

Mather, F.C., 'The General Strike of 1842', in Quinault, R., and Stevenson, J., (Eds), *Popular Protest and Public Order*, London, 1974.

McCord, N., '1871 Strike - Prelimineries', in Allen, E., Clarke, J.F., McCord, N., and Rowe, D.J., (Eds), *The Strikes of the North-East Engineers in 1871: The Nine Hours League*, Newcastle, 1971.

McKendrick, N., 'Josiah Wedgewood and Cost Accounting in the Industrial Revolution', *Economic History Review*, 2nd Series, Vol. XXIII (1970).

McKinlay, Alan, and Zeitlin, Jonathan, 'The Meanings of Managerial Prerogative: Industrial Relations and the Organisation of Work in British Engineering, 1880-1939', *Business History*, Vol. 31, No. 2, 1989, pp 32-47.

McNeil, Ian, 'Hydraulic Power Transmission: The First 350 Years', *Transactions of the Newcomen Society*, Vol. 47 (1974-76), pp 149-159.

Measom, G., *The Official Illustrated Guide to the London & North-Western Railway*, London, (n.d. but c.1854).

Millar, John, *William Heap and His Company 1866*, Hoylake, 1976.

Milward, Alan S., and Saul, S.B., *The Development of the Economies of Continental Europe 1850-1914*, London, 1977.

Mokyr, Joel, 'Evolution and Technological Change: A New Metaphor for Economic History?', in Fox, Robert, (Ed), *Technological Change: Methods and Themes in the History of Technology*, Studies in the History of Science, Technology and Medicine, Harwood Academic, 1996.

More, C., *Skill and the English Working Class, 1870-1914*, London, 1980.

Morton, John, *Thomas Bolton & Sons Ltd. 1783-1983*, Ashbourne, 1983.

Moss, Michael S., and Hume, John R., *Workshop of the British Empire*, London, 1977.

Mott, R.A., 'Dry and Wet Puddling', *Transactions of the Newcomen Society*, Vol. 49, 1977-78, pp 153-8.

Musson, A.E., 'James Nasmyth and the Early Growth of Mechanical Engineering', *Economic History Review*, 2nd Series, Vol. 10, No. 1, 1957-58.

Musson, A.E., 'The 'Great Depression' in Britain, 1873-1896: A Reappraisal', *Economic History Review*, Vol. 12 (1959-60), pp 199-228.

Musson, A.E., 'Joseph Whitworth and the Growth of Mass-Production Engineering', *Business History*, Vol. XVII, No. 2 (1975), pp 109-149.

Musson, A.E., 'British Origins', in Mayr, O., and Post, R.C., (Eds) *Yankee Enterprise: the Rise of the American System of Manufactures*, Washington DC, 1982.

Nicolson, Murdoch, and O'Neill, Mark, (Eds), *Glasgow Locomotive Builder to the World*, Glasgow, 1987.

O'Brien, P., (Ed), *Railways and the Economic Development of Western Europe 1830-1914*, Oxford, 1983.

O'Brien, Patrick; Griffiths, Trevor; and Hunt, Philip, 'Technological Change During the First Industrial Revolution: The Paradigm Case of Textiles, 1688-1851', in Fox, Robert (Ed) *Technological Change: Methods and Themes in the History of Technology*, Studies in the History of Science, Technology and Medicine, Harwood Academic, 1996.

Orcutt, H.F.L., 'Machine Shop Management in Europe and America', paper to the Institution of Mechanical Engineers, *Engineering Magazine*, January-March 1902.

De Pambour, Chev. F.M.G., *A Practical Treatise on Locomotive Engines Upon Railways*, London, 1836.

De Pambour, Comte F.M.G., *A Practical Treatise on Locomotive Engines*, London, 1840.

Payen, Jacques, *La Machine Locomotive en France*, Lyon, 1986.

Payne, P.L., 'The Emergence of the Large-Scale Company in Great Britain', *Economic History Review,* Vol. 20 (1967).

Payne, P.L., 'Industrial Entrepreneurship and Management in Great Britain', in Mathias, P., and Postan, M.M., (Eds), *The Cambridge Economic History of Europe VII*, Part I, 1978.

Payne, P.L., *British Entrepreneurship in the Nineteenth Century*, London, 1988.

Picon, Antoine, 'Towards a History of Technological Thought', in Fox, Robert, (Ed) *Technological Change: Methods and Themes in the History of Technology*, Studies in the History of Science, Technology and Medicine, Harwood Academic, 1996.

Pole, William, (Ed), *The Life of Sir William Fairbairn, Bart.*, London, 1877.

Pollard, Sidney, 'Factory Discipline in the Industrial Revolution', *Economic History Review*, 2nd Series, Vol. 16 (1963-64), pp 254-271.

Pollard, Sidney, *The Genesis of Modern Management: A Study of the Industrial Revolution in Great Britain*, Cambridge, Mass, 1965.

Pollard, S., 'Fixed Capital in the Industrial Revolution in Britain', in Creuzet, F., *Capital Formation in the Industrial Revolution*, London, 1972.

Prais, S.J., *The Evolution of Giant Firms in Britain*, Cambridge, 1976.

Prior, Ann, and Kirby, Maurice, 'The Society of Friends and the Family Firm, 1700-1830', *Business History*, Vol. 35, 1993, pp 66-85.

Redman, Ronald N., *The Railway Foundry Leeds: 1839-1969*, Norwich, 1972.

Reed, Brian, 'Norris Locomotives', *Locomotives In Profile*, Windsor, 1971, Vol. 1, No. 11.

Reed, Brian, *150 Years of British Steam Locomotives*, Newton Abbot, 1975.

Reed, Brian, *Crewe Locomotive Works and its Men*, Newton Abbot, 1982.

Revill, George, 'Railway Paternalism and Corporate Culture, 'Railway Derby' and the Formation of the ASRS', in Divall, Colin, (compiler), *Workshops, Identity and Labour*, Working Papers in Railway Studies, Number three, Institute of Railway Studies, York, 1998.

Riden, Philip, *The Butterley Company 1790-1830*, Chesterfield, 1973.

Rippy, J. Fred., *British Investments in Latin America, 1822-1949*, Minneapolis, 1959.

Roll, E., *An Early Experiment in Industrial Organisation: Being a History of the Firm of Boulton & Watt, 1775-1805*, London, 1930.

Rolt, L.T.C., *A Hunslet Hundred*, Dawlish, 1964.

Rolt, L.T.C., *Tools For The Job*, Dawlish, 1965.

Rosenberg, N., 'Economic Development and the Transfer of Technology: Some Historical Perspectives', *Technology and Culture*, Vol. 11 (1970).

Rowe, D.J., 'Trade Unions and Strike Action in the North-East', in Allen, E., Clarke, J.F., McCord, N., and Rowe, D.J., (Eds), *The Strikes of the North-East Engineers in 1871: The Nine Hours League*, Newcastle, 1971.

W.D. Rubinstein, *Capitalism, Culture and Decline in Britain, 1750-1990*, London, 1993.

Saul, S.B., *Studies in British Overseas Trade 1870-1914*, Liverpool, 1960.

Saul, S.B., 'The Engineering Industry', in Derek H Aldcroft (Ed), *The Development of British Industry and Foreign Competition 1875-1914*, London, 1968.

Saul, S.B., 'The Market and the Development of the Mechanical Engineering Industries in Britain, 1860-1914', in Saul, S.B., (Ed), *Technological Change: The United States and Britain in the Nineteenth Century*, London, 1970.

Scranton, Philip, and Licht, Walter, *Work Sights: Industrial Philadelphia, 1890-1950*, Philadelphia, 1986.

Scranton, Philip, *Endless Novelty*, Princeton, 1997.

Simon, H.A., *Models of Bounded Rationality*, 2 Vols., Cambridge, USA, 1982.

Simon, H.A., 'Rationality in Psychology and Economics', *Journal of Business*, Vol. 59 (1986).

Slezak, J.O., *Die Lokomotivfabriken Europas, Internationales Archiv Für Lokomotivgeschichte*, Wien, 1962.

Smiles, Samuel, *The Life of George Stephenson*, London, 1857

Smiles, Samuel, *Industrial Biography*, 1863.

Smiles, Samuel, (ed), *James Nasmyth Engineer An Autobiography*, London, 1883.

Smith, R.T., 'John Gray and His Expansion Valve Gear', *Transactions of the Newcomen Society*, Vol. 50 (1979/80), pp 139-154.

Solomons, D., 'The Historical Development of Costing', in Solomons, D., (Ed), *Studies in Cost Analysis*, 2nd Ed., 1968.

Southall, Humphrey, 'Industrial Protest: 1850-1900', in Charlesworth, Andrew, *et al*, *An Atlas of Industrial Protest in Britain 1750-1990*, Basingstoke, 1996.

Steeds, W., *A History of Machine Tools 1700-1910*, Oxford, 1969.

Sullivan, Richard J., 'The Revolution of Ideas: Widespread Patenting and Invention During the English Industrial Revolution', *Journal of Economic History*, Vol. L., No. 2, (1990), pp 349-361.

Supple, B., *Essays in British Business History*, Oxford, 1977.

Taksa, Lucy, 'Political and Industrial Mobilization, Workplace Culture and Citizenship at the New South Wales Railways and Tramways Department Workshops 1880-1932', in Divall, Colin, (compiler), *Workshops, Identity and Labour*, Working Papers in Railway Studies, Number Three, Institute of Railway Studies, York, 1998, pp 1-24.

Tann, Jennifer, 'Marketing Methods in the International Steam Engine Market: The Case of Boulton and Watt', *Journal of Economic History*, Vol. 38, 1978, pp 363-391.

Thomas, John, *The Springburn Story*, Dawlish, 1964.

Thomas, R.H.G., *The Liverpool & Manchester Railway*, London, 1980.

Usselman, Steven Walter, 'Air Brakes for Freight Trains: Technological Innovation in the American Railroad Industry, 1869-1900', *Business History Review*, Vol. 58 (Spring 1984), pp 30-50.

Usselman, Steven Walter, *Running The Machine: The Management of Technological Innovation on American Railroads, 1860-1910*, unpublished PhD. thesis, University of Delaware, 1985.

Vamplew, Wray, 'Scottish Railways and the Development of Scottish Locomotive Building in the Nineteenth Century', *Business History Review*, Vol. 46, No. 3, 1972, pp 320-338.

van-Helten, J.J., and Cassis, Y., (Eds), *Capitalism in a Mature Economy: Financial Institutions, Capital Exports and British Industry 1870-1939*, London, 1990.

van Riemsdijk, J.T., 'The Compound Locomotive, Part I 1876-1901', *Transactions of the Newcomen Society*, Vol. XLIII (1970-71), pp 1-17.

van Riemsdijk, J.T., *Compound Locomotives*, Penryn, 1994.

Walker, Charles, *Thomas Brassey Railway Builder*, London, 1969.

Wardley, P., 'The Anatomy of Big Business: Aspects of Corporate Development in the Twentieth Century', *Business History*, Vol. 33, No. 2 (1991).

Warren, J.G.H., *A Century of Locomotive Building By Robert Stephenson & Co 1823/1923*, Newcastle on Tyne, 1923.

Wear, Russell, 'The Locomotive Builders of Kilmarnock', *Industrial Railway Record*, No. 69, (1977), pp 325-408.

Westwood, J.N., *Locomotive Designers in the Age of Steam*, London, 1977.

White, John H. Jr., *A History of the American Locomotive*, Baltimore, 1968, later editions New York, 1979 and Baltimore, 1997.

White, John H. Jr., *A Short History of American Locomotive Builders in the Steam Era*, Washington D.C., 1982.

Wiener, M.J., *English Culture and the Decline of the Industrial Spirit, 1850-1980*, Cambridge, 1981.

Wigham, Eric, *The Power to Manage: A History of the Engineering Employers' Federation*, London, 1973.

Wilkins, M., 'The Free-Standing Company, 1870-1914: An Important Type of British Foreign Direct Investment', *Economic History Review*, 2nd Series, Vol. 41 (1988), pp 259-282.

Williamson, O.E., *The Economic Institutions of Capitalism: Firms, Markets, Relational Contracting*, New York, 1985.

Williamson, O.E., *Economic Organisation: Firms, Markets and Policy Control*, London, 1986.

Wilson, John F., *British Business History, 1720-1994*, Manchester, 1995.

Wilson, R.B., (Ed), *D. Gooch Memoirs and Diary*, Newton Abbott, 1972.

Zeitlin, Jonathan, 'Between Flexibility and Mass Production: Strategic Ambiguity and Selective Adaptation in the British Engineering Industry, 1830-1914', in Sabel, Charles F., and Zeitlin Jonathan, (Eds), *World of Possibilities, Flexibility and Mass Production in Western Industrialization*, Cambridge, 1997.

List of Archive collections

Beyer Peacock collection
Manchester, Science & Industry Museum, ref. YA 1966.24.

Bidder collection
Science Museum library and Archives, Wroughton, Wiltshire, ref. GPB.

Hodgkin collection
Durham County Record Office, ref. D/HO.

Jones & Potts collection
Institution of Mechanical Engineers Library and Archive, ref. IMS.

Nasmyths Gaskell & Nasmyth Wilson collection
City of Salford Archives, ref. U 268.

Livesey & Henderson collection
Institution of Mechanical Engineers Library and Archive.

Neilson Reid & Co. collection
University of Glasgow Archive Services, ref. GB 248 UGD 010.

Pease-Stephenson collection
Durham County Record Office, ref. D/PS.

Robert Stephenson & Co. collection
Search Engine, National Railway Museum, York, ref. ROB.

Starbuck collection
Tyne & Wear Record Office, ref. 131.

Vulcan Foundry collection
Merseyside Maritime Museum, Liverpool: Maritime Archives and Library, ref. B/VF, DX/603.

Index

Locomotive manufacturing sites are listed by company name rather than by site.
Page references in **bold** include illustrations.

A

Aberdeen Railway 167
Abt, Carl Roman (innovator, rack railway system) 77
Accidents 70
Accounting procedures 134, 160-170, 174, 190
Adams, William (Engineer) 79, 87
Adamson, Daniel (Engineer) 84
African locomotive market 30, 32, 45, 93
Agents, (representative and commission) 41-4, 57-8, 186
    (*See also* Edward F. Starbuck)
Alexandra (Newport & South Wales) Docks and Railway 168, **173**
Allen, Everitt & Sons (boiler tube manufacturer) 108
Amalgamation of manufacturers 179-180
American Locomotive Company (ALCo) 35, 180, 191/2
American locomotive industry
    (commercial) 24, 32-5, 42, 44-9, 53, 185, 187/8
    (design & manufacturing) 69, 74/5, 81, 84, 89, 91-3, 100, 102, 117, 125/6
    (organisational) 14/15, 19/20, 35, 148, 166, 180, 190-2
American Locomotive Manufacturers Association 180
Anglo Chilean Nitrate & Railway Co. **43**, 79, **172**
Annual locomotion production 18
Anti-friction rings for locomotives 88
Antofagasta (Chile) & Bolivia Railway 79
Apprenticeships 132, 142/3, 146, **147**, 148
Argentinian locomotive market 32, **33**, 34, **125**
Armstrong, Sir William G. 153/4
Articulated locomotive designs 78
Australian locomotive market
    (New South Wales) 19, 24/5, 27/8, 32-4, 42-3, 45, 57/8, **75**, **194**
    (Queensland) 19, 24/5, 27/8, **29**, 32-4, 43, 45, 57/8, 77, **194**
    (South Australia) 19, 24/5, 27/8, 32-4, 43, 45, 54, 56-8
    (Victoria) 19, 24/5, 27/8, **32**, 33/4, 43, **45**, 57/8, 93
Austrian locomotive market 26, 46, 61, 81
Avignon Marseille Railway (France) **86**
Avonside Engine Co. (See also Stothert Slaughter & Co. and Slaughter Grüning & Co.) 26, **54**, 134, 154, **160**, 178
Axle technology 82/3, 95, 107, 110, 114

B

Balance weights 76
Baldwin Locomotive Works, USA
    (commercial) 19, 35, 44, 46, 48/9, 53, 59,
    (design and manufacture) 81, 84, 89, 100, 102, 104, 113, 117/8, 120, 126
    (organisation) 15, 138/9, 145, 147/8, 166, 172

Bangor, Piscataquis Canal & Railroad (USA) **20**
Barings Bank 20, 33
Barclay, Andrew & Son, Barclay, Andrew, Sons & Co. 26, 106, 115, 136, 165, 177, 180
Barclays & Co. 180
Barr Morrison & Co. 49, **50**
Batch production 13-5, 17, 22, 27, 34, 39, 42, 58, 69, 75, 89-90, 100-2, 118-26, 145, 186, 188
Bayonne & Biarritz Railway (France) 81
Beattie, Joseph (engineer) 80/1, 111
Bedlington Iron Co. 105-7, 135, 148/9, **151**, 162, 190
Belgian locomotive industry 19, 23/4, 52/3, 55, 86, 94, 191
Belgian locomotive market 21/2, 26, 46, 55, 80
Benet, L. (French manufacturer) 122, 177
Bennett, Robert (of Beyer Peacock & Co.) 42
Bessemer Steel 82/3, 106/7, 109
Beyer Peacock & Co.
    (administration) 145, 149, 164, 167, 170
    (commercial) **25**, 40-3, 48, **49**, 52, **55**, 56-60, 62
    (design and manufacturing) 77-81, 83, 106/7, 111-5, 117, 126,
    (organisation and strategy) 13, 122, 134, 137, 141, 172, 174, **175**, 177, **178**
Beyer, Charles (of Beyer Peacock & Co.) 40/1, 55, 141, 170, 174/5
Birley family (investors) 174/5
Birmingham Battery Co. (boiler tube manufacturer) 108
Black Hawthorn & Co. 44, **153**, 174
Blastpipe technology 80
Bodmer, John (engineer) 76, 113, 119, 122
Boilers (design and manufacture) 81-3, 105/6, 111, 115, 139, 150
Bolton, Thomas & Sons (copper plate supplier) 84, 108
Bombay & Baroda Railway (India) **127**
Borsig locomotive works (Germany) 35, 41, 56
Bottomley, G. (engineer) 87
Bought-in locomotive components 106-9,
Boulton, Matthew, and Watt, James (engineers) 39, 57, 161
Bourdon, Eugène (French engineer) 108
Bowling Iron Works (Yorkshire) 105/6
Braking systems 87/8
Brassey, Thomas (contractor) 24, 28, 175
Brighton locomotive workshops 145
Bristol & Exeter Railway **142**
British Empire locomotive market 19, 24, 30
Broughton Copper Co. (boiler tube manufacturer) 108
Brown & Sharpe (tool-maker, USA) 113
Browne, Benjamin (of R. & W. Hawthorn Leslie & Co.) 155, 168, 173, 175, 193,
Bruce, George (engineer) 110
Brunel, Isambard K. (engineer) 55
Brunel, Marc (engineer) 100

Brunswick Iron Works 107
Budden, W.H. (of R. Stephenson & Co.) 135
Buddicom, William (engineer) 177
Buenos Aires Great Southern Railway (Argentina) **33**, 92, **93**
Burma Railway **78**
Burnett, Robert (of Beyer Peacock & Co.) 137
Burrell Foundry 104/5
Bury, Edward & Co. 39-40, 46, 71-2, 85, 150, 176
Bury Curtis & Kennedy 23, 47, 51, **70**, 72, **73**, 171, 180
Business plans for locomotive manufacturers 174
Butterley Co. 39

C

Caledonian Railway **41**, 56, **57**, **72**, **167**
Cambrian Railways **24**
Cammell, Charles & Co. (spring-maker) 109
Canadian locomotive market 27, 93, **119**, 177
Capital for railway investment/exports 17-21, 24, 26, 30/1, 36, 186
Capital investment and equipment for manufacturers 21, 35, 99-100, 102, 109-118, 154, 168, 188, 190
Central Argentine Railway **34**
Cheadle Copper & Brass Co. 107
Clark, Daniel Kinnear, (engineering author) 89
Clement, Joseph (manufacturer) 109
Cleminson, James (engineer) 79
Clark, Edwin Kitson (of Kitson & Co.) 145/6, 150, 154, 156, 176
Clayton, William (of Hudswell Clark & Rodgers) 137
Clerical duties for locomotive manufacturers 134-6, 162, 166, 170
Cockerill, John & Co. (Belgian manufacturer) 52, 55/6, 177
Colliery locomotive market 33, **42**, 46, 53, **137**, **189**
Cologne Minden Railway (Germany) 56
Companies Act
   (1856) 12, 14, 161, 178
   (1862) 12, 14, 27, 134, 168
   (1900) 161
Competing Designs in the Overseas Market 92-6
Compound locomotives 81, 87
Condensing locomotives 78
Conner, Benjamin (Caledonian R'way Loco. Super.) 41, 57, 72
Consulting engineers 72, **73**, 74/5, 89/90, 119, 187/8, 191
Contracting industry 24-6, 28/9
Cook, Edward (of R. Stephenson & Co.) 135
Copiapo & Caldera Railway (Chile) **73**
Copper and brass supply industry **84**
Copper fireboxes 57
Coulthard, Ralph (of R. Coulthard & Co.) **174**
Craft locomotive firms 14/5, 34, 79, 99, 102, 112, 116, 120, 123/4, 132, 142, 161, 166, 173/4, 178, 188-193
Crampton, Thomas R. (engineer) 47, **51**, 76, 86, **87**
Cranes (see lifting equipment)
Crank axles 70
Crédit Mobilier (French bank) 26/7
Crewe locomotive works (LNWR)
   (employment) 132/3, 138, 141, 145, 154, 190
   (administration) 23/4, 27, 61/2, 83/4, 89, 107, 113, 116, 126
'Crewe'-type locomotives **72**, 76
Cross, William (of R. & W. Hawthorn Leslie & Co.) 125, 134, 145, 165, 173
Crow, George (of R. Stephenson & Co.) **116**, 124, 136, 138, 149/50

Crown Agents for the Colonies 31, 45
Cudworth, James (South Eastern R'way Loco. Super.) 80

D

Darlington locomotive works (North Eastern Railway) 145
Data cards for locomotives 48, **49**, 91
Davies & Metcalfe (railway equipment manufacturer) 108
Delivery and commissioning of locomotives 38, 60/1, 118, 122, 126/7, 176, 186
Depreciation 161, 168-70, 190
Derby locomotive works (Midland Railway) 24, 138, 154
Design, evolution of locomotive 72-9, 92, 100/1, 118, 126, 146
Development of:
   thermo-dynamic technology 80/1
   material technology 81-4
Dewrance, John (railway engineer) 80, 108
Dick Kerr & Co. 45, 49, 53, 178
Dickinson, Harris (of R. Stephenson & Co.) 135
Diversification 171-4
Dodds & Son **142**
Doncaster locomotive works (Great Northern Railway) 145
Double-Fairlie locomotive type **43**, **78**
Douglas, George K. (of R. Stephenson & Co.) 134
Draughtsmen, drawings and tracings 46, 55, 69, 79, 88/9, **102-4**, 110, 119-21, 143
Dredge, James (engineering journalist) 88
Drills and drilling machines 113, **116**
Dublin & Drogheda Railway **49**
Dübs & Co.
   (commercial) **35**, 53, 62, **77**, **91**, 107, **175**
   (organisational) 14, 94, 123, 134/5, 138, 173, 175, 177, 180, 191/2
Dübs, Henry (of Dübs & Co.) 134, 137, 175, 177, 180
Dudley, Sedgley & Wolverhampton Tramways **172**
Dunaburg Vitebsk Railway (Russia) 29
Dundalk & Enniskillen Railway 55, 59
Dundee, Perth & Aberdeen Junction Railway **87**

E

Eastern Counties Railway 76, 167
Eastern Railway (France) 81
East India Railway 52
East Lancashire Railway **14**, **108**
Eastwick & Harrison (USA locomotive manufacturer) 20
Ebbw Vale Co. **26**
Egyptian locomotive market 41, 55/6, **57**
Electrification of workshops 117
Employers' Federations 150, 152-6, 190
Employers' Liability Act 149
Employment 13/4, 19, 35, 102, 109, 117, 131-156, 188-90,
England, George & Co. 51
Evans, W.W. (American engineer) 93
Exhibitions, International 42, 46, 51-4, 81, 109, 113, 116, 119, 187

F

Factory Acts Extension Act (1867) 142, 145, 153
Factory system 9, 100, 120, 131-3, 139, 189/90
Failures of locomotive manufacturers 14, 17, 19, 35, 161, 165-7,

171, 174, 180/1
Fairbairn, Kennedy & Naylor (manufacturer) 115, **116**
Fairbairn, Peter (manufacturer) 177
Fairbairn, Thomas (of William Fairbairn & Sons) 58
Fairbairn, (Sir) William (engineer & manufacturer) 39, 41, 110, 139, 150, 177
Fairbairn, William & Sons/Fairbairn & Co.
 (commercial) **26**, 39, 41-3, 48, 51, 55, 58, 108
 (organisational) 10, 13, 88, 110-2, 138/9, 150-2, 177
Fairlie, Robert (locomotive engineer) **43**, 53, **78**
Fairlie Engine & Steam Carriage Co. 78
Falcon Engine & Car Works 44
Family firms 9-10, 12, 46, 175
Far East locomotive market 19, 32-5, 44/5, 52/3
Farnley Iron Co. (iron works) 105/6
Feed pumps (for boilers) 87
Fenton, Murray & Jackson 22, 120
Fell, John, (innovator, centre-rail adhesion system) 77
Firebox design and manufacture 57, 70/1, 80-3, **84**, 87, 93, 115
Fire-tube manufacture 70, 84, 101, 107, **108**, 109
Fives-Lille Cie, (French manufacturer) 172, 178
Fletcher Jennings & Co. 13, **26**, 48, 178
Foremen, works 11, 61, 132, 134-38, 145, 149, 189
Forging 70, 83, 100, 102, 106/7, 114, **115**, 123, 141, 177
Forrester, George & Co. 150
Fossick & Hackworth 144
Fowler, John & Co. 41, 53, 57, 155, 171, **172**
Fox, Sir Charles (consulting engineer) 28, 68, 92
Fox, James (manufacturer) 109
Fox Walker & Co. (See also Peckett & Sons) 48, 53, **119**
Frame design and manufacture 70-2, 75/6, 83, 86, 93, 105/6, 115, 121, 123
Franco-Prussian War 26, 31, 165
French & Donnison (spring-manufacturer) 109
French locomotive industry 19, 23/4, 26, 34, 52/3, 81, 85, 89, 94, 191
French locomotive market 21/2, **26**, 33, 46, 61, 75, 82, **86**, 177
Friedmann, Alexander (Austrian engineer) 79
Frustrated orders 57

G

Garforth, W.J. & J. 171
Garrett, G.H. (of R. Stephenson & Co.) 181
George & Co. (Rotherham) (iron works) 82, 109
German locomotive industry
 (commercial) 19-24, 29-35, 41, 52/3, 63, 69, 75, 78, 86, 95, 124
 (organisational) 11, 20, 81-3, 89, 91/2, 94, 138, 177/8, 184, 188-91
German locomotive market 20-2, 24, 26, 41, 46, 61, 75, 82, 86,
Giffard, Henri, (innovator, steam injector) 87, 108
Gilkes Wilson & Co. **22**, 56
Glasgow Paisley & Greenock Railway 23
Globe locomotive works (Boston, USA) 166
Glyn & Co. (bank) 59
Gooch, Daniel (of the Great Western Railway) 80, 89, **118**, 164
Gooch, William Frederick (of Vulcan Foundry Co.) 134/5
Gordon & Biddulph (ironworks owners) 105
Grand Junction Railway 23, 39, 70, 76, 176
Grand Trunk Railway (Canada) 177
Grant Ritchie & Co. 49

Gray, John (locomotive engineer) 86
Great Central Railway 34
Great Depression (1873) 31, 34, 190, 193
Great Eastern Railway 52, 57, 62, 81, 109
Great Exhibition (1851) **51**, 52, 82, 152
Great Indian Peninsula Railway **28**, **58**
Great Northern Railway 34, **91**, 178
Great North of England Railway **56**
Great Southern & Western Railway (Ireland) **73**
Great Southern of India Railway **25**
Great Western Railway 19, 23/4, 27, 55, 62, **70**, 76-8, 80, 89, **118**, 120, **160**
Green, Charles (tube manufacturer) 107
Gresham & Craven (component manufacturers) 108
Gryaze-Tsaritsin Railway (Russia) 81

H

Hackworth, Timothy/Hackworth & Downing 10, **40**, 71, **105**, 107
Hackworth-type locomotives **71**
Hall, Joseph (of R. Stephenson & Co.) 138
Hambros Bank 20, 44
Hamilton, Lord George (Secretary of State for India) 95
Hancock's inspirator (pump + injector) 79
Harrison, T.E. (railway engineer) 164
Hartmann locomotive works (Germany) 78
Haswell, John (of William Fairbairn & Sons) 138
Hawkshaw, Sir John (consulting engineer) 28, 68, 92
Hawthorn, William (of R. & W. Hawthorn) 175/6
Hawthorn, R. & W.
 (administrative) 103, 107, 112/3, 137, **140**, 143, 145/6, 148, **153**, 154, 164, 166, 170,
 (commercial) **19**, 43, 47, 51, 85, 123, 176
 (organisational) 10, 115, 172, 177, 179/80
Hawthorn, R. & W., Leslie & Co. 14, 54, 113, 115, 125, 134, 145, 155, 165, **168**, **173**, 174/5, 178/9, 192/3
Hawthorns (of Leith) 176
Henschel & Sohn (Germany) 81
Hick Hargreaves & Co. 171
Hoadley, Joseph (American industrialist) 180
Holt, Francis (of Beyer Peacock & Co.) 137
Horwich locomotive works (Lancashire & Yorkshire Railway) 83
Housing for employees of locomotive workshops 133, **141**, 142
Howe, William (of R. Stephenson & Co.) 86
Hudswell, William (of Hudswell Clark & Rodgers) 137
Hudswell & Clarke/ Hudswell Clark & Rodgers 26, **79**, 109, 114, 119, 124, **137**, 145, 173, 179
Hughes, Henry, & Co. 53, 172
Hull & Barnsley Railway **91**
Hunslet Engine Co. **15**, 26, **33**, **42**, 62, 105, 145, 173, 179, **189**
Hutchinson, William (of R. Stephenson & Co.) 136, 150, 170

I

Imperial Railways (Japan) **187**
Incorporation of manufacturing companies 13, 160/1, 171, 174, 178-81, 191/2
Indian locomotive market
 (commercial) 19, 24, **25**, 27, **28**, **30**, **32**, 33, 45, 52, **58**, 60, **95**, 127

Indian Locomotive Market *continued*
    (standardisation of designs) 90, **91**, 187
Indian State Railways **32**, 60, **91**
India Office 45, 90, 95
Industrial locomotive market
    (commercial) 25, **26**, 27, **33**, 34, **42**, 44-6, 48/9, 53/4, 59, **76**, **79**, 124, **125**, **137**, **172/3**, 176
    (design and manufacture) 109, **119**, 120, 172, **189**
Industrial relations in the locomotive industry 8, 14, 131-3, 136, 142-5, 149-156, 190
Injector design and manufacture 79, 87, 108
Institution of Civil Engineers 41, 45, 154
Institution of Engineers (Scotland) 92
Institution of Mechanical Engineers 81, 119
Inverness & Aberdeen Junction Railway 55
Inter-Colonial Railway (Canada) **175**
Investment and status for locomotive manufacturers 160/1, 168, 174-7
Irish Midland Railway 57
Iron foundries 104/5, 114, 123
Iron plate supplies 82, **105**, 106, 111, 177
Italian locomotive market 22, 24, 26, 77

J

Japanese locomotive market **Front cover**, 33/4, 44, 46, 92, **125**, **187**
Joint Stock Companies Act (1856) 161, 178
Jones, John (of Jones Turner & Evans and Jones & Potts) 134, 156
Jones & Potts 41, 76, 109, **118**, 122, 134, 141, 152, 166, **167**
Jones Turner & Evans **82**
Joy, David, (railway engineer) 79/80

K

Kessler (Maschinenfabrik Esslingen, Germany) 52
Kinmond Hutton & Steel 177
Kirkup, Lawrence (of R. Stephenson & Co.) 150
Kirtley & Co. **47**, 171
Kitching, W. & A. **71**
Kitson, J. Hawthorn 146, 155, 175
Kitson & Co.
    (commercial) **9**, **27/8**, **30**, **34**, **45**, **52**, **58**, 61/2, **67**, 77, **79**, **91**, **155**, **172**
    (organisational) 12, 61/2, 78, 114, 145-7, 150, 154-6, 172, 174-6, 179
Todd Kitson & Laird/Kitson Laird & Co./Kitson Thompson & Hewitson/Kitson & Hewitson 22, 51, 60, **71**, **73/4**, 76, 85, 164, 177
Kitson-Meyer locomotive type **79**
Krupp, Alfred (German steel manufacturer) 82, **83**, 84, 109

L

Labour, skilled and un-skilled 13, 19, 21/2, 27, 34/5, 90, 100, 109-13, 117, 124, 131-56, 185, 188-90
Lancashire & Yorkshire Railway 58, 61/2, 83, 90, **91**, **186**
Lange, Herman (of Beyer Peacock & Co.) 77
Licences for patent use 68, 70, 75, 77/8, 86, 88
Lifting equipment (shear-legs, cranes & traversers) 110, 113-5, 117, 121/2, 125/6, 172

Lindheim, Hermann (German manufacturer) 177
Lilleshall Engineering Co. 53
Link motion (Stephenson) **80**, 86
Liverpool & Manchester Railway **39**, **69**, **71**, 80, 107, 111, 120, 176
Livesey, James (consulting engineer) 77, 79, 92
Llangollen Railway 56
Llynvi & Ogmore Railway **153**
'Lock-out' industrial dispute (1897) 34, 142, 145, 155/6
Locomotive Builders Association (USA) 61
Locomotive Manufacturers Association 60-3, 90, 95, 124, 188, 192
Lombardy-Venetian Railway (Italy) **55**, **Back cover**
London & Birmingham Railway 22, 39/40, 46, **164**
London & Blackwall Railway 167
London Brighton & South Coast Railway **89**
London Chatham & Dover Railway 41
London, City of, locomotive market 26, 28, 31, 36, 39, 42, 44-6, 52/3, 63, 91, 184, 187/8
London & North Western Railway 24, 27, 51, 61/2, **74**, 80, 82-4, 86/7, 89/90, 107, **164**, 167
London & South Western Railway 23, **35**, 80, 111
London Tilbury & Southend Railway 62, **88**
'Long Boiler' locomotive design **56**, **75**, **76**, 85, **86**, 88, **103**, **194**
Longridge, Michael (of R. Stephenson & Co., Bedlington Ironworks and R.B. Longridge & Co.) 105-7, 135, 148-51, 156, 162, 169, 176
Longridge, R.B., & Co. 43, **75**, 105, 143, 145, 149, **151**
'Long Swing' overseas investments 18, 33
Lower Silesian Railway (Germany) 87
Low Moor Iron Co. (Bradford) 82, **105**, 106
Lynton & Barnstaple Railway 34

M

Machine tools
    (application) 47/8, 90, 99, **110**, 112-4, 115, 117/8, 120-6, 139-40, 148, 152
    (design and development) 13, 21, 100, **101**, 109, **111**, **115/6**, 117, 132, 141/2, 172, **189**
Mackenzie, William (contractor) 24
Madras Railway (India) 59, **155**
Maffei, Joseph von (German railway engineer) 138
Mallet, Anatole (Swiss railway engineer) 78, 81
Management and supervision in workshops 39, 131-56, 160-81, 189-93
Managerial capitalism 11/2, 191/2
Manby, Charles (of R. Stephenson & Co.) 41
Manchester & Milford Railway **168**
Manchester Sheffield & Lincolnshire Railway **9**, **76**, **89**
'Mania' years (1844-47) 22/3
Manning Wardle & Co. 26, 62, **76**, **89**, 105, **119**, 124, **125**, 146, **168**, 173, 179
Mapplebeck & Co. (tube manufacturer) 108
Market evolution and development for locomotives 19-35, 125
Marketing methods 46-54
Marseille-Avignon Railway (France) **80**, 86, 122, 177
Marshall, Francis (of R. & W. Hawthorn Leslie & Co.) 155
Marshall William Prime (R. Stephenson associate) 46
Mass production (American system) 100, 107, 117/8, 120, 124, 139, 188

INDEX

Master Car-Builders Association (USA) 88
Material development 69/70, 81-4
Mather Dixon & Co. 180
Maudslay, Sons & Field (manufacturer) 100, 109, **113**, 118
McConnell, James (Caledonian Railway) 74, **75**, 80
McCulloch, Thomas & Sons 45
Melbourne & Hobsons Bay Railway (Australia) **32**
Metropolitan Railway 77/8
Meyer, Jean (French consulting engineer) 78
Middle East locomotive market 19, 32, 68
Midland Railway 24, **27**, 34, **78**, 81, 90, 106-8, 123-5, 181
Midland Railway (India) **95**
Mining companies 44, 53, 99, 111, **137**, 172, 176, 178, 180
Molyneaux, Thomas (of Beyer Peacock & Co.) 149
Monk Bridge Iron Co. 105, 177
Mont Cenis Railway (Italy) 77
Motion components 83, 106/7, 114
Mountain railways 77
Mount Lyell Railway & Mining Co. (Australia) **77**
Munich-Augsburg Railway (Germany) 138
Murdoch Aitken & Co. **11**

N

Narrow gauge locomotive designs **48**, **77**, 78-9
Nasmyth, James (of Nasmyths Gaskell & Co. & J. Nasmyth & Co. 107, **110/1**, 112, 118/9, 137/8, 141/2, 146, 148, 151, 156, 175
Nasmyths Gaskell & Co. 48, 107, **114**, 120, **121**, 147, 173/4, 188
Nasmyth Wilson & Co. **12**, 13, 62, 111, 118, 122, **137**, 139, **151/2**, **169**, 187
Naylor & Vickers (iron works) 82/3, 107, 109
Neath Abbey Iron Works 22
Neilson & Mitchell 22
Neilson & Co.
(commercial) **29**, 40, **41**, **47-9**, 56, **57**, **72**, **84**, **91-3**, 94, 123, **194**
(organisational) 62, 75, 92, **99**, 123, **131**, 134, 138, 141, **143**, 149, 154, 156, 165, 175, 177, 192
(design and manufacturing) 108, 112/3, 115, **116**, 117, 173
Neilson Reid & Co. 35, 53, 94, **104**, 117, **127**, 134, **143**, 155, 180, 191
Neilson, Walter (of Neilson & Mitchell, Neilson & Co. and Clyde Locomotive Co.) 40, 92, 141, 175, 180, 192, 193
Netherlands locomotive market 22, 24, 30, 41, 85/6
New South Wales Government Railways (Australia) 42, 137, **194**
Newcastle & Berwick Railway 57, **137**, 149, **152**
Newcastle & Carlisle Railway 47, **136**, **163**
New Zealand Government Railway **93**, 94
New Zealand locomotive market 32-5, 93
Nine hours movement (industrial relations) 153/4
Nippon Railway (Japan) **125**
Non-ferrous components (copper, brass, gun-metal) 84, 106, 121, 123, 164, 171
Norris locomotive works (USA) 20, 92
North British Iron Hematite Co. 177
North British Locomotive Company 35, 174, **179**, 180, 191-3
North Eastern Railway 22, **49**, **56**, 70, **136/7**, 150, **152**, 168, **173**, 192/3
North Eastern Railway (Victoria, Australia) 45
Northern Railway (France) **26**, **127**
Norwegian Government Railway 56, 58, 77
Nova Scotia Railways (Canada) **84**, **92**, **105**, **119**

O

Odessa-Balta Railway (Ukraine, Russia Empire) **30**
Oil-burning locomotive designs 81
Origins of main line locomotives 69-72
Overend Gurney Bank failure (1866) 28, 167
Overseas ventures of locomotive manufacturers 177/8

P

Pambour, Chev. Guyonneau de (French engineer) 47
Paris-Orleans Railway (France) **31**, 85
Paris-Rouen Railway (France) 177
Partnership capitalism 11-5, 30, 134, 191, 193
Paternalism in the locomotive industry 131-3, 148-50
Patent Exhaust Steam Injector Co. (Davies & Metcalfe Ltd) 87
Patents (locomotives and components) 39, 42/3, 51, 56, 68, 70-2, 75-81, 84-9, 103, 107/8, 111, 114, 116/7, 179
PATENTEE locomotive type 46, **71**
Patent Shaft & Axletree Co. (wheel & axle maker) **109**
Peacock, Richard (of Beyer Peacock & Co.) 174/5
Pease, Edward (of R. Stephenson & Co.) 70, 136, 139, 176, 181, 193
Pease, Joseph (1799-1872) (of R. Stephenson & Co.) 43, 136, 146, 181, 193
Pease, Sir Joseph (1828-1903) (of R. Stephenson & Co.) 181, 193
Peckett & Sons (See also Fox Walker & Co.) 48, 51, 146
Pennsylvania Steel Co. (USA) 84,
Peto, Brassey & Betts (contracting consortium) 28, 122, 149, 171, 177
Peto, Samuel (contractor) 24, 28
Piecework and piece-mastering 132, 138, 142, 145, 152, 156, 165, 189
PLANET-type locomotive design **19**, **20**, **39**, **69**, 80, **102**
Portuguese locomotive market 21, 24
Poti-Tiflis Railway (Georgia, Russian Empire) **191**
Potts, Arthur (of Jones & Potts) 134, 166/7
Pricing of locomotives 44, 54, 57-9, 188
Private Limited companies 12, 14, 134, 191
Production Engineering practices 96, 120-6
Profit & Loss statements 161, **163/4**, 165, 168/9, **170**, 171
Progressive locomotive manufacturers
(equipment) 13, 25, 30, 34, 99/100, **101**, 117/8, 120, 122-5, 161
(labour) 14-15, 27, 35, 132, 135, 141/2, 156, 161, 189
(organisational) 58, 79, 90, 102, 112, 166, 173-5, 178, 180, 185, 188, 190-3
Proprietorial contact 30, 39-42, 56
Public Limited companies 12, 14, 134, 191

Q

Quaker dynasties in locomotive manufacture 10
Queensland Government Railways (Australia) **29**, **77**, **194**

R

Radial axles for locomotives 79, 87, **88**
Railway-owned locomotive workshops 23, 25, 34, 68, 72, 133, 138, 145/6, 149, 185/6, 188/9
Rainhill locomotive trials (1829) 39, **69**
Ramsbottom, John (London & North Western Railway) 83, 87

Raw material supply to locomotive manufacturers 14, 54, 57 59-61, 81, 101, 104-7, 133, 161-6, 180, 190
Receipts, payments and working capital 161, 166-8
Recession (1842-3) **21**, 22, 152, 166, 180
Recession (1847-50) 23, 25, 43, 122, 167, 171, 175
Recession (1893-7) 155, 173, 180
Reid, Hugh (of Neilson Reid & Co.) 155
Reid, James (of Dübs & Co.) 175, 180
Rendell, Sir Alexander (consulting engineer) 28, 68, 90, 92
Rennie, G. & J. 22
Rennie, Robert (of Bedlington Iron Co.) 107
Rhenish Railway (Netherlands) 55
Rhymney Railway **169**
Ridley, J.H. (of R. & W. Hawthorn) 175
Riveting of locomotive components 110-3, 139
Roberts, Richard (of Sharp Roberts & Co.) 110-2, 118/9, 151
Robertson, Henry (of Beyer Peacock & Co.) 170, 175
ROCKET locomotive (Liverpool & Manchester Railway) 51, **69**, **162**
Rothschilds Bank 20, 59
Rothwell & Co. **142**
Royalties on patents 84-7, 89, 122
Russian locomotive industry 19, 31, 35, 53
Russian locomotive market 20/1, 24/5, 27, **29/30**, 31, 41, 53, **61**, 62, **191**
Ruston Proctor & Co. 53
Rybinsk-Bologoye Railway (Russia) **29**

S

Sacré, Alfred (of Yorkshire Engine Co. and Avonside Engine Co.) 134
Sacré, Edward (of Yorkshire Engine Co.) 62
Sales practices 54-7
Salter, George & Co. (spring balance manufacturer) 48, 108
Saxon Bavarian Railway (Germany) 59
Scandinavian locomotive market 19, 22, 24, 27, 30, 41, 61
Schaken Caillet et Cie. (French locomotive maker) 178
Schenectady locomotive works (USA) 46
Schmidt, Wilhelm (German engineer) 81
Schneider et Cie. (French locomotive manufacturer) 52
Schröeders Bank 20
Scinde Railway (India) 75
Scottish Central Railway **167**
Second-hand locomotive market 25
Sharp Roberts & Co. **21**, 42, 76, **111**, 120, 137, 151/2
Sharp Brothers 12, 40, 43, 47, 74, **76**, 86, 111, 114, 119, 152, 175
Sharp Stewart & Co.
 (commercial) **17**, **24**, **29**, **31**, **48**, 53, 59, 62, 87, **88**, 108, **125**, **185**
 (design and manufacturing) 91, **101**, 106, 112/3, **115**, 116/7, 188
 (administrative) 35, 94, 112, 149, 155, 178, **179**, 180, 191/2
Shear-legs (see lifting equipment)
Shrewsbury & Chester Railway **75**
Shop system of manufacturing 121/2
Slaughter Grüning & Co. 41, 77, 122, 134, **135**
Slaughter, Edward (of Stothert & Slaughter, Slaughter Grüning & Co. and Avonside Engine Co.) 134, 154
Smith, Sydney (pressure gauge manufacturer) 108
Smyrna-Aiden Railway (Turkey) **59**
Snowball, Edward (of R. Stephenson & Co. & Neilson & Co.) 75, 146, 150

Societe Alsacienne de Constructions Mécaniques (SACM - French locomotive manufacturer) 81
Sole proprietor (engineering manufacturer) 9/10, 39/40
Somerset & Dorset Joint Railway 90
South American locomotive market 19, 24/5, 27/8, 30, 32, **33/4**, 35, **43**, 44/5, 52/3, 68, 77, **79**, 92/3, **125**, 186
South Devon Railway **160**
South Eastern Railway 43, **51**, 80, 108/9
South Indian Railway 30
Spanish locomotive market 24, 26, 172
Spares, locomotive components 119
Spencer, John & Sons (spring manufacturer) 109
Spring technology 52, 109
Staffordshire iron supplies 82, 105-7
Standardisation
 (of components) 68, 76, 90, 99, 102, 118-22, 125, 148, 180, 184, 186/7
 (of designs) 68/9, 73/4, 89-90, **91**, 95, 102, 126, 188, 193
Starbuck, Edward (locomotive agent) 41, 43-5, 56, 59/60, 85/6
Staveley Iron Works **168**
Steam hammers 83, 107, **114/5**, 141
Steam pressures 83
Steel technology for locomotive components 82, **83**, 104, 106-9, 115, 177
Stephenson, George (railway engineer) 39/40, 69-71, 104, 107/8, 134/5, 146, 176
Stephenson George Robert (of R. Stephenson & Co.) 134, 149, 154, 156
Stephenson, Robert (of R. Stephenson & Co.) 39-41, 43, 45, 51, 69, 71, 85, 109, 134-6, 146, **147**, 149/50, 176/7
Stephenson, Robert Co.
 (administrative) 138/9, 143, 146, **147**, 148-50, 153/4, 156, **162/3**, **165**, 166, **169/70**, 171, **173**
 (commercial) 11, 20, 22/3, **32**, **39**, 40-3, 46, **50/1**, 55-8, **59**, 60, **61**, 62, **75**, 85, **86**, 113, 124/5, **136**, 172, **194**
 (design & manufacturing) 56, **69/70**, 75, 77, 79, 80, 84/5, **86**, 87, 91, **102/3**, 104-7, 109/10, **112**, **115-17**, 120, **122**, **124**, **126**, 188
 (organisational) 10, 14, 45, 61/2, 75, 122, 134-6, 176-80, **181**, 192/3
Stettin Maschinenbau A.G. (German locomotive maker) 81
Stewart, Charles (of Sharp Stewart & Co.) 175
Stock locomotives 122, 172/3
Stockton & Darlington Railway **19**, **22**, 51, 56, **70/1**, 166
Stone, J. & Co. (feed-pump manufacturer) 48
Sub-contracting of locomotives and components 81, 101/2, 122
Summerlee Ironworks (Scotland) 177
Summers Groves & Day 171
Superheating development 80/1
Swedish Government Railway **175**
Swedish locomotive industry 35
Swindon locomotive workshops (GWR) 8, 23/4, 27, 134, 138, 141, 145
Swiss locomotive market 24
Sydney & Goulburn Railway (Australia) 42, **75**
Systemisation of locomotive manufacturing 100, 186

T

Taff Vale Railway **38**, **103**, **181**
Talabot, Paulin (French engineer) 55, 177
Taltal Railway (Chile) 79

INDEX

Tank locomotives **26**, 44, 48, 50/1, 53/4, 56, **76-9**, 106, **169**, **181**, 189

Tariffs on international trade 21, 23/4, 31, 36, 55, 82, 187

Tarragona, Barcelona & France Railway (Spain) **135**

Tayleur, Charles & Co.(See also Vulcan Foundry Co.) **19**, 57, 76, 134, 137, **141**, **164**, **176**

Tayleur, Charles Sr. and Jr. (of Charles Tayleur & Co.) 176

Tayleur, Edward (of Charles Tayleur & Co.) 134

Taylor Brothers & Co. (Iron works) 105-7

Telegraph, electric (company communication) 55, 57-8

Templates and gauges 79, 103, 107, 118, 120, 126, 187

Tender invitations and submissions 40, 54-7, 60, 62, 79, 84, 117

Textile industry 9, 46, 61, 100, 139, 145, 168, 171

Textile machinery 9, 13, 19, 42, 85, 99/100, 111/2, 118, 120, 122, 176

Trade fairs 46, 53/4, 189

Trades unions 132/3, 145, 150-56, 190

Trinidad Government Railways **155**

Tsarskoe-Selo Railway (Russia) **61**

Tulk & Ley **87**

Turton, Thomas & Sons (spring manufacturer) 109

Tweddell, Ralph Hart (of R. & W. Hawthorn) 113

Tyre technology for locomotives 82, 95, 106/7, 109

U

United Kingdom locomotive market **18**

United States of America locomotive market 19-21, 46, 72, 74

V

Vacuum Brake Co. 88, 108/9

Valve gears and variable cut-off for locomotives 80, 86

Vauclain locomotive type 81, 89

Victorian Railways (Australia) **32**, 43, **45**, 93

Vienna financial crash (1873) 31

Vienna-Raab Railway (Austria) **21**, 138

Vulcan Foundry Co.
    (administration) 12, **106**, 134/5, **143**, **164**, 178/9, 191
    (commercial) **38**, 39, 52, 62, **78**, 84, **91**, 94, **95**, **103**, **127**, 138, 165, **184**, **186**

W

Wages for locomotive manufactory employees 112, 138/9, 142-5, 147, 150, 154/5, 164, 190

Walker & Brother **14**

Walker, John (of R. Stephenson & Co.) 181

Walschaerts, Egide (Belgian railway engineer) 80

Wardropper, Thomas (of R. Stephenson & Co.) 61

Waterford & Kilkenny Railway (Ireland) 57

Waterford & Limerick Railway (Ireland) **67**, **184**

Watt, James (Scottish engineer) 39

Weatherburn, Robert (of Kitson & Co.) 61, 124

Webb, Francis (LNWR Chief Mechanical Engineer) 79, 83, 87, 116

Wedgewood, Josiah (industrialist) 161

West Lancashire Railway **185**

West Midland Railway 56

Westinghouse, George, Brake Co. 87/8, 108

Wheel design and development 70/1, 76, 106/7, 109/10, 115/6

Wheel smiths 141

Whitworth, Joseph (manufacturer) **110**, 111/2, 118/9

Wilkes, John, & Co. (boiler tube manufacturer) **108**

Willis, Robert (of Nasmyth & Co.) 137

Wilson, E.B., & Co. 40, 42, 51, 76, 86, **89**, 91, 112-5, 119, 122, **173**, 175, 180, 188

Windsor & Annapolis Railway (Nova Scotia, Canada) **119**

Wöhler, August (Prussian railway engineer) 87

Wolverhampton locomotive workshops (GWR) 24

Wolverton locomotive workshops (LNWR) 74, 138, 145

Woods, Edward (consulting engineer) **73**, 79

Worcester Engine Co. 180, **181**

Working capital 11, 58, 105, 161, 164, 166-169, 172/3, 180, 189

Working hours for manufactory employees 140, 143, **144**, 150, 153/4

Workmen's Compensation Act (1880) 149

Workshop lighting 126

Workshops adopted by locomotive manufacturers 123-6

Y

Yates, William (Lancashire & Yorkshire Railway engineer) 186

York, Newcastle & Berwick Railway 57, **137**, **152**

York & North Midland Railway 56

Yorkshire Engine Co. 12, 14, **30**, **32/3**, 62, **93**, **178**, **191**

Yorkshire iron supplies 82, 105